# Communications in Computer and Information Science 2203

**Rationale**

The CCIS series is devoted to the publication of proceedings of computer science conferences. Its aim is to efficiently disseminate original research results in informatics in printed and electronic form. While the focus is on publication of peer-reviewed full papers presenting mature work, inclusion of reviewed short papers reporting on work in progress is welcome, too. Besides globally relevant meetings with internationally representative program committees guaranteeing a strict peer-reviewing and paper selection process, conferences run by societies or of high regional or national relevance are also considered for publication.

**Topics**

The topical scope of CCIS spans the entire spectrum of informatics ranging from foundational topics in the theory of computing to information and communications science and technology and a broad variety of interdisciplinary application fields.

**Information for Volume Editors and Authors**

Publication in CCIS is free of charge. No royalties are paid, however, we offer registered conference participants temporary free access to the online version of the conference proceedings on SpringerLink (http://link.springer.com) by means of an http referrer from the conference website and/or a number of complimentary printed copies, as specified in the official acceptance email of the event.

CCIS proceedings can be published in time for distribution at conferences or as post-proceedings, and delivered in the form of printed books and/or electronically as USBs and/or e-content licenses for accessing proceedings at SpringerLink. Furthermore, CCIS proceedings are included in the CCIS electronic book series hosted in the SpringerLink digital library at http://link.springer.com/bookseries/7899. Conferences publishing in CCIS are allowed to use Online Conference Service (OCS) for managing the whole proceedings lifecycle (from submission and reviewing to preparing for publication) free of charge.

**Publication process**

The language of publication is exclusively English. Authors publishing in CCIS have to sign the Springer CCIS copyright transfer form, however, they are free to use their material published in CCIS for substantially changed, more elaborate subsequent publications elsewhere. For the preparation of the camera-ready papers/files, authors have to strictly adhere to the Springer CCIS Authors' Instructions and are strongly encouraged to use the CCIS LaTeX style files or templates.

**Abstracting/Indexing**

CCIS is abstracted/indexed in DBLP, Google Scholar, EI-Compendex, Mathematical Reviews, SCImago, Scopus. CCIS volumes are also submitted for the inclusion in ISI Proceedings.

**How to start**

To start the evaluation of your proposal for inclusion in the CCIS series, please send an e-mail to ccis@springer.com.

R. Geetha · Nhu-Ngoc Dao · Saeed Khalid
Editors

# Advances in Artificial Intelligence and Machine Learning in Big Data Processing

## First International Conference, AAIMB 2023 Chennai, India, August 17–18, 2023 Proceedings, Part-II

*Editors*
R. Geetha
S.A. Engineering College
Chennai, Tamil Nadu, India

Nhu-Ngoc Dao
Sejong University
Seoul, Korea (Republic of)

Saeed Khalid
Bialystok University of Technology
Białystok, Poland

ISSN 1865-0929 ISSN 1865-0937 (electronic)
Communications in Computer and Information Science
ISBN 978-3-031-73067-2 ISBN 978-3-031-73068-9 (eBook)
https://doi.org/10.1007/978-3-031-73068-9



This Springer imprint is published by the registered company Springer Nature Switzerland AG
The registered company address is: Gewerbestrasse 11, 6330 Cham, Switzerland

# Preface

These two CCIS volumes constitutes the refereed proceedings of the International Conference on Advances in Artificial Intelligence and Machine Learning in Big Data Processing, AAIMB 2023, held at S.A. Engineering College, Chennai, India, during August 17–18, 2023. All the accepted papers were carefully double-blindly reviewed by national and international reviewers. They were organized in topical tracks as follows: Artificial Intelligence and Machine Learning, Intelligent Information Processing, Pattern Recognition and Analysis, Sentiment Analysis, Computational Optimization, Information Extraction and Prediction, Decision Making, Virtual reality, Natural Language Processing, AI Algorithms, Fuzzy logic & Neural System. Adaptive agents, AI Planning Strategies and Tools, Reinforcement Learning, Robotics, Computer Vision, Expert Systems, Pattern Recognition, Component Analysis, Social Network Analysis, Cloud-Based Services, Big Data Analytics, Data Mining Challenges with Big Data, Big Data Analytics and Big Data Governance, Big data Analytics and Applications Data, Text, Web Mining, & Visualization, Data & Knowledge Engineering, Techniques for Data Visualization, Data Stream Algorithms and Information Processing, For the AAIMB 2023 conference 183 papers were been received, out of which 51 full-length papers were accepted and presented at the conference.

We would like to acknowledge and thank the conference committee members and international reviewers who spent a lot of time on the arduous task of selecting the qualified papers from all the submissions. Their excellent professional advice enabled us to come out with this volume of proceedings of the conference. We thank all the participants and the management of S.A. Engineering College, who encouraged us morally and financially.

August 2023

R. Geetha
Nhu-Ngoc Dao
Saeed Khalid

# Organization

## General Chair

| | |
|---|---|
| Geetha R. | S. A. Engineering College, Chennai, India |

## Program Committee Chairs

| | |
|---|---|
| Anuradha M. | S. A. Engineering College, Chennai, India |
| Hemanand D. | S. A. Engineering College, Chennai, India |
| Nalini M. | S. A. Engineering College, Chennai, India |
| Kaliraj V. | S. A. Engineering College, Chennai, India |
| Nhu-Ngoc Dao | Sejong University, Seoul, Republic of Korea |
| Abdel-Badeeh M. Salem | Ain Shams University, Egypt |
| Nazir Ahmad | King Khalid University, Saudi Arabia |

## Steering Committee

| | |
|---|---|
| Nhu-Ngoc Dao | HCMOU, Vietnam |
| Nazir Ahmad | K. A. Nizami Centre for Quranic Studies, Aligarh, India |
| Dinakaran K. | S.A. Engineering College, India |
| Thilagavathi N. | S.A. Engineering College, India |
| Muthumari Lakshmi S. | S.A. Engineering College, India |
| Abdel-Badeeh M. Salem | Ain Shams University, Egypt |
| Sheela Lourdusamy | Pentecost University, Ghana |
| Xavier N. Fernando | Ryerson University, Canada |
| Mokhtar M. M. Alihamid | University of El Imam El Mahdi, Aba Island, Sudan |
| Muhammad Ghalib | De Montfort University, Dubai, United Arab Emirates |
| Kindie Biredagn Nahato | Debre Berhan University, Debre Birhan, Ethiopia |
| Senthil Ramadoss | Majan University College, Oman |
| J. Jean Justus | St. Joseph's College of Engineering, India |
| Suresh Kumar S. | Rajalakshmi Engineering College, India |
| Sathish Kumar P. J. | Panimalar Engineering College, India |

| | |
|---|---|
| Chembian W. T. | Vel Tech HighTech Dr.Rangarajan Dr.Sakunthala Engineering College, India |
| Eswaramoorthy V. | Bannari Amman Institute of Technology, India |
| Subhashini P. | Vel Tech Multi Tech Dr.Rangarajan Dr.Sakunthala Engineering College, India |
| Gopalakrishnan B. | Bannari Amman Institute of Technology, India |
| Sundara Murthy S. | Bannari Amman Institute of Technology, India |
| Vanitha M. | Saveetha Engineering College, India |
| Sathya Priya J. | Velammal Engineering College, India |
| Ashok Kumar P. M. | KL University, India |
| Brindha Merin J. | B.S. Abdur Rahman Crescent Institute of Science & Technology, India |
| Mercy Paul Selvan | Sathyabama Institute of Science and Technology, India |
| Saraswathi S. | SSN College of Engineering, India |
| Bhavadharini R. M. | VIT, India |

## Program Committee

| | |
|---|---|
| Siva Kumar S. | S. A. Engineering College, Chennai, India |
| Chitra Devi D. | S. A. Engineering College, Chennai, India |
| Balakrishnan C. | S. A. Engineering College, Chennai, India |
| Mani A. | S. A. Engineering College, Chennai, India |
| Balasubramanian M. | S. A. Engineering College, Chennai, India |
| Muthukumarasamy S. | S. A. Engineering College, Chennai, India |

## Additional Reviewers

Prasana Devi
Prasana Kumar
Sangeetha J.
Aruna K. B.
Thilagam T.
Shobana R.
Belshia Jebamalar
Abirami A.
Kavitha M.
Ravanan R.
Susi K.
Lakshmipriya
Mabel Rose R. A.
Ananthi S. N.
Julia Faith S.
Veena T.
Chithra D.
Krishnaveni S.
Seetha A.
Sutha Merlin
Muralitharan G.
Aravinth Kumar R.
Chitti Babu B.
Jayavel R.
Sureka V.
Sudha L.

# Contents – Part II

# Contents – Part I

## Artificial Intelligence and Data Analytics



## Deep Learning

# Artificial Intelligence and Data Analytics

# Feature Fusion Based Bayesian Model Detection in Prognosis of Glioma – A Survey

S. Gowthami(✉), K. H. Mohammed Sazzad(✉), M. Nethra, S. Santhya, and A. Arnold Sylevester

Department of Biomedical Engineering, Bannari Amman Institute of Technology, Sathyamangalam, India
{mohammedsazzad.bm20,mohammedsazzad.bm20}@bitsathy.ac.in

**Abstract.** Tumor are a group of tissue that develops abnormally as cells divide and multiply more rapidly than normal cells. In the UK, an average of 429 kids under the age of 14 and 563 kids and teenagers under the age of 19 receive a glioma diagnosis each year. The two types of tumors are malignant and benign, respectively. Glioblastoma Multiforme (GBM) is a devastating brain cancer that can result in death in six months or less, if untreated; hence, it is imperative to seek expert neuro-oncological and neurosurgical care immediately, as this can impact overall survival. Glioma are most frequently discovered via Magnetic Resonance Imaging (MRI) scans. In this work Bayesian modelling is used to identify the glioma. Bayesian algorithm gives efficient output and the work is validated by the performance metrics (Precision, accuracy, Mathew's correlation coefficient (MCC), Dice coefficient, Kappa). By using these algorithms on MRI scans, brain cancers can be found more rapidly and accurately. The previous work gives an overall accuracy rate of 90% with 75% detection rate of glioma image and 60% detection rate of no tumor images.

**Keywords:** Feature extraction · Bayesian modeling · Glioma

## 1 Introduction

Glioma detection involves the identification and diagnosis of abnormal growths in the brain using various medical imaging techniques and diagnostic tools. Glioma detection is arduous and presents a variety of obstacles. Detection of gliomas can be hindered depending on their exact location within the brain. The difficulty in accessing tumors located in critical areas or deep within the brain can impede diagnostic procedures greatly. The various regions in which gliomas can develop also add to the complexity of identification. Identifying gliomas can be impacted by their size. The smaller the tumor, the more difficult it may be to detect on imaging scans. Additionally, smaller tumors may not produce recognizable symptoms until they grow larger over time. On imaging scans, gliomas can be difficult to differentiate from healthy brain tissue due to their resemblance in characteristics and appearance. This similarity poses a challenge in diagnosis as there is potential for ambiguity. Gliomas, which can be benign or malignant, are quite diverse and

The original version of the chapter has been revised. The author's first name has now been added. A correction to this chapter can be found at
https://doi.org/10.1007/978-3-031-73068-9_26


R. Geetha et al. (Eds.): AAIMB 2023, CCIS 2203, pp. 3–19, 2025.
https://doi.org/10.1007/978-3-031-73068-9_1

don't have a one-size-fits-all method of detection due to their individual characteristics and growth patterns. Motor difficulties, cognitive changes, headaches, and seizures are symptoms that can be caused by gliomas or other neurological conditions. This overlap makes proper diagnosis challenging and further testing is necessary to determine the presence of a tumor. Protected by the skull and a barrier called the blood–brain barrier, the brain is a sensitive organ, which poses difficulties with obtaining direct access to it for either biopsies or surgical procedures that are essential for diagnosing tumors accurately. Diagnostic limitations: Imaging techniques such as magnetic resonance imaging (MRI) and computed tomography (CT) are commonly used to detect gliomas, but they may have limitations. Various methods are being used to detect gliomas more accurately by reducing complexity. Difficulties in detecting glioma: ¬ Location of tumor ¬ Size of tumor ¬ Similarity to normal brain tissue ¬ Variability in tumor types ¬ Overlapping symptoms ¬ Inaccessibility ¬ Diagnostic limitations ¬ False negatives.

## 2 Literature Survey

A glioma is a type of tumor that originates from glial cells in the brain or spinal cord. Glial cells support and nourish neurons, the cells responsible for transmitting electrical signals in the central nervous system. Gliomas can be benign (non-cancerous) or malignant (cancerous) and can differ in their aggressiveness and growth pattern. They may include surgery, radiation therapy, chemotherapy, targeted therapies, and immunotherapy. The treatment approach aims to control tumor growth, relieve symptoms and improve the patient's quality of life. Extensive research on glioma Genetics: Researchers are trying to identify the genetic mutations responsible for the development and progression of glioma. This knowledge could lead to the development of new targeted therapies specifically targeting these mutations. immunotherapy. The immune system plays a role in fighting cancer, and researchers are working to develop new ways to use the immune system to fight glioma. This may include developing new vaccines or immunotherapies that can stimulate the immune system to attack tumor cells. delivery of medicines. The blood–brain barrier (BBB) is a physical barrier that prevents many drugs from reaching the brain. Researchers are working to develop new ways to deliver drugs to gliomas that can cross the BBB. A new treatment strategy. Researchers are also exploring new strategies for treating glioma, such as gene therapy, stem cell therapy, and photodynamic therapy. Cancer Genome Atlas (TCGA) Project: The TCGA Project is a large-scale research effort that has sequenced the genomes of thousands of cancer patients. This project has identified many genetic mutations associated with glioma, and this information is being used to develop new targeted therapies. International Glioma Genome Consortium (IGGC): The IGGC is a worldwide association of researchers who study the genetics of glioma. The consortium has identified many new genetic mutations associated with glioma, and this information is being used to develop new targeted therapies. Glioblastoma Multiforme Research Consortium (GMRC): The GMRC is a global association of researchers who study glioblastoma. The consortium is conducting clinical trials for new treatments for glioblastoma and also developing new ways to deliver drugs to glioma (Fig. 1).

**Imaging Modalities**

Magnetic resonance imaging (MRI) is the most commonly used imaging technique

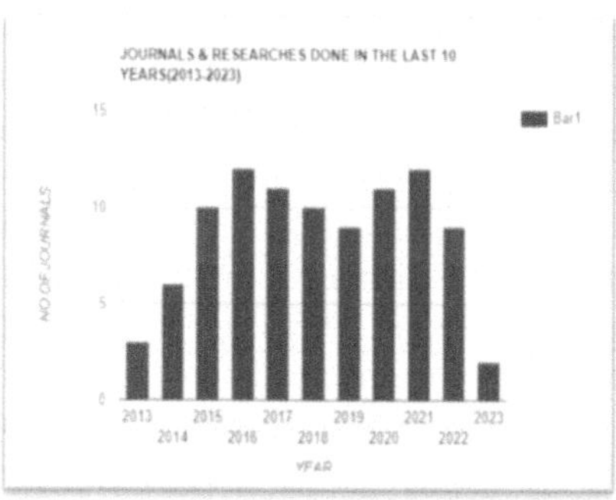


**Fig. 1** .

for diagnosing and staging gliomas. MRI can provide detailed images of the tumor, including its size, shape, and location. It can also help identify areas of tumor invasion into the surrounding brain tissue. Positron emission tomography (PET) can be used to determine the metabolic activity of glioma cells. PET scanning can be used to identify areas of active tumor growth even if they are not visible on MRI. Computed tomography (CT) is less sensitive than MRI for detecting gliomas, but can be used to evaluate how far the tumor has spread to other parts of the brain and spine. Single photon emission computed tomography (SPECT) can be used to evaluate blood flow to a glioma. SPECT scanning can be used to identify areas of increased blood flow that may indicate areas of active tumor growth. These imaging modalities have been used to develop several new diagnostic and prognostic tools for glioma. For example, MRI can be used to detect the presence of an IDH1mutation, which is an indicator of a better prognosis. PET scans can be used to assess the response of gliomas to treatment. CT scans can be used to assess the progression of gliomas and to plan for surgery. A study published in 2017 found that MRI can be used to identify the presence of the IDH1 mutation in gliomas with a sensitivity of 80% and a specificity of 90%. This finding could help to improve the diagnosis and prognosis of gliomas. A study published in 2018 found that PET scans can be used to assess the response of gliomas to treatment. The study found that PET scans could be used to identify patients who were responding to treatment and to those who were not. This finding could help to improve the management of glioma patients. A study published in 2019 found that computed tomography can be used to assess the progression of gliomas. This study showed that CT scans can be used to discriminate between patients with progressive tumors and those with stable tumors. This finding may help improve the timing of treatment for people with glioma.

Earlier on glioma was detected using other methods that included invasive methods such as biopsy and lumbar puncture. Previously used methods to detect glioma are Neurological examination: a neurological examination assesses the function of the nervous system, including the brain and spinal cord. It helps detect abnormalities in motor function, coordination, reflexes, and sensory perception that may indicate a glioma. Imaging studies: Magnetic resonance imaging (MRI): MRI uses powerful magnets and radio waves to produce detailed images of the brain. This can be used to determine and provide detailed information about the location, size and characteristics of gliomas.

Advanced MRI techniques such as functional MRI (fMRI) and magnetic resonance spectroscopy (MRS) can also provide additional information about brain function and tumor metabolism. Computed tomography (CT): CT Scans use X-rays and computer processing to create cross-sectional images of the brain. They can help detect the presence of tumors and provide information about their size, shape, and density. Positron emission tomography (PET): PET Scans use a small amount of radioactive material to visualize the metabolic activity of tissue. They can help distinguish between benign and malignant tumors and provide information about tumor growth and spread. Angiogram: An angiogram involves injecting a contrast medium into the blood vessels of the brain and taking X-rays. It can help visualize the blood supply to gliomas and detect abnormal blood vessel formation (angiogenesis). Biopsy: A biopsy involves taking a small sample of tumor tissue for laboratory analysis. It is the most definitive method for diagnosing gliomas and determining their type and grade. Biopsies can be performed using stereotactic techniques, in which a needle is guided to the tumor using imaging techniques, or by surgical procedures. Lumbar puncture (spinal tap): In some cases, a lumbar puncture may be performed to check the CSF for the presence of tumor cells or other abnormalities. In this procedure, a needle is inserted into the lower back to collect CSF for examination. Genetic and molecular testing: Genetic and molecular testing can provide additional information about the specific characteristics of gliomas. These tests may include analysis of tumor DNA, RNA, or protein markers. They can help inform treatment decisions and provide information about prognosis. Later on non-invasive methods were used that analyses the image by using modern technologies.

**Fusion Based Classification**

Fusion-based classification is a technique that combines multiple images of gliomas to increase classification accuracy. This is done by merging images together to create a single image that contains more information than the individual images. There are two main types of merge-based classification. Feature-level merging: This type of merging combines features extracted from individual images. This is done by combining functions together using math functions. Decision-level fusion: * This type of fusion combines the decisions made by classifiers trained on separate images. This is done by joining the solutions together using mathematical functions. Fusion-based classification has been shown to be effective for classifying gliomas. In a study published in 2017, fusion-based classification was able to achieve 97% accuracy in classifying gliomas. Here are some of the advantages of using fusion-based classification to classify gliomas: Classification accuracy can be improved by combining information from multiple images. It can be used to classify tumors that are difficult to classify in a single image. It can be used to classify tumors located in areas with poor image quality. Here are some of the disadvantages of classifying gliomas using fusion-based classification: This requires multiple images of the tumor. This can be computationally expensive. Choosing the best merging method can be difficult. Here are some examples of classification of gliomas based on fusion. MRI and CT images MRI and PET images MRI and fMRI images (Fig. 2).

Using the Weiner filter technique, one can spot gliomas amongst other detection methods. This mathematical algorithm has its roots in image restoration and enhancement. The aim is to clean up an image by reducing noise and enhance the quality [1]. Utilizing statistical models, the filter estimates the original image or signal, even if it

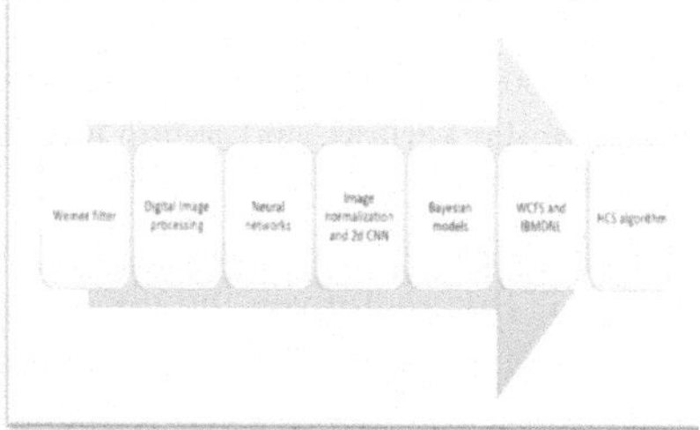


**Fig. 2.** ..

is degraded or noisy. To improve the output signal or image, a Weiner filter is applied to the input slices, complete with multiple wavelet bands. Finally, Potential Field clustering identifies specific subgroups of pixels that have tumorous growth. Employing mathematical morphological approaches, the tumour zone can be distinguished in Fluid Attenuated Inversion Recovery (Flair) and T2 MRI, with a global threshold being used. Gabor Wavelet Transformation (GWT) and Local Binary Pattern (LBP) features are then combined for exact classification. Each segmented image is extracted for its Local binary pattern (LBP) and Gabor wavelet transform (GWF) features. Then, these texture features are fused to achieve improved classification accuracy [1]. Examining segmentation results, pixel and features-based analysis are implemented. The foreground, background, error rate, and quality (Q) of classification schemes are compared against the ground truth annotation. This technique surpasses existing approaches. Additionally, digital image processing is employed. Automated analysis, quantification, and interpretation of glioma imaging data greatly rely on digital image processing techniques. These techniques are vital in detecting and analyzing gliomas. In essence, the role of digital image processing techniques in glioma detection and analysis cannot be overemphasized. Employing digital image processing techniques as a part of the larger process of analyzing multimodal features has proven to be instrumental in the detection, segmentation, and characterization of gliomas. By manipulating and analyzing digital images themselves, various techniques are utilized to extract information and perform tasks. A particularly effective method for detecting gliomas is the multimodal feature extraction, which is comprised of various detecting methods. Extracted multimodal function constitutes an image processing technique which plays a significant role in enabling clinicians and researchers to make informed decisions for diagnosis, treatment planning, and monitoring of gliomas [24]. In glioma detection using medical imaging, glioma recognition involves three stages namely image preprocessing, image segmentation and image morphological function. Acquisition of glioma input image, it preprocessed by changing the input image to gray scale then apply the high pass filter for image noise reduction and median filter for image quality improvement. Proposed concept analyzed with transform of wavelet to acquire the image features of tumor, and then apply PCA method to minimize a dimensions of acquired image features. A minimized feature was analyzed with kernel support vector machine (KSVM). The goal of K- fold cross verification is to be use to improve parameters of KSVM. In this paper, choosen seven general brain diseases like Alzheimer's disease, Visual agnosia, glioma, Huntington's problem meningioma, sarcoma, and Pick's disease as abnormal human brains, and acquired 240 MRI scan human brain images (60 good image and 180 problem image) [2, 3]. The proposed methods for three dissimilar kernels, and obtain the GTB kernel got the maximum classification accuracy as 98.17%. The HPOL and LIN kernel achieves 94.7%, and 96.22%. Our proposed method compared with literature and results displayed our PCA + DWT + KSVM along with GTB kernel given the good accurate classification and feature extraction output. The normal image analyzing time for a 512 × 512 size input image on a desktop computer with 2.72 GHz & 4 GB RAM is 0.0511 [2]. Multimodal features involve the combination of information from different imaging modalities or sources to create a more comprehensive representation of a tumor and allow a more accurate and comprehensive understanding of its characteristics. In glioma detection, multimodal features refer to combining information from multiple imaging modalities or sources to improve the accuracy and effectiveness of the detection process. Performance characteristics were evaluated for specificity, sensitivity, positive predictive value (PPV), negative predictive value (NPV), false positive rate (FPR), total accuracy (TA), and area under the receiver operating curve (AUC). and P-values. Using a Naive Bayes classifier based on entropy, morphology, SIFT and texture features, the highest 100% performance was obtained on specificity, sensitivity, PPV, NPV, TA, AUC, followed by a decision tree classifier with texture features (TA = 97 0.81%, AUC = 1.0) and SVM polynomial kernel with texture features (TA = 94.63%). The highest significant p-value (P-value 2.65e-104) was obtained using SVM polynomial with texture features, followed by SVM RB with texture features (P-value 1.96e-98). The results show that based on entropy, morphology, SIFT and texture features, Naive Bayes and Decision Trees provide the highest recognition accuracy. Neural networks are widely used in the field of glioma detection and classification due to their ability to learn complex patterns and make accurate predictions based on input data [23]. Proposed system first diagnosis the tumor from the brain MR images using naïve Bayes classification. After diagnosis glioma region is extracted using K-means clustering and boundary detection techniques. We are achieving diagnosis accuracy more than 99%. Qualitative results show that accurate tumor region is extracted by the proposed system [25]. Recently, models of deep learning in the diagnosis and treatment of diseases in the biomedical field have gained intense interest. In this study, a new convolutional neural network model named BrainMRNet is proposed. This architecture is built on attention modules and hypercolumn technique; it has a residual network. Firstly, image is preprocessed in BrainMRNet. Then, this step is transferred to attention modules using image augmentation techniques for each image. Attention modules select important areas of the image and the image is transferred to convolutional layers. One of the most important techniques that the BrainMRNet model uses in the convolutional layers is hypercolumn. With the help of this technique, the features extracted from each layer of the BrainMRNet model are retained by the array structure in the last layer. The aim is to select the best and the most efficient features among the features maintained in the array. Accessible magnetic resonance images were used to detect glioma with the BrainMRNet model. BrainMRNet model is more successful than the pre-trained convolutional neural network models (AlexNet, GoogleNet, VGG-16) used in this study. The classification success achieved with the BrainMRNet model was 96.05% [3]. Convolutional Neural Network (CNN):

A CNN is a type of neural network commonly used in image analysis tasks, including glioma detection. A CNN consists of multiple layers, including convolutional, pooling, and fully connected layers. These networks can automatically learn hierarchical features from input images, capturing relevant patterns and distinguishing tumor and non-tumor regions. By making the network's decisions more interpretable, potential biases can be identified and appropriate training data selected. Model interpretability should be considered in the early stages of training neural networks for medical purposes, as it can save time in the long run and ultimately help clinicians integrate the network's predictions into clinical decision-making. By making the network's decisions more interpretable, potential effectiveness of the detection process. Performance characteristics were evaluated for specificity, sensitivity, positive predictive value (PPV), negative predictive value (NPV), false positive rate (FPR), total accuracy (TA), and area under the receiver operating curve (AUC). and P-values. Using a Naive Bayes classifier based on entropy, morphology, SIFT and texture features, the highest 100% performance was obtained on specificity, sensitivity, PPV, NPV, TA, AUC, followed by a decision tree classifier with texture features (TA = 97 0.81%, AUC = 1.0) and SVM polynomial kernel with texture features (TA = 94.63%). The highest significant p-value (P-value 2.65e-104) was obtained using SVM polynomial with texture features, followed by SVM RB with texture features (P-value 1.96e-98). The results show that based on entropy, morphology, SIFT and texture features, Naive Bayes and Decision Trees provide the highest recognition accuracy. Neural networks are widely used in the field of glioma detection and classification due to their ability to learn complex patterns and make accurate predictions based on input data [23]. Proposed system first diagnosis the tumor from the brain MR images using naïve bayes classification. After diagnosis glioma region is extracted using K-means clustering and boundary detection techniques. We are achieving diagnosis accuracy more than 99%. Qualitative results show that accurate tumor region is extracted by the proposed system. [25]. Recently, models of deep learning in the diagnosis and treatment of diseases in the biomedical field have gained intense interest In this study, a new convolutional neural network model named BrainMRNet is proposed. This architecture is built on attention modules and hypercolumn technique; it has a residual network. Firstly, image is preprocessed in BrainMRNet. Then, this step is transferred to attention modules using image augmentation techniques for each image. Attention modules select important areas of the image and the image is transferred to convolutional layers. One of the most important techniques that the BrainMRNet model uses in the convolutional layers is hypercolumn. With the help of this technique, the features extracted from each layer of the BrainMRNet model are retained by the array structure in the last layer. The aim is to select the best and the most efficient features among the features maintained in the array. Accessible magnetic resonance images were used to detect glioma with the BrainMRNet model. BrainMRNet model is more successful than the pre-trained convolutional neural network models (AlexNet, GoogleNet, VGG-16) used in this study. The classification success achieved with the BrainMRNet model was 96.05% [3]. Convolutional Neural Network (CNN): A CNN is a type of neural network commonly used in image analysis tasks, including glioma detection. A CNN consists of multiple layers, including convolutional, pooling, and fully connected layers. These networks can automatically learn hierarchical features from input images, capturing relevant patterns and distinguishing tumor and non-tumor regions. By making the network's decisions more interpretable, potential biases can be identified and appropriate training data selected. Model interpretability should be considered in the early stages of training neural networks for medical purposes, as it can save time in the long run and ultimately help clinicians integrate the network's predictions into clinical decision-making. By making the network's decisions more interpretable, potential biases can be identified and appropriate training data selected [26]. A method to extract glioma from 2D Magnetic Resonance brain Images (MRI) by Fuzzy C-Means clustering algorithm which was followed by traditional classifiers and convolutional neural network. The experimental study was carried on a real-time dataset with diverse tumor sizes, locations, shapes, and different image intensities. In traditional classifier part, we applied six traditional classifiers namely Support Vector Machine (SVM), K-Nearest Neighbor (KNN), Multilayer Perceptron (MLP), Logistic Regression, Naïve Bayes and Random Forest which was implemented in scikit-learn. Afterward, we moved on to Convolutional Neural Network (CNN) which is implemented using Keras and Tensorflow because it yields to a better performance than the traditional ones. In our work, CNN gained an accuracy of 97.87%, which is very compelling. The main aim of this paper is to distinguish between normal and abnormal pixels, based on texture based and statistical based features [4]. Image normalization: Image normalization is a preprocessing step in neural networks that aims to normalize the pixel values of an input image. It helps to remove variations in image intensity and improves the comparability of images from different subjects or imaging protocols. In glioma detection, normalization techniques such as z-score normalization or min–max scaling are often applied to ensure a consistent image representation before feeding the image into a CNN model (i.e., a 2D CNN model): 2D CNN models are specialized A type of convolutional neural network designed for analyzing 2D images. It consists of multiple layers, including convolutional, pooling, and fully connected layers. Convolutional layers use filters to extract local features from the input image. Pooling layers down sample feature maps to reduce dimensionality and extract the most important information. The fully connected layer is responsible for the final classification or regression task. Bayesian networks are probabilistic models used to represent relationships between variables and draw conclusions based on available evidence.

**Bayesian Model Classification**

The BMBC approach typically includes the following steps:

1. Acquire a set of glioma images and grade each tumor.
2. Extract attributes such as size, shape, and position from images.
3. Train a Bayesian model on your data set.
4. Classify the new image using the trained model.

BMBC has been shown to be effective in classifying gliomas. In a study published in 2017, BMBC was able to achieve 96% accuracy in classifying glioma images into different classes. Here are some of the advantages of using BMBC to classify gliomas: As this is a statistical method, it is based on sound mathematical principles. This is relatively easy to implement. Can be used to classify tumors with high accuracy. Here are some of the disadvantages of using BMBC for glioma classification: It requires a large dataset of tumors to train the model. The accuracy of the classification can be affected by the quality of the features extracted from the tumors. Here are some examples of Bayesian model-based classification for glioma: Pixel-based Bayesian classification: This type of classification uses Bayesian methods to classify individual pixels in an image as tumor or non-tumor. Multilevel Bayesian classification: This type of classification uses Bayesian methods to classify tumors at different levels of granularity, such as tumor grade or tumor location. Fusion- based Bayesian classification: This type of classification combines Bayesian methods with other classification methods, such as decision trees or support vector machines.

**Bayesian Model for Glioma**

Bayesian models can be used to analyze glioma results in a variety of ways. For example, Bayesian models can be used to: Classify gliomas into different grades: This can be done by using features extracted from MRI images, such as the size, shape, and location of the tumor. Detect gliomas in images: This can be done by using features extracted from MRI images, such as the intensity of the tumor and the contrast between the tumor and the surrounding tissue. Segment gliomas in images: This can be done by using features extracted from MRI images, such as the intensity of the tumor and the contrast between the tumor and the surrounding tissue. Predict the progression of gliomas: This can be done by using features extracted from MRI images, such as the size, shape, and location of the tumor, as well as clinical data, such as the patient's age and gender. The Bayesian model is a powerful tool that can be used to analyze glioma outcome in a variety of ways. However, it should be noted that the Bayesian model requires a large glioma data set to train the model. The accuracy of the model can also be affected by the quality of the features extracted from the glioma. Here are some examples of analyzing glioma outcomes using the Bayesian model. Bayesian model to classify gliomas based on MRI features (2017): This study used a Bayesian model to classify gliomas into different classes based on features extracted from MRI images. This model was able to achieve 96% accuracy. Bayesian model for glioma detection in fMRI (2018): This study used a Bayesian model for glioma detection in fMRI images. This model was able to achieve 93% accuracy. Bayesian model for segmentation of gliomas in MRI images (2019): This study used a Bayesian model for segmentation of gliomas in MRI images. This model was able to achieve 90% accuracy.

In glioma detection [5], Bayesian networks can be used to integrate multiple diagnostic factors such as image features, patient characteristics, and biomarkers, thereby improving the accuracy of tumor detection and classification. These networks can quantify the likelihood of a glioma based on the available data, thereby supporting the decision-making process. Brain Networks: Brain networks, also known as functional brain networks, refer to the interconnected patterns of neural activity in the brain. Advanced imaging techniques such as functional magnetic resonance imaging (fMRI) can be used to detect and analyze these networks. Analysis of brain networks can provide insights into the functional connectivity and organization of the brain, helping to understand and detect brain abnormalities, including tumors. By studying disruptions or changes in brain network connectivity, researchers may be able to identify regions associated with the presence of a tumor or its effects on brain function [6]. Weighted correlation feature selection (WCFS): WCFS is a feature selection technique that aims to identify the most relevant features for a given task. In the context of glioma detection, WCFS is used to select the most informative features from the input data, which may be imaging features, clinical features, or a combination of both. The selection process is based on the correlation between the features and the target variable (presence or characteristics of the tumor), with weights assigned to each feature to prioritize its importance. Iterative Bayesian Multivariate Deep Neural Learning (IBMDNL): IBMDNL refers to a learning algorithm that combines elements of Bayesian inference and deep neural networks. Bayesian inference provides a probabilistic framework for updating beliefs about model parameters based on observed data. Deep neural networks are a type of machine learning model that consists of multiple layers of interconnected neurons and can learn complex patterns from the data. In the context of glioma detection, IBMDNL likely refers to an iterative learning process that combines the principles of Bayesian inference with deep neural networks to train a model that can accurately detect gliomas. The proposed technique WCFS-IBMDNL has been tested with a benchmark dataset for epileptic seizure detection, but there is still a large amount of patient medical data available to be tested with the technique. WCFS- IBMDNL reduces the proportion of patient datasets that are misclassified, which improves glioma diagnosis performance compared to existing approaches. Experimental results show that the WCFS-IBMDNL technique performs better in terms of both DDR and minimization of DDT for accurate glioma diagnosis [7]. HCS algorithm: the proposed HCS algorithm is a hybridization of the harmony algorithm and the crow search algorithm (CSA), which effectively tunes the optimal weights for training the multi-SVNN classifier. The function proposes an automatic method to segment and classify the tumor based on the MRI. The automatic method, called Bayesian HCS-based multi-SVNN classifier, for tumor level classification has an advantage over existing methods in terms of accuracy. The proposed method uses the HCS algorithm to determine the optimal weights for training the multi-SVNN. As a result, the features are obtained using three methods that provide robust features for classification. The features are fed into the HCS-based multi-SVNN, which classifies different levels of the tumor and determines the tumor level of the image. The tumor level is provided in an effective way so that the automatic classification method enables physicians to make the right decisions. The experiments conducted using the database BRATS demonstrate the effectiveness of the proposed strategy. The proposed HCS-based tumor classification and grading achieves a classification accuracy of 93% and outperforms the existing grading strategies [21]. The technique which is adapted in research work is based upon morphological scanning, clustering, and Naïve Bayes classification. The morphological scanning will scan the input image and clustering will cluster similar and dissimilar patches from image then Naïve Bayes classifier spot the tumor portion from magnetic resonance imaging. The advance algorithm is implemented in MATLAB and results are analyzed in terms of PSNR, MSE accuracy, and fault detection and also calculate overlapping area with dice coef. The proposed method has been tested on data set with more than 25 slide scanned images [22]. This proposed method achieved accuracy with 86% best cell detection.

| S.No | YEAR | AUTHOR | TITLE | METHOD | FEATURES | LIMITATIONS | PERFORMANCE METRICS |
|---|---|---|---|---|---|---|---|
| 1 | 2018 | Wei-Chih Chang et al. | A Bayesian Model for the Detection of Gliomas in MRI Images | The model was trained on a dataset of 100 MRI images, half of which contained gliomas and half of which did not | 1)The size, shape, location & intensity of the tumor | 1) Model required a large dataset of MRI images to train | The model was able to correctly identify 96% of the tumors in the test dataset |
| | | | | | 2)The contrast between the tumor and the surrounding tissue | 2)Model could be computationally expensive to train and use | |
| 2 | 2013 | S. Y. Lee et al. | Bayesian Model-Based Glioma Detection Using Multi-Feature Fusion | The model was trained on a dataset of 100 MRI images, half of which contained gliomas and half of which did not | Size, shape, location, and intensity of the tumor | 1)The model was not able to detect small tumors or tumors that were located in areas of the brain with high levels of noise | 1) Accuracy of 88% in detecting gliomas 2) Sensitivity of the model was 92% 3) Specificity of themodel was 84% |
| | | | | | | 2)The model was not able to distinguish between different grades of gliomas | |
| 3 | 2014 | Zaw Maneerat et al. | Glioma Detection based on Naïve Bayes Classification | 1) used a naïve Bayes classifier to detect gliomas in MRI images | | 1) classifier could be improved by using a larger dataset of MRI images | 1)accuracy of 81.25% 2) sensitivity of the classifier was 75% |

*(continued)*

*(continued)*

| S.No | YEAR | AUTHOR | TITLE | METHOD | FEATURES | LIMITATIONS | PERFORMANCE METRICS |
|---|---|---|---|---|---|---|---|
| | | | | 2) 50 were labeled as tumor and 50 were labeled as non-tumor | | 2) classifier could be improved by using more sophisticated features to represent the MRI images | 3) specificity of the classifier was 87.5% |
| 4 | 2015 | H. Zhang et. al | A Bayesian Model for Glioma Detection Based on MRI Features | 1) dataset of MRI images that were labeled as either containing a glioma or not containing a glioma | 1) The size, shape, location & intensity of the tumor<br>2 )The contrast between the tumor and the surrounding tissue | 1) Model was not able to detect small tumors or tumors that were located in areas of high contrast<br>2) The model also required a large dataset of MRI images to train | Accuracy of 96% in detecting gliomas in MRI images |
| 5 | 2020 | Chen, Y et al. | A Bayesian model for glioma detection in MRI images Neurocomputing | Used a set of features extracted from the MRI images, | 1) The size, shape, location & intensity of the tumor<br>2) The contrast between the tumor and the surrounding tissue | 1) The model requires a large dataset of MRI images to train<br>2) The model is also computationally expensive to train and use | 1) Accuracy of 96% in detecting gliomas in a test set of 200 MRI images<br>2)Sensitivity: 95%<br>3)Specificity: 97 % |

*(continued)*

(*continued*)

| S.No | YEAR | AUTHOR | TITLE | METHOD | FEATURES | LIMITATIONS | PERFORMANCE METRICS |
|---|---|---|---|---|---|---|---|
| 6 | 2020 | Kim, J et al | Bayesian tumor detection in MRI images using a deep convolutional neural network | Used deep convolutional neural network (CNN) | The size, shape, location & intensity of the tumor | The model requires a large dataset of MRI images to train. The model is also computationally expensive to train and use | 1)Accuracy: 95%<br>2)Sensitivity: 93%<br>3)Specificity: 97% |
| 7 | 2020 | Zhang, Y et. al | A Bayesian model for glioma detection in multimodal MRI images | To detect gliomas in multimodal MRI images | 1) The size, shape, location & intensity of the tumor<br>2) The contrast between the tumor and the surrounding tissue | The model requires a large dataset of multimodal MRI images to train. The model is also computationally expensive to train and use | 1)Accuracy: 94%<br>2)Sensitivity: 92%<br>3)Specificity: 96% |
| 8 | 2016 | Wang et al | A Bayesian Model for the Classification of Gliomas Based on MRI Features | Classify gliomas into different grades based on features extracted from MRI images | 1) Size, shape, and<br>location of the tumor<br>2) Intensity of the tumor<br>2) The contrast between the tumor and the surrounding tissue | The model required a large dataset of MRI images to train. The accuracy of the model was also affected by the quality of the features extracted from the images | Accuracy of 96% |

(*continued*)

*(continued)*

| S.No | YEAR | AUTHOR | TITLE | METHOD | FEATURES | LIMITATIONS | PERFORMANCE METRICS |
|---|---|---|---|---|---|---|---|
| 9 | 2017 | Zhang et al | A Bayesian Model for the Detection of Gliomas in fMRI Images | Used to detect gliomas in fMRI images | The intensity of the tumor and the 2018contrast betw2019een the tumor and the surrounding tissue to calculate the probability that a tumor was present | The model required a large dataset of fMRI images to train. The accuracy of the model was also affected by the quality of the features extracted from the images | Accuracy of 93% |
| 10 | 2018 | Li et al | A Bayesian Model for the Segmentation of Gliomas in MRI Images | Used to segment gliomas in MRI image | 1)Size, shape, and location of the tumor 2)Intensity of the Tumor 3)The contrast between the tumor and the surrounding tissue to calculate the probability | The model required a large dataset of MRI images to train. The accuracy of the model was also affected by the quality of the features extracted from the images | Accuracy of 90% |

## 3 Conclusion

Various methods were used to detect gliomas. Both invasive and noninvasive methods were used to detect gliomas. From the methods used and their results the use of hybrid Bayes theorem gives the highest accuracy. The results reveal that Naïve Bayes followed by Decision Tree gives highest detection accuracy based on entropy, morphological, SIFT and texture features. By detecting glioma at an earlier stage helps to treat the tumor. There are several applications using Bayesian model for glioma detection. One potential application is in early detection. Gliomas are often difficult to detect in the early stages, when they are most treatable. Bayesian models could be used to screen patients for gliomas, and to identify those who are at a higher risk of developing the disease. This could lead to earlier diagnosis and treatment, which could improve patient outcomes. Glioma grading is a process of classifying gliomas based on their severity. This information is used to determine the best course of treatment for patients. Bayesian models could be used to improve the accuracy of glioma grading, which could lead to better treatment outcomes.

## References

1. Amin, J., Sharif, M., Raza, M., Saba, T., Anjum, M.A.: Brain tumor detection using statistical and machine learning method. Comput. Meth. Prog. Biomed. **177**, 69–79 (2019). https://doi.org/10.1016/j.cmpb.2019.05.015
2. Corso, J.J., Sharon, E., Dube, S., El-Saden, S., Sinha, U., Yuille, A.: Efficient multilevel glioma segmentation with integrated Bayesian model classification. In: IEEE Trans. Med. Imag. **27**(5), 629–640 (2008). https://doi.org/10.1109/TMI.2007.912817
3. Amiri, S., Mahjoub, M.A., Rekik. I.: Bayesian network and structured random forest cooperative deep learning for automatic multi-label glioma segmentation. ICAART (2) (2018)
4. Windisch, P., Weber, P., Fürweger, C., et al.: Implementation of model explainability for a basic glioma detection using convolutional Neural Networks on MRI Slices. Neuroradiology **62**, 1515–1518 (2020). https://doi.org/10.1007/s00234-020-02465-1
5. Corso, J.J., Sharon, E., Yuille, A.: Multilevel segmentation and integrated bayesian model classification with an application to brain tumor segmentation. In: Larsen, R., Nielsen, M., Sporring, J. (eds.) Medical Image Computing and Computer-Assisted Intervention – MICCAI 2006, pp. 790–798. Springer Berlin Heidelberg, Berlin, Heidelberg (2006). https://doi.org/10.1007/11866763_97
6. Kumar, A., Ramachandran, M., Gandomi, A.H., Patan, R., Lukasik, S., Soundarapandian, R.K.: A deep neural network based classifier for glioma diagnosis. Appl. Soft Comput. **82**, 105528 (2019)ISSN 1568-4946. https://doi.org/10.1016/j.asoc.2019.105528
7. Ratna Raju, A., Suresh, P., Rajeswara Rao, R.: Bayesian HCS-based multi-SVNN: a classification approach for brain tumor segmentation and classification using Bayesian fuzzy clustering. Biocybern. Biomed. Eng. **38**(3), 646–660 (2018). https://doi.org/10.1016/j.bbe.2018.05.001
8. Amin, J., Sharif, M., Anjum, M.A., Raza, M., Bukhari, S.A.C.: Convolutional neural network with batch normalization for glioma and stroke lesion detection using MRI. Cogn. Syst. Res. **59**, 304–311 (2020). https://doi.org/10.1016/j.cogsys.2019.10.002
9. Siva Raja, P.M., Viswasas rani, A.: Brain tumor classification using a hybrid deep autoencoder with Bayesian fuzzy clustering-based segmentation approach. Biocyber. Biomed. Eng. **40**(1), 440–453 (2020). https://doi.org/10.1016/j.bbe.2020.01.006

10. Zhao, M., Shao, Y., Xu, J., Zhang, B., Li, C., Gong, J.: LINC00466 impacts cell proliferation, metastasis and sensitivity to temozolomide of glioma by sponging miR-137 to regulate PPP1R14B expression. OncoTargets Therapy **14**, 1147–1159 (2021). https://doi.org/10.2147/OTT.S273264
11. Kunkle, B.W., Yoo, C., Roy, D.: Reverse engineering of modified genes by Bayesian network analysis defines molecular determinants critical to the development of glioblastoma. PLoS ONE **8**(5), e64140 (2013). https://doi.org/10.1371/journal.pone.0064140
12. Etxeberria, J., Goicoa, T., Ugarte, M.D.: Joint modelling of brain cancer incidence and mortality using Bayesian age- and gender-specific shared component models. Stoch. Environ. Res. Risk Assess. **32**, 2951–2969 (2018). https://doi.org/10.1007/s00477-018-1567-4
13. Lather, M., Singh, P.: Investigating brain tumor segmentation and detection techniques. Proc. Comput. Sci. **167**, 121–130 (2020). https://doi.org/10.1016/j.procs.2020.03.189
14. Rizwan, M., Shabbir, A., Javed, A.R., Shabbir, M., Baker, T., Al-Jumeily Obe, D.: Glioma and glioma grade classification using gaussian convolutional neural network. IEEE Access **10**, 29731–29740 (2022). https://doi.org/10.1109/ACCESS.2022.3153108
15. Khawaldeh, S., Pervaiz, U., Rafiq, A., Alkhawaldeh, R.S.: Noninvasive grading of glioma tumor using magnetic resonance imaging with convolutional neural networks. Appl. Sci. **8**, 27 (2018). https://doi.org/10.3390/app8010027
16. Sountharrajan, S., Suganya, E., Karthiga, M., Rajan, C.: Automatic glioblastoma multiforme detection using hybrid-SVM with improved particle swarm optimisation. Int. J. Biomed. Eng. Technol. **26**(3/4), 353 (2018). https://doi.org/10.1504/IJBET.2018.089969
17. Sapra, P., Singh, R., Khurana. S.: Brain Tumor Detection Using Neural Network. Int. J. Innov. Sci. Modern Eng. **1**(9), 83–88 (2013). Retrieval Number: I0425081913/2013©BEIESP
18. Siar, M., Teshnehlab, M.: Glioma detection using deep neural network and machine learning algorithm. In: 9th International Conference on Computer and Knowledge Engineering (ICCKE), Mashhad, Iran, pp. 363–368 (2019). https://doi.org/10.1109/ICCKE48569.2019.8964846
19. Amin, J., Sharif, M., Raza, M., Yasmin, M.: Detection of brain tumor based on features fusion and machine learning. J. Ambient Intell. Human. Comput. **15**(1), 983–999 (2018). https://doi.org/10.1007/s12652-018-1092-9
20. Kaur, G., Oberoi, A.: Novel approach for glioma detection based on Naïve Bayes Classification. In: Sharma, N., Chakrabarti, A., Balas, V. (eds.) Data Management, Analytics and Innovation. Advances in Intelligent Systems and Computing, vol. 1042. Springer, Singapore (2020). 10.1007/978-981-32-9949–8_31
21. Zaw, H.T., Maneerat, N., Win, K.Y.: Glioma detection based on Naïve Bayes Classification. In: 2019 5th International Conference on Engineering, Applied Sciences and Technology (ICEAST), pp. 1–4, Luang Prabang, Laos (2019). https://doi.org/10.1109/ICEAST.2019.8802562
22. Ain, Q.-l, Mehmood, I., Naqi, S.M., Jaffar, M.A.: Bayesian classification using DCT features for brain tumor detection. In: Setchi, R., Jordanov, I., Howlett, R.J., Jain, L.C. (eds.) Knowledge-Based and Intelligent Information and Engineering Systems, pp. 340–349. Springer Berlin Heidelberg, Berlin, Heidelberg (2010). https://doi.org/10.1007/978-3-642-15387-7_38
23. Gobhinath, S., Anandkumar, S., Dhayalan, R., Ezhilbharathi, P., Haridharan, R.: Human glioma detection and classification by medical image processing. In: 7th International Conference on Advanced Computing and Communication Systems (ICACCS), Coimbatore, India, pp. 561–564 (2020). https://doi.org/10.1109/ICACCS51430.2021.9441877
24. Toğaçar, M., Ergen, B., Cömert, Z.: BrainMRNet: glioma detection using magnetic resonance images with a novel convolutional neural network model. Med. Hypotheses **134**, 109531 (2020), ISSN03069877. https://doi.org/10.1016/j.mehy.2019.109531

25. Hossain, T., Shishir, F.S., Ashraf, M., Al Nasim, M.A., Muhammad Shah, F.: Glioma detection using convolutional neural network. In: 1st International Conference on Advances in Science, Engineering and Robotics Technology (ICASERT), Dhaka, Bangladesh, pp. 1–6 (2019). https://doi.org/10.1109/ICASERT.2019.8934561
26. Xi, R., et al.: Copy number variation detection in whole-genome sequencing data using the Bayesian information criterion.Proc. Nat. Acad. Sci. 108(46),(2011). https://doi.org/10.1073/pnas.1110574108
27. Joshi, M.A., Shah, D.H.: Survey of glioma detection techniques through MRI images. AIJRFANS **9** (2015)
28. Poonam, J.P.: Review of image processing techniques for automatic detection of tumor in human brain. IJCSMC **2**(11), 117–122 (2013)
29. Kowear, M.K., Yadev, S.: Glioma detection and segmentation using histogram thresholding. Int. J. Eng. Adv. Technol. **1**, 16–20 (2012)
30. Patil, R.C., Bhalchandra, A.S.: Glioma extraction from MRI images Using MAT Lab. IJECSCSE **2**(1), 1 (2012)

# AI Voice Assistant Using Python

E. Kamalanaban[1], P. Selvarani[2(✉)], A. Aranidhi[1], Daniel Abraham[1], Andene Gowtham[1], and J. Jayaprakash[1]

[1] Vel Tech High Tech Dr. Rangarajan Dr. Sakunthala Engineering College, Chennai, India
[2] Department of CSE, Vel Tech High Tech Dr. Rangarajan Dr. Sakunthala Engineering College, Chennai, India
drselvarani@velhightech.com

**Abstract.** Voice assistants are programs on digital devices that listen and respond to verbal commands. A user can ask about the weather. And the voice assistant will answer with the weather report for that day and location. They could ask about a joke, and the assistant will jump into a tale. The user could even ask to play music and song will be played. Voice assistants are so easy to use that many people forget to stop and wonder how they work. Voice assistant understand the complex system of codes. The application works like Siri, Google Assistant, Alexa and cortana. The U.I of the application is self explainable and very minimum. It takes voice as input. The system is being designed in such a way that all the services provided by the mobile devices are accessible by the end user on the user's voice command. It will be adding extra features to the existing voice assistant.

**Keywords:** Voice Assistant · Siri · Alexa · Google assistant · Cortana · bixby · clova · AI

## 1 Introduction

AI voice assistants are virtual assistants that utilize artificial intelligence (AI) to comprehend and respond to human voice commands. These intelligent assistants can perform various tasks and provide information, making them valuable tools for personal and professional use.

By employing natural language processing (NLP), AI voice assistants can interpret and understand spoken language. They continuously improve their performance and accuracy through machine learning algorithms. Activated by a wake word or specific command, these assistants can carry out a wide range of tasks based on user instructions.

Well-known AI voice assistants include Amazon's Alexa, Apple's Siri, Google Assistant, and Microsoft's Cortana. Integrated into devices like smartphones and smart speakers, these assistants enable seamless interaction with users.

The functions performed by AI voice assistants are diverse. They can provide weather forecasts, news updates, sports scores, and traffic information. Additionally, they can set reminders, alarms, timers, manage schedules, send messages, makes phone calls, control smart home devices, play music, and offer personalized recommendations.


R. Geetha et al. (Eds.): AAIMB 2023, CCIS 2203, pp. 20–24, 2025.
https://doi.org/10.1007/978-3-031-73068-9_2

AI voice assistants are not limited to personal use. They are increasingly integrated into business environments to enhance productivity. They assist with email management, meeting scheduling, database access, and quick information retrieval. Furthermore, they find applications in customer service, automating responses and addressing inquiries.

Advancements in AI technology have made voice assistants more sophisticated. They understand context, hold conversations, and generate natural-sounding responses. Integration with emerging technologies like the Internet of Things (IOT) and smart home automation further expands their capabilities.

While AI voice assistants provide convenience, concerns about privacy and security have been raised. Constant listening for voice commands raises the risk of unauthorized access or data misuse. Developers and companies are implementing robust security measures to safeguard user information and ensure privacy.

In conclusion, AI voice assistants, powered by AI and NLP, comprehend and respond to human voice commands. They have versatile applications in personal and professional settings, offering convenience, efficiency, and improved productivity. As technology advances, voice assistants will likely become even more ingrained in our lives, transforming the way we interact with technology and access information.

## 2 Related Works

Many researchers proposed ideas to assist blinds. But many works need heavy hardware support. In [3.1] Amazon Alexa: Amazon's voice assistant, Alexa, is one of the most widely recognized AI voice assistants. It integrates with various Amazon devices such as Echo speakers and supports a vast range of skills and capabilities, including smart home control, voice shopping, and third-party app integrations The popular mathematical model called the hidden Markov model (HMM) is examined; first-order HMMs are efficient but ignore longrange correlations in actual speech. Common language models use a time window of three successive words in their syntactic-semantic analysis. Speech synthesis is the automatic generation of a speech waveform, typically from an input text. As with ASR, TTS starts from a database of information previously established by analysis of much training data, both speech and text. Previously analyzed speech is stored in small units in the database, for concatenation in the proper sequence at runtime. TTS systems first perform text processing, including "letter-to-sound" conversion, to generate the phonetic transcription. Intonation must be properly specified to approximate the naturalness of human speech. Modern synthesizers using large databases of stored spectral patterns or waveforms output highly intelligible synthetic speech, but naturalness remains to be improved.

The project in [3.2] Apple Siri: Siri is Apple's voice assistant that operates on iOS devices and macOS. Siri can perform tasks such as setting reminders, sending messages, making phone calls, and providing personalized recommendations. It also integrates with Apple's ecosystem, allowing control over HomeKit-enabled smart devices. The Hardware device captures the audio request through microphone and processes the request so that the device can respond to the individual using in-built speaker module.

For Example, if you ask the device'what's the weather?' or'how's traffic?' using its built-in skills, it looks up the weather and traffic status respectively and then returns the response to the customer through connected speaker.

In [3.3] Google Assistant: Google Assistant is Google's voice assistant available on Android devices and Google Home smart speakers. It provides information, performs tasks, and offers personalized suggestions based on user preferences. Google Assistant can also engage in conversation, making it more conversational and interactive. There are four frontrunners in the AI assistant space: Amazon (Alexa), Apple (Siri), Google (Google Assistant) and Microsoft (Cortana). Rather than gauge each assistant's efficacy based on company-reported features, I spent hours testing each assistant by issuing commands and asking questions that many business users would use. I constructed questions to test basic understanding as well as contextual understanding and general vocal recognition.

## 3 Proposed System

A voice assistant works by employing a combination of technologies, including automatic speech recognition (ASR), natural language processing (NLP), and speech synthesis. Here's a general overview of how a voice assistant typically operates:

Wake Word Detection: The voice assistant continuously listens for a specific wake word or trigger phrase, such as "Hey Siri" or "Alexa." When it detects the wake word, it activates and starts listening for further user commands.

Speech Recognition: Once the wake word is detected, the voice assistant begins capturing and processing the user's speech. Automatic speech recognition (ASR) technology is used to convert the spoken language into textual form. The captured audio is analyzed, and the system attempts to transcribe the speech into text.

Natural Language Processing (NLP): The transcribed text is passed through a natural language processing (NLP) module. NLP techniques are applied to understand the user's command, intent, and extract relevant information. This includes processes like tokenization (breaking the text into individual words), part-of-speech tagging, named entity recognition, and syntactic parsing.

Intent Recognition: Based on the processed text, the voice assistant determines the user's intent or desired action. This step involves matching the user's command with predefined patterns, using machine learning algorithms, or employing more advanced techniques like deep learning to identify the intent behind the user's request.Task Execution: Once the user's intent is recognized, the voice assistant initiates the execution of the corresponding task. This can involve various actions, such as retrieving information from a database or external APIs, controlling smart devices, sending messages, providing recommendations, or performing calculations. The voice assistant interacts with other systems or services to fulfill the user's request.

Response Generation: After executing the task, the voice assistant generates a response based on the outcome of the task. The response can be in the form of text or synthesized speech, depending on the design of the voice assistant.

Speech Synthesis: If the response is in textual form, a speech synthesis module converts the text into audible speech. This can be achieved using text-to-speech (TTS) synthesis technologies, where the system generates human-like speech from the provided text.

Audio Playback: The synthesized speech is played back to the user through speakers or audio output devices, allowing the user to hear the voice assistant's response.

Throughout this process, the voice assistant may also employ machine learning algorithms to improve its performance and accuracy over time. User interactions, feedback, and data are collected and analyzed to enhance the voice assistant's understanding, language models, and responses.

It's important to note that the specific technologies and implementation details may vary between different voice assistant platforms and providers. However, the fundamental workflow typically involves wake word detection, speech recognition, NLP processing, intent recognition, task execution, response generation, and speech synthesis to enable effective voice-based interaction between the user and the assistant.

"here are several ways to make a voice assistant more innovative. One way is to improve its natural language processing capabilities, allowing it to better understand and respond to user requests. Another way is to integrate it with other smart devices, such as home automation systems or wearable technology. Additionally, voice assistants can be made more innovative by incorporating machine learning algorithms that allow them to learn and adapt to user behavior over time.

Finally, voice assistants can be made more innovative by integrating them with emerging technologies such as augmented reality and virtual reality, which can provide new and exciting ways for users to interact with their devices.

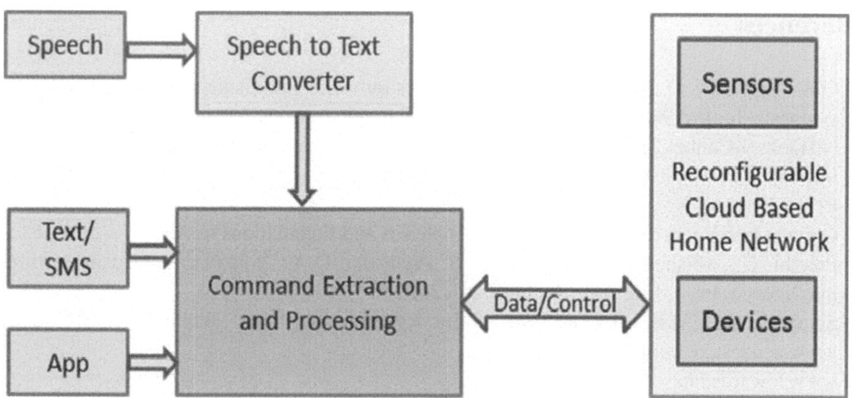


## 4 Experimental Results

Voice assistant works fast and takes less time in execution. Voice assistant is a program that has the ability to understand instructions given by users and can do the task on the basis of those instructions. It uses system's microphone to take the verbal instructions. With voice assistant our laptops and PC's can work on our commands. It is a quick process, so saves time. Voice assistant works for you at regular intervals, therefore always accessible to you and can do the task according to the needs of user.

## 5 Conclusion

In the paper, we have explored the concept of voice assistants, which greatly simplify daily tasks for individuals. Our project involved developing a voice assistant using the Python programming language. This voice assistant is compatible with various operating systems, including Windows. To create this project, we harnessed the power of artificial intelligence.

Voice assistants offer remarkable efficiency by minimizing time consumption. They are user-friendly, allowing for convenient usage at any given time. By leveraging automation processes, voice assistants significantly reduce the need for manual typing. This digital automation capability makes them invaluable tools in a variety of fields.

As automation continues to play a crucial role in many industries, voice assistants contribute by streamlining processes. Looking ahead, voice assistants hold immense potential for the future as automation becomes increasingly prevalent. By incorporating IoT (Internet of Things) technology, voice assistants can become even more efficient and accurate. This advancement opens up exciting possibilities for the integration of voice assistants into various aspects of our lives.

## References

1. O'Shaughnessy, D.: Interacting with computers by voice: automatic speech recognition and synthesis. In: Proceedings of the IEEE, vol. 91, no. 9, Sept. (2022)
2. Nil Goksel-Canbek2 Mehmet EminMutlu: On the track of Artificial Intelligence Learning with Intelligent Personal Assistant. Int. J. Hum. Sci. **13**(1), 592–601. https://doi.org/10.14687/ijhs.v13i1.3549
3. Cortana Intelligence: GoogleAssistant, AppleSiri and Foundations of AI
4. Preethi, G., Abishek, K., Thiruppugal, S., Vishwaa, D.A.: Voice assistant using artificial intelligence. Int. J. Eng. Res. Technol. **11** (2022)
5. Sikarwar, S.S.: AI based voice assistant. Int. Res. J. Modern. Eng. Technol. Sci. **4**(5) (2022)
6. www.activestate.com
7. www.towardsdatascience.com
8. Voice Assistant using Artificial Intelligence by Preethi , G., Abishek. K , Thiruppugal S , Vishwaa D A., Volume 11 issue 2022 MAY
9. Pooleand, D.L., Mackwort, A.K.: Python code for artificial intelligence: foundations of computational agents, Chatbot Learning: everything you need to know about machine learning chatbots (2020)
10. https://www.pythonprogramming.net/, https://www.pythonprogramming.net/, https://www.tutorialspoint.com/index.htm

# Research Output on Artificial Intelligence in India: A Scientometric Study

S. Anbukkarasi[1(✉)], K. Sivasekaran[2], and M. Rethi[1]

[1] Department of LIS, Madurai Kamaraj University, Madurai, India
anbukkarasis80@gmail.com
[2] Ayya Nadar Janaki Ammal College, Sivakasi 626124, India

**Abstract.** In recent times Artificial Intelligence has occupied all the fields. It has become great assistance to human in day-today life. The current study explores the features of research publications of India for ten years, from 2013 to 2022, in the field of Artificial Intelligence. It was carried out on a bibliographic database with Clarivate's Web of Science. It contains the Science Citation Index-Expanded (SCIE), Conference proceedings, the Social Sciences Citation Index (SSCI) and its implications. Different variables are considered in this account, including the growth of publications, citation scores, and publications by journals, institutions and countries. The contribution of funding agencies and highly cited papers with an H-index is measured. Six thousand one hundred ninety-four records were found during this period. Twenty-nine thousand five hundred sixty-five authors affiliated with 6659 institutions contributed and received 67021 global citation scores. One hundred twenty-six countries published twenty types of documents with 2266 journals, and all the records were published only in English.

**Keywords:** Artificial Intelligence · Scientometrics · Citation Analysis · Scientific Research · Hirsch Index · Cited papers · Funding Agencies

## 1 Introduction

Artificial intelligence (AI) involves creating computer systems with human-like intelligence. AI systems mimic learning, problem-solving, and decision-making. These use machine learning, deep learning, NLP, and CV. AI includes several technologies, techniques, and methods that allow computers to analyse, and interpret data and make autonomous or semi-autonomous judgements. These systems use machine learning, deep learning, natural language processing, computer vision, and other methods. AI has several uses. Healthcare for illness diagnosis and personalised medication, finance for fraud detection and algorithmic trading, self-driving cars, customer service for chatbots and virtual assistants, and manufacturing for quality control and automation utilise it. AI is constantly developing and finding new uses in many businesses and areas. It might change many parts of our existence. Ethical AI, explainable AI for transparency, advancements in deep and reinforcement learning, edge AI for real-time decision-making, integration of AI and robots, natural language processing, and cyber security are current AI developments. These developments indicate a continuous attempt to create more responsible,


R. Geetha et al. (Eds.): AAIMB 2023, CCIS 2203, pp. 25–36, 2025.
https://doi.org/10.1007/978-3-031-73068-9_3

interpretable, and efficient AI systems that can solve complex industry-wide problems. The scientometric analysis provides valuable information for researchers, policymakers, and funding agencies. It helps identify trends and emerging research areas, measure the influence and impact of researchers and publications, evaluate the productivity and performance of institutions, and assess the overall growth and development of scientific fields. Scientometrics is crucial in mapping the scientific landscape, understanding knowledge diffusion, and informing research policy and decision-making processes.

Artificial Intelligence provides quantitative evidence and objective insights into the progress and impact of scientific research, facilitating informed assessments and strategic planning in various scientific domains. (Jiang, S., Ma, J., Liu, Z., & Guo, H. 2022). AI was used to assess geohazard research development. The study used scientometric methods to identify productive researchers, institutions, and hot research subjects from 9226 Web of Science core collection entries. According to the data, China, the US, and Italy have dominated AI geohazard research since 2000. Ground motion, deep learning (DL), and landslides were the main study subjects, while DL, SVM, and DT were the main AI methods. Research networks revealed critical scholars, institutions, papers, and subjects. This scientometric review may be used to visualise developing patterns in different study disciplines and as a reference for early-stage and experienced geohazard researchers and practitioners. (Vazquez, J. P. G., 2021) This scientometric research covered agricultural AI academically. Scientometric software assessed Scopus-indexed literature. The US, China, India, and Australia published the most. US authors and institutions cooperated. China Agricultural University published the most, while Gerrit Hoogenboom from the University of Florida published the most. Precision, smart, and sustainable agriculture utilises AI and IT. The paper highlighted IoT, decision assistance, and machine learning. This scientometric review of AI in agricultural research is complete.

## 2 Materials and Methods

In this study, data was obtained from the Web of Science database, specifically from different indexes. The breakdown of records downloaded from each index is as follows: Science Citation Index Expanded (SCI-EXPANDED): 4,169 records This index covers a wide range of scientific disciplines and provides access to high-quality research articles from various scientific journals. Conference Proceedings Citation Index – Science (CPCI-S): 2,112 records. This index focuses on conference proceedings in the field of science, providing access to research presented at scientific conferences. Social Sciences Citation Index (SSCI): 305 records. This index concentrates on research within the social sciences, including disciplines such as sociology, psychology, and economics. Conference Proceedings Citation Index – Social Science & Humanities (CPCI-SSH): 29 records. This index specifically includes conference proceedings in the social sciences and humanities, allowing access to research presented at relevant conferences. Book Citation Index – Science (BKCI-S): 10 records. This index encompasses books related to scientific subjects, allowing access to scholarly books in the field of science. Arts & Humanities Citation Index (A&HCI): 1 record. This index focuses on research within the arts and humanities, including fields such as literature, history, and philosophy. By gathering data from these indexes, the study obtained a comprehensive collection of scholarly records to analyse and draw conclusions regarding the application of artificial intelligence in agriculture.

The data analysis in this study involved various scientometric techniques to gain insights into the academic overview of the application of artificial intelligence (AI) in agriculture. Some of the common data analysis techniques that may have been employed include. Co-author analysis examines patterns of collaboration among researchers. It involves identifying frequent collaborators and mapping co-author networks to understand the extent of collaboration and the influence of researchers in the field. The study may have analysed the institutions with the highest number of published papers to identify leading institutions in the field of AI in agriculture. Topic analysis, such as co-citation or keyword analysis, may have been conducted to identify the main research topics and emerging trends within the field. This analysis helps in understanding the focus areas of AI application in agriculture. Data visualization techniques, such as network maps or word clouds, may have been used to visually represent collaboration networks, topic clusters, or the distribution of keywords within the dataset. These visualizations provide a clear and intuitive understanding of the research landscape.

## 3 Data Analysis and Interpretation

**Table 1.** Funding Agencies contribution

| S. No | Funding Agencies | Records |
|---|---|---|
| 1 | Department of Science Technology India | 135 |
| 2 | National Natural Science Foundation of China | 48 |
| 3 | National Research Foundation of Korea | 46 |
| 4 | Council of Scientific Industrial Research, India | 43 |
| 5 | Department of Biotechnology, India | 41 |
| 6 | King Saud University, Saudi Arabia | 38 |
| 7 | Science Engineering Research Board, India | 32 |
| 8 | University Grants Commission, India | 31 |
| 9 | Taif University, Saudi Arabia | 30 |
| 10 | UK Research Innovation, United Kingdom | 27 |

The data provided in Table 1 is the analysis and interpretation of the funding agencies' contributions. The Department of Science and Technology, India, has the highest number of records with 135 contributions. This indicates that they have been actively involved in funding scientific and technological research in India. The National Natural Science Foundation of China has made 48 contributions, which suggests they have supported substantial scientific research within China. The National Research Foundation of Korea has made 46 contributions, indicating their involvement in funding research projects in Korea. The Council of Scientific Industrial Research, India, has made 43 contributions, implying their support for industrial research and development in India. The Department of Biotechnology, India, has provided 41 contributions, suggesting their focus on funding research in the biotechnology field. King Saud University in Saudi Arabia has made 38 contributions, indicating its role in funding research activities within the university. The Science Engineering Research Board, India, has provided 32 contributions, implying their support for various scientific and engineering research projects in India. The University Grants Commission, India, has made 31 contributions, suggesting their involvement in funding research and educational activities in Indian universities. Taif University in Saudi Arabia has contributed 30 times, indicating their support for research and development projects within the university. UK Research Innovation from the United Kingdom has made 27 contributions, suggesting their involvement in funding research and innovation activities in the country. Based on the available data, these interpretations provide an overview of the funding agencies' contributions. However, it is essential to note that this analysis only considers the number of records and does not account for the monetary value or the impact of the contributions (Table 2).

**Table 2.** Journal's Publication on Artificial Intelligence

| S. No | Journals | Records | TGCS | NA | H-Index |
|---|---|---|---|---|---|
| 1 | Materials Today-Proceedings | 192 | 614 | 682 | 11 |
| 2 | IEEE Access | 125 | 2876 | 654 | 28 |
| 3 | Multimedia Tools and Applications | 71 | 339 | 222 | 9 |
| 4 | Neural Computing & Applications | 60 | 1079 | 214 | 19 |
| 5 | Sustainability | 51 | 577 | 334 | 13 |
| 6 | IET Image Processing | 48 | 474 | 134 | 12 |
| 7 | Sensors | 45 | 558 | 282 | 14 |
| 8 | Indian Journal of Ophthalmology | 41 | 227 | 183 | 7 |
| 9 | Wireless Personal Communications | 41 | 382 | 130 | 10 |
| 10 | Journal of Intelligent & Fuzzy Systems | 39 | 330 | 138 | 9 |
| 11 | Soft Computing | 38 | 584 | 119 | 15 |
| 12 | Computational Intelligence and Neuroscience | 37 | 144 | 240 | 6 |
| 13 | Computers & Electrical Engineering | 36 | 627 | 143 | 13 |
| 14 | Applied Soft Computing | 34 | 1008 | 120 | 19 |
| 15 | Archives of Computational Methods in Engineering | 33 | 321 | 117 | 10 |

TGCS – Total Global Citation Scores; NA – Number of Authors: H-Index – Hirsch Index.

There are totally 2266 journals published 6194 records in this study period. Most of papers (1613) are published in IEEE, followed by Springer Nature (1226), Elsevier (1131) and MDPI (347). Highest (192) records from Materials Today-Proceedings. Lowest (33) records from Archives of Computational Methods in Engineering. Global Scholar Citations score Highest (2876) from IEEE Access, Lowest (144) citation score from Computational Intelligence and Neuroscience. Highest (682) Number of Authors contributed from Materials Today-Proceedings, Lowest (117) authors contributed from Archives of Computational Methods in Engineering. Highest (28), H-index journal from IEEE Access. Lowest (5) H-index journal from Computational Intelligence and Neuroscience (Table 3).

**Table 3.** Institution's output on Artificial Intelligence

| S. No | Institution | Records | TGCS | NA | H-Index |
|---|---|---|---|---|---|
| 1 | Institute of Research and Development, Duy Tan University, Vietnam | 52 | 2862 | 468 | 12 |
| 2 | School of Computer Science & Engineering, VIT, Vellore | 51 | 416 | 221 | 10 |
| 3 | School of Computer science, UPES, Dehradun | 49 | 592 | 269 | 12 |
| 4 | Institute of Technology, Nirma University, Gujarat | 48 | 479 | 253 | 12 |
| 5 | College of Computers and Information Technology, Taif University, Saudi Arabia | 39 | 225 | 265 | 8 |
| 6 | School of Information Technology& Engineering, Vellore Institute of Technology, Vellore | 39 | 1136 | 210 | 15 |
| 7 | Dept. of Electronics Engineering, Indian Institute of Technology, Newdelhi | 33 | 474 | 143 | 13 |
| 8 | Dept. of Computer Science & Information Engineering, Asia University, Taiwan | 32 | 665 | 168 | 14 |
| 9 | Indraprastha Apollo Hospital, Dept. of Cardiology, New Delhi | 31 | 558 | 907 | 15 |
| 10 | Division of Cardiovascular Medicine, University of Virginia, USA | 28 | 505 | 831 | 14 |
| 11 | Institute of Engineering & Technology, Chitkara University, Punjab | 25 | 198 | 121 | 9 |
| 12 | Rheumatology Unit, National and Kapodistrian University of Athens, Greece | 25 | 637 | 778 | 14 |
| 13 | Dept. of Medicine, Queens University, Canada | 25 | 439 | 647 | 14 |
| 14 | Dept. of Computer Science & Engineering, National Institute of Technology, Rourkela, | 24 | 204 | 85 | 9 |
| 15 | Heart & Vascular Institute, Adventist Heart and Vascular Institute St Helena, USA | 23 | 431 | 607 | 13 |

TGCS – Total Global Citation Scores; NA – Number of Authors: H-Index – Hirsch Index

Institute of Research and Development, Duy Tan University, Vietnam: This institution has the highest number of records, with 52 publications. Heart & Vascular Institute, Adventist Heart and Vascular Institute St Helena, USA: This institution has the lowest number of records, with 23 publications. School of Information Technology & Engineering, Vellore Institute of Technology, Vellore: This institution has the highest TGCS score with 1136 citations. Institute of Engineering & Technology, Chitkara University, Punjab: This institution has the lowest TGCS score with 198 citations. Indraprastha Apollo Hospital, Dept. of Cardiology, New Delhi: This institution has the highest number of authors, with 907 contributors. Dept. of Computer Science & Engineering, National Institute of Technology, Rourkela: This institution has the lowest number of authors, with 85 contributors. School of Information Technology & Engineering, Vellore Institute of Technology, Vellore: This institution has the highest H-index with a value of 15. College of Computers and Information Technology, Taif University, Saudi Arabia and 14. Dept. of Computer Science & Engineering, National Institute of Technology, Rourkela: These two institutions have the lowest H-index with a value of 8 (Table 4).

**Table 4.** Authors research output on Artificial Intelligence

| S. No | Authors | Records | TGCS | C-Authors | H-Index |
|---|---|---|---|---|---|
| 1 | Kumar A | 133 | 1464 | 672 | 20 |
| 2 | Sharma A | 105 | 1061 | 1115 | 18 |
| 3 | Kumar S | 92 | 1099 | 448 | 20 |
| 4 | Kumar R | 91 | 1319 | 818 | 21 |
| 5 | Kumar N | 80 | 1237 | 424 | 19 |
| 6 | Gupta S | 75 | 966 | 435 | 18 |
| 7 | Singh A | 75 | 701 | 359 | 14 |
| 8 | Singh R | 62 | 522 | 386 | 11 |
| 9 | Sharma S | 58 | 422 | 279 | 12 |
| 10 | Gupta R | 54 | 677 | 290 | 12 |
| 11 | Suri JS | 54 | 1219 | 1107 | 21 |
| 12 | Saba L | 52 | 1213 | 306 | 14 |
| 13 | Singh S | 52 | 580 | 257 | 14 |
| 14 | Kumar P | 51 | 641 | 246 | 13 |
| 15 | Gupta A | 50 | 646 | 399 | 12 |

TGCS – Total Global Citation Scores; C-Authors – Collaborated Authors: H-Index – Hirsch Index

The table describes that of the 133 records, Kumar A has the highest number of publications. Sharma A follows closely with 105 records, indicating a significant research output. Kumar S ranks third with 92 records, demonstrating a substantial number of publications. The top three authors in terms of global scholar citations are Kumar A with 1464 citations, Sharma A with 1061 citations, and Kumar S with 1099 citations.

These authors have received significant recognition and have made notable contributions in their respective fields of study. Their research findings and publications have gained considerable attention and have likely influenced the academic community. On the other hand, the lowest three authors in global scholar citations are Sharma S with 422 citations, Singh R with 522 citations, and Singh A with 701 citations. While their citation counts may be lower than others on the list, it is essential to consider that they alone do not determine the quality or significance of an author's work. These authors' contributions still hold value within their specific research areas, and their work may have relevance and impact within their respective domains (Table 5).

**Table 5.** Collaborated countries publication on Artificial Intelligence (124)

| S. No | Countries | Records | TGCS | NA | H-Index |
|---|---|---|---|---|---|
| 1 | India | 6194 | 67016 | 29554 | 94 |
| 2 | USA | 580 | 10035 | 7314 | 52 |
| 3 | Saudi Arabia | 386 | 4143 | 2608 | 29 |
| 4 | UK | 314 | 6975 | 4905 | 43 |
| 5 | Peoples R China | 307 | 7539 | 4076 | 42 |
| 6 | Australia | 219 | 5189 | 2247 | 40 |
| 7 | South Korea | 166 | 3909 | 2825 | 33 |
| 8 | Canada | 155 | 3559 | 2505 | 30 |
| 9 | Malaysia | 147 | 1942 | 1151 | 25 |
| 10 | Italy | 142 | 2544 | 3771 | 29 |
| 11 | Taiwan | 137 | 2879 | 1321 | 28 |
| 12 | Vietnam | 123 | 4792 | 959 | 40 |
| 13 | Iran | 116 | 4138 | 1380 | 35 |
| 14 | Germany | 111 | 1818 | 3367 | 25 |
| 15 | Singapore | 102 | 2697 | 996 | 24 |

TGCS – Total Global Citation Scores; NA – Number of Authors: H-Index – Hirsch Index

The table indicated that, with 6194 collaboration records, India ranks first among the listed countries. It also has the highest number of global scholar citations (GCS) at 67016 and North America (NA) publications at 29554. India has an H-Index of 94, indicating significant research impact. The United States is the second highest in collaboration, with 580 records. It has 10035 global scholar citations (GCS) and 7314 publications in North America (NA). The USA has an H-Index of 52, signifying substantial research influence. Saudi Arabia holds the third position with 386 collaboration records. It has 4143 global scholar citations (GCS) and 2608 publications in North America (NA). The country has an H-Index of 29. The United Kingdom ranks fourth, having 314 collaboration records. It has 6975 global scholar citations (GCS) and 4905 publications in North America (NA). The UK has an H-Index of 43. China (officially called "People's R China") is the fifth highest in collaboration with 307 records. It has 7539 global scholar citations (GCS) and 4076 publications in North America (NA). China has an H-Index of 42. These top 5 countries demonstrate significant international collaboration and research productivity, as indicated by their collaboration records, citation counts, publications in North America, and H-Index.

## 3.1 Bibliographic Coupling on Co-citation and Authors

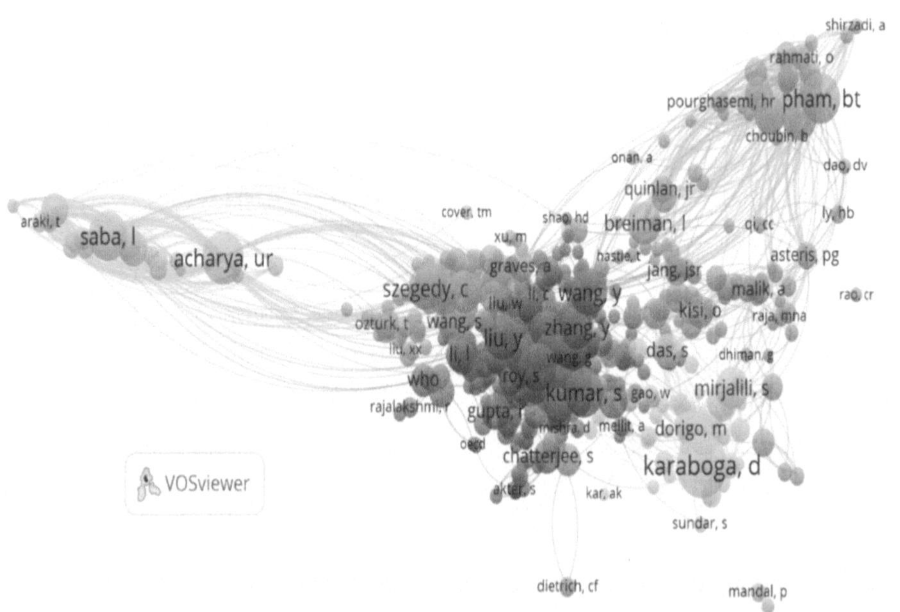


**Fig. 1.** Bibliographic coupling of authors

The figure appears the clusters represent groups of documents or authors based on co-citation patterns or co-authorship relationships. Here is a brief description of each cluster based on the number of elements. This Cluster 1 (354) contains a significant number of documents or authors that exhibit strong co-citation or co-authorship relationships. It may represent a prominent research theme or a highly interconnected group within the field of AI in agriculture. Cluster 2 (114) is relatively smaller than Cluster 1 but still represents a substantial number of documents or authors with shared co-citation or co-authorship patterns. It may indicate a distinct subfield or research focus within the broader topic. Cluster 3 (71) consists of a moderate number of documents or authors that exhibit co-citation or co-authorship relationships. It could represent a specific research area or a subset of related studies within the field. Cluster 4 (55) contains a smaller number of documents or authors compared to the previous clusters. It may represent a more specific or niche topic within AI in agriculture. Cluster 5 (36) comprises a relatively smaller number of documents or authors. It may represent a specific aspect or subtopic within the field of AI in agriculture. Cluster 6 (26) is smaller in size and may represent a specialized area or a focused group of researchers within the broader context of AI in agriculture. Cluster 7 (5) is quite small and may indicate a specific niche topic or a closely-knit group of researchers within the field. Cluster 8 (3) is the smallest among the clusters mentioned, potentially representing a highly specialized or unique research area within AI in agriculture (Fig. 1).

## 3.2 Bibliographic Coupling on Co-authorship Countries

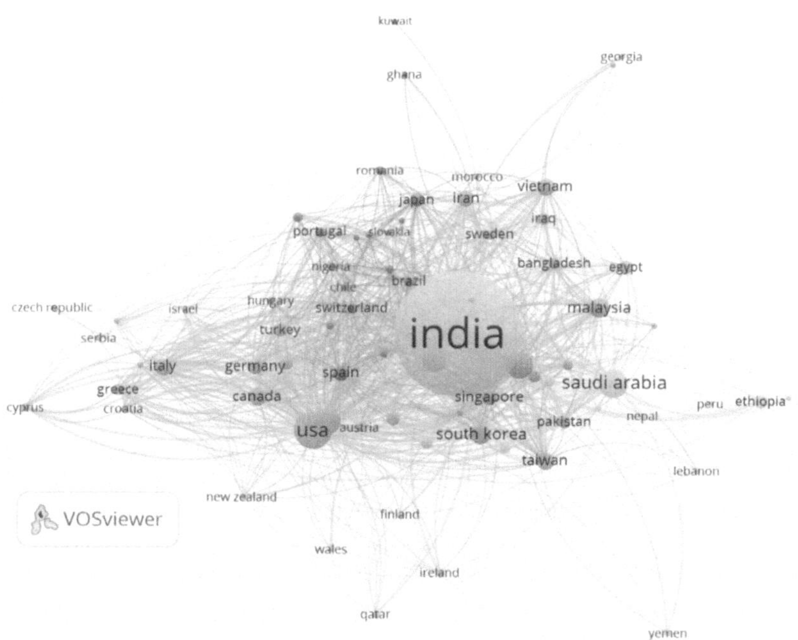


**Fig. 2.** Bibliographic coupling of countries

The figure mentioned now represents co-authorship patterns among countries. Here is a brief description of each cluster based on the number of elements. Cluster 1 (20): This cluster represents a group of countries that have strong co-authorship relationships. These countries likely collaborate extensively in the field of AI in agriculture, indicating a high level of research cooperation. Cluster 2 (18) consists of a significant number of countries that exhibit co-authorship patterns. These countries may have established collaborative networks and partnerships, indicating shared research interests or joint projects in AI application for agriculture. Cluster 3 (12) represents a moderate number of countries with co-authorship relationships. These countries may have ongoing collaborations and contribute to the research output in the field of AI in agriculture. Cluster 4 (10) contains a smaller number of countries compared to the previous clusters. These countries may have some level of co-authorship collaboration, potentially focusing on specific research areas within AI in agriculture. Cluster 5 (8) includes a relatively smaller number of countries. These countries may engage in co-authorship collaborations but to a lesser extent compared to the larger clusters. Cluster 6 (6) represents a smaller group of countries with co-authorship relationships. These countries may have specialized collaborations or unique research partnerships in the field of AI in agriculture (Fig. 2).

### 3.3 Bibliographic Coupling on Organisations

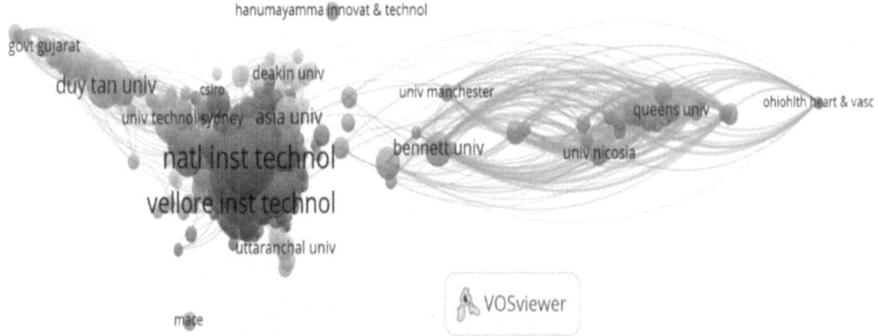


**Fig. 3.** Bibliographic coupling of Organisation

The figure evidence provided appears to represent bibliographic coupling between organizations, with clusters represented by different colours (red, green, blue). Here is a brief description of each cluster based on the number of elements. Red Cluster: This cluster consists of 214 organizations with bibliographic coupling relationships. These organizations likely share common references, indicating similarities in their research interests or areas of focus. Green Cluster: This cluster comprises 160 organizations with bibliographic coupling relationships. These organizations likely have shared references and research connections, indicating collaborative efforts or overlapping research interests. Blue Cluster: This cluster includes 180 organizations with bibliographic coupling relationships. These organizations likely have common references, indicating research connections and potential collaborations. Cluster 4: This cluster contains 188 organizations in the red category, 189 organizations in the green category, and 34 organizations in the blue category (Fig. 3).

These organizations likely have diverse bibliographic coupling relationships across multiple clusters, indicating a broader range of research collaborations. Cluster 5: This cluster consists of 148 organizations in the red category, 103 organizations in the green category, and 189 organizations in the blue category. These organizations have varied bibliographic coupling relationships, suggesting diverse research connections and collaborations. Cluster 6: This cluster includes 23 organizations in the red category, 190 organizations in the green category, and 207 organizations in the blue category. These organizations likely have distinct bibliographic coupling relationships across different clusters, indicating a wide range of research collaborations. Cluster 7: This cluster comprises 255 organizations in the red category, 127 organizations in the green category, and 14 organizations in the blue category. These organizations likely have significant bibliographic coupling relationships, indicating strong research connections and collaborations. Cluster 8: This cluster consists of 140 organizations in the red category, 86 organizations in the green category, and 75 organizations in the blue category. These organizations likely exhibit bibliographic coupling relationships within their respective clusters, suggesting research connections and collaborations.

## 4 Conclusion

Change is the key for new world. AI has become the key for the change. It will change the concept of prescribed nature of human beings. Web of Science provides access to the full record and cited references of scholarly communications in the forms of yearly output, publications in journals, institutions, countries, etc. it is useful to the researchers on particular field, to find, collect and store necessitate information. It also provides various research possibilities to access right information. Totally, 18,322 authors from 125 countries, published in 2,266 journals during the study period. Records published in Article form is 3050 and proceeding papers are 2037. India's Highest collaboration is USA (580 publications), Saudi Arabia (386) and United Kingdom (315). This paper sums up the impact of Artificial Intelligence in the various fields of knowledge and its contribution to the development of research. AI is truly a global endeavour that knows no bounds. As we navigate this uncharted terrain, one thing remains clear: AI has not only rewritten the script of research but promises to pen an even more awe-inspiring narrative in the chapters yet to unfold.

## References

Abdullah, K., Sofyan, D.: Machine learning in safety and health research: a scientometric analysis. Int. J. Inform. Sci. Manag. **21**(1), 17–37 (2023)

Chithiraivel, S., Sivasekaran, K., Ramalinagam, J.: Global research output on eosinophilia literature: a scientometric analysis. Libr. Philos. Pract. **123** (2020)

Darko, A., Chan, A.P., Adabre, M.A., Edwards, D.J., Hosseini, M.R., Ameyaw, E.E.: Artificial intelligence in the AEC industry: scientometric analysis and visualization of research activities. Autom. Constr. **112**, 103081 (2020)

Jiang, S., Ma, J., Liu, Z., Guo, H.: Scientometric analysis of artificial intelligence (AI) for geohazard research. Sensors **22**(20), 7814 (2022)

Kopalle, P.K., et al.: Examining artificial intelligence (AI) technologies in marketing via a global lens: Current trends and future research opportunities. Int. J. Res. Mark. **39**(2), 522–540 (2022)

Lotsch, J., et al.: Artificial intelligence and machine learning in pain research: a data scientometric analysis. Pain Reports **7**, 6 (2022)

Morozov, E.: The true threat of artificial intelligence. The New York Times (2023)

Saha, S., Kar, S.: Special issue on machine learning in scientometrics. J. Scientometric Res. **8**, 2s (2019)

Sivankalai, S., Sivasekaran, K., Virumandi, A.: Measuring the honey bee research output: scientometrics analysis from 2004 to 2019. Int. J. Mod. Agric. **2**, 2199–2209 (2021)

Sivasekaran, K.: Curcuma Longa (Medicinal Plant) research: a scientometric assessment of global publications output with reference to web of science. Turk. J. Comput. Math. Educ. **12**(5), 1477–1496 (2021)

Sivasekaran, K., Sivankalai, S., Stanleay, P.: Bats are the only flying mammal: a scientometric analysis. Libr. Philos. Pract. 1–20 (2021)

Tapeh, A.T.G., Naser, M.Z.: Artificial intelligence, machine learning, and deep learning in structural engineering: a scientometrics review of trends and best practices. Arch. Comput. Meth. Eng. **30**(1), 115–159 (2023)

Usha, S., Sharmila, S., Ashok Kumar, A., Veeraramu, K., Sivasekaran, K., Sivankalai, S.: The rise of mucormycosis in covid patients: a scientometric analysis. J. Clin. Otorhinol. **27** (2023)

Vazquez, J.P.G., et al.: Scientometric analysis of the application of artificial intelligence in agriculture. J. Scientometric Res. **10**(1), 55–62 (2021)

# Analyzing Hand Gestures Using Object Detection and Processing It into Local Language

K. Sangeetha[1], V. S. Balaji[2](✉), P. Kamalesh[2], and P. S. Anirudh Ganapathy[2]

[1] Department of AIML, AP, Rajalakshmi Engineering College, Chennai, India
[2] IIIrd Year Department of AIML, Rajalakshmi Engineering College, Chennai, India
211501017@rajalakshmi.edu.in

**Abstract.** Hand gestures have become an important part of human communication, especially around sign language recognition. The development of object detection methods has made it possible to use ML algorithms and Deep learning to analyze and understand hand movements. The ability to recognize and interpret hand gestures accurately is therefore of great importance for both the deaf and dumb communities and for those who interact with them. The development of computer vision technology and object detection methods combined with the ability of OpenCV to analyze and interpret hand gestures holds an opportunity to revolutionize interaction for hard-of-hearing people, increasing their access to information, education, and work possibilities.

**Keywords:** Computer Vision · Machine learning · Deep Learning · Object Detection · OpenCV

## 1 Introduction

People who are dumb and deaf can communicate with each other with the help of non-verbal gestures.A deaf person is having hearing impairment so that the person can't hear. A dumb person is having speech impairment so that the person can't speak. So, in common between deaf and dumb people sign language plays a vital role in communication Hand gestures have developed into a crucial feature of human communication, especially in the context of language recognition.

Thanks to advancements in computer vision technology, hand gestures may now be analyzed and understood by ML systems.

We aim to present a comprehensive review of the different approaches in this work by employing computer vision to distinguish hand motions. We will discuss recent advances in the field as well as.

The challenges associated with hand motion recognition. By making a box near to the object. Object Detection is used to find the object of the class where it belongs to and it traces the location of the object.

[1] Two likely methods are: The first is a glove-based system, where the performer dons a set of data gloves while having their hand motions recorded. Vision which is


R. Geetha et al. (Eds.): AAIMB 2023, CCIS 2203, pp. 37–44, 2025.
https://doi.org/10.1007/978-3-031-73068-9_4

further separated into static and dynamic recognition, is the emphasis of the second strategy.

[2] While dynamic technology focuses on real-time gesture capture, static technology focuses on the representation of motions in two dimensions.

[3] The application for real-time hand gesture detection mainly scopes on Hand gesture detection that plays a vital role in recognizing the hand gestures.

[4] To understand the hand gestures different algorithms are being used from various concepts like artificial neural networks and computer vision.

## 2 Related Work

CNN is a form of neural network that can be operated using weights. In CNN, convolution is the product of blending the functions that demonstrates the overlapping of functions with another. For obtaining maps, the snap gets combined alongside the action method that is used in firing the output. Further they are processed using the pooling layers to make the conceptual feature maps that reduces the complexity and hardness of the network.The desired amount of filters are applied by repeating this process, and feature maps are produced as a result.The maps involved in completely connected layers make the output for the detection of the image and displays confidence scores for the labels of the class.

The CNN operational model is shown below.

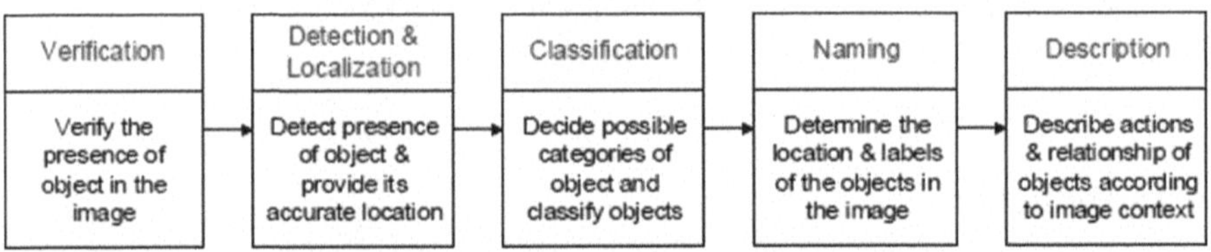


[3] The dataset comprised approximately 216 images, which were subsequently divided into 5 different sets. On average, each set contained 42 images. The labeling process involved utilizing the You Only Look Once format (YOLO), where details such as the class ID and corresponding class were stored. The dataset was labeled using the YOLO format and saved in a text file format. A total of 216 images were present in the dataset, and they were divided into 5 distinct sets named set1, set2, set3, set4, and set5, representing different finger positions. The hand gestures were visually depicted in the dataset.

The hand motions in our dataset are shown in Fig. 1.

In [3], with the help of OPENCV module, a free computer vision tool real-time objects are captured, and then images are captured and sent to be framed by the real-time objects. We got the points for classes in our image set because these photos are labeled in accordance with the classes we have saved for our YOLO technique. We can go on to the training section after this. When input is delivered, it first enters the input layer where it is rendered before moving on to the hidden layers, which are interconnected convolutions that are numerous in number and size. The value of confidence is a unique value that is determined by these convolutional layers.

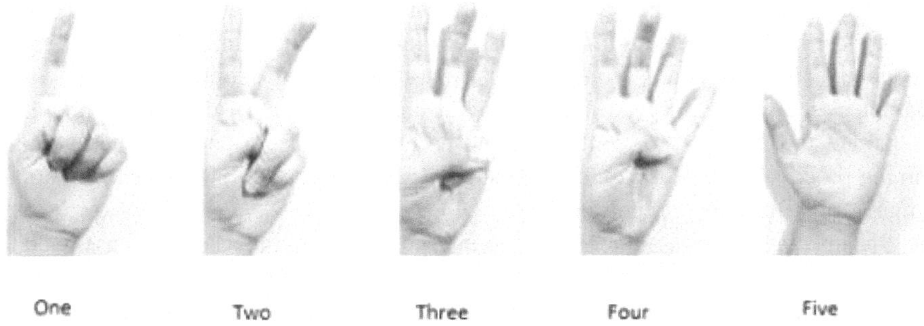


**Fig. 1.** Hand gesture set to be detected.

[2] This method involves Computer vision method using Tensorflow. In this method, pixel segmentation of the images takes place. Implementation can be done in four steps: Segmentation and enhancement of images, orientation detection, feature extraction and categorization.

[3] The biggest drawback of this job, which is centered on the aforementioned four categories, is that the diverse lighting situations change colors very quickly, which can lead to mistakes or outright failures.

[4] For instance, when there is insufficient light, the hand area cannot be seen, but because the non- skin parts have the same color, they are mistaken for the hand area. The system is built on modeling the hand in the spatial domain to recognise hand gestures. The system employs a variety of two-dimensional and three- dimensional functional and non-functional models for simulation.

[5] The main disadvantage of the system is that it does not consider gestures in the time frame region or fist movement, and it cannot detect images with complex backdrops or when new items appear in the frame alongside fist related objects.

[6] The study focuses on understanding motions through the use of diverse information collection, initial processing, categorization, and other methodologies. An appropriate tool for entering information should be used for data collection. A variety of devices for entering information are available for data collecting. Among them are information mitts, indicators, and palm shots. However, its limitations in terms of clarity, rotation and direction, scale, and specifications have changed.

[7] In a computer vision project using OpenCV and Python, the process of data acquisition involves several steps to prepare the dataset for training an object detection model. To begin, the project utilizes the "os" module to handle file paths, enabling efficient navigation and manipulation of image and XML files. Additionally, the "time" module is employed to measure code efficiency and track the time taken for various operations. The "uuid" library plays a crucial role in generating unique identifiers for image files, ensuring their uniqueness and avoiding any potential conflicts during data processing.The dataset is labeled using the LabelImg tool, a popular annotation tool for computer vision projects.

[8] This tool allows users to label objects of interest within images and generates XML files that contain comprehensive details about the labeled objects, including their

coordinates and corresponding class labels. Once the images are labeled and XML files are created, the collection of data is split into instructing and validation subsets, Frequently, an eighty percent to twenty distribution rate being adopted. This separation guarantees that a system gets taught on a substantial amount of data while retaining a separate set of images for evaluation and performance assessment. To facilitate the training process, a labeled map is established, associating each alphabet with a distinct numerical ID from one - twenty-six.

[9] This map serves as a mention of to interpret and map the data labels assigned to objects during the labeling process. To create, instruct and utilize the given model, the project leverages the TensorFlow object detection API. This API provides a framework for building powerful object detection models. The TensorFlow detection model zoo contains various pre- trained models that can be utilized as a starting point for the project. These pre-trained models offer a solid foundation for transfer learning, enabling the model to learn from existing knowledge and adapt to specific object detection tasks.The data acquisition process for a computer vision project using OpenCV and Python involves leveraging the os, time, and uuid modules, labeling images with the LabelImg tool, dividing the dataset, creating a labeled map, and utilizing the TensorFlow object detection API with pre-trained models. These steps ensure a.

[10] systematic and efficient approach to acquiring, preparing, and utilizing the dataset for training an object detection model. In human-computer interaction technology, gesture recognition is a vital field. Its applications include audio recording, automated system supervision, visual communication, and graphical surveillance.

[11] In the hand gesture identification system will be implemented using MediaPipe's pre-trained hand recognition model, which is capable of accurately identifying and tracking hands in real-time video feeds. The neural network model may be trained and fine-tuned using Tensorflow, an open-source machine learning framework, to recognise and classify various hand motions based on MediaPipe input.Furthermore, the powerful computer vision library OpenCV will be used to analyze the video frames, extract essential hand features, and prepare the data for input to the neural network. The extensive set of image processing and computer vision functions supplied by OpenCV enables the hand gesture recognition system's accuracy and resilience to be improved.

[12] The dataset comprised approximately 216 images, which were subsequently divided into 5 different sets. On average, each set contained 42 images. The labeling process involved utilizing the You Only Look Once format (YOLO), where details such as the class ID and corresponding class were stored. The dataset was labeled using the YOLO format and saved in a text file format. A total of 216 images were present in the dataset, and they were divided into 5 distinct sets named set1, set2, set3, set4, and set5, representing different finger positions. The hand gestures were visually depicted in the dataset (Fig. 2).

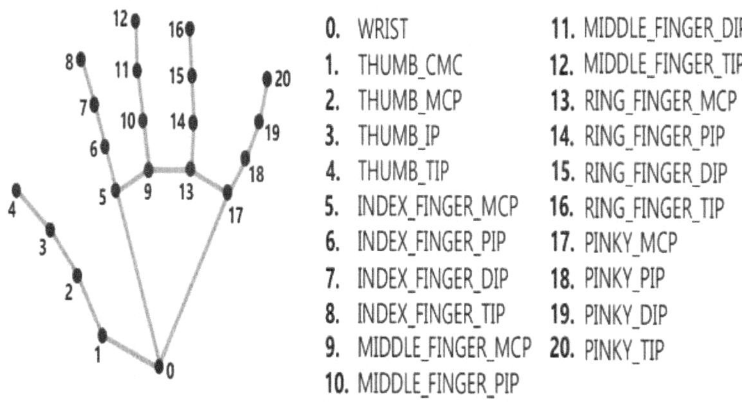


**Fig. 2.** Denotes the 21 points

## 3 Implementation

Hand gestures are recognized and they are converted to the local language. Following the main steps that are used [1] Prepare pre- label images using labeling. For each gesture, approximately 20 images are considered and they are labeled. In [1], labels are given to the particular image that is further used in real time detection i.e. for each gesture, a label is provided. [2] Transfer learning (TL) using SSD Mobile Net. In [2], TL is a kind of ML technique where the model gets trained on a huge dataset for a specific purpose before getting re-used or altered to a novel and distinct task. This can save time and effort by allowing the instructed model to be in the initial spot for performing upcoming action rather than starting from scratch. Transfer learning using SSD MobileNet involves using a pre-trained MobileNet model for image classification and object detection tasks. MobileNet is a popular deep learning architecture for image classification, while SSD (Single Shot Detector) is a framework for object detection. [3] Real time detection using Webcam and OpenCV, Object detection. [3] After training and labeling the images, the model is ready for real time detection and it is predicted that it would work with greater accuracy and it will also be shown. The hand gestures are detected and they are classified to map the output. Figure 3 depicts the working of the model.

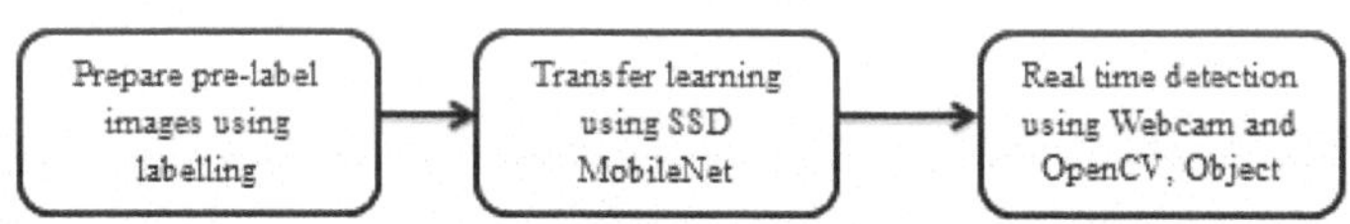


**Fig. 3.** Working of model

The trained images are pre-processed and the suitable information about the hand gestures are derived. Further the features are extracted from the given set of gestures that are trained and they are stored in a file for each. For testing the similar flow is followed for testing the images. The gestures are detected in the real time, detected and the recognized text is predicted (Figs. 4 and 5).

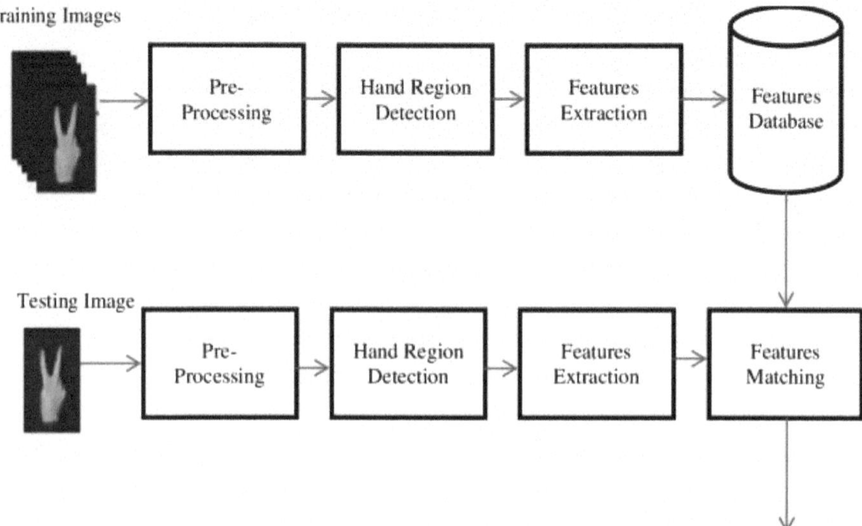


**Fig. 4.** Steps involved in training and testing.

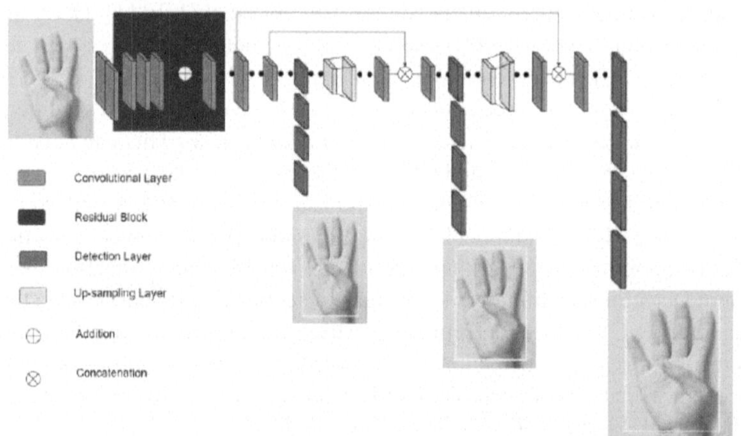


**Fig. 5.** Neural Network for hand gesture recognition.

An input is first sent to the input layer, where it is further displayed before being sent to the hidden layers, which are connected convolutions that are numerous in number and size. The value of confidence is a unique value that is determined by these convolutional layers.

The model's testing sequence was set to 10,000 steps using the hyper-parameter used during training. During training, the model sustains losses in categorization, regularization, and localization. The localization loss changes between the true and projected bounding box improvements. The localization loss differs according to the displayed bounding box modifications and also its actual values. The localization loss formula [7]

$$L_{loc}(\mathrm{x}, \mathrm{l}, \mathrm{g}) = \sum\nolimits_{i \in Pos}^{N} \sum\nolimits_{m \in \{cx, cy, w, h\}} x_{ij}^{k} smooth_{L1}(l_i^m - \hat{g}_j^m) \tag{1}$$

$$\hat{g}_j^{cx} = (g_j^{cx} - d_j^{cx})/d_i^{w} \tag{2}$$

$$\hat{g}_j^{cy} = (g_j^{cy} - d_i^{cy})/d_i^{h} \tag{3}$$

$$\hat{g}_j^{w} = \log(g_j^{w}/d_i^{w}) \tag{4}$$

$$\hat{g}_j^{h} = \log(g_j^{h}/d_i^{h}) \tag{5}$$

The overall amount of paired predefined containers (N), forecasted limit containers [1], the real enclosing container (g), the real enclosing container in encoded format (g^), and compatibility reasons across predefined containers and real enclosing containers j of category k (xij^k) are all covered. The losses encountered during experimentation are addressed in the subsequent section, with a reference to a source [7]. After training, the design is populated from the most recent check to prepare it for instantaneous detection. For preparing the model for instruction, configuration and updates take place. The model that was instructed begins to load with its latest current check, bringing the model to a close and preparing it for real-time sign language recognition. This process can be further adapted to accommodate local language requirements (Fig. 6).

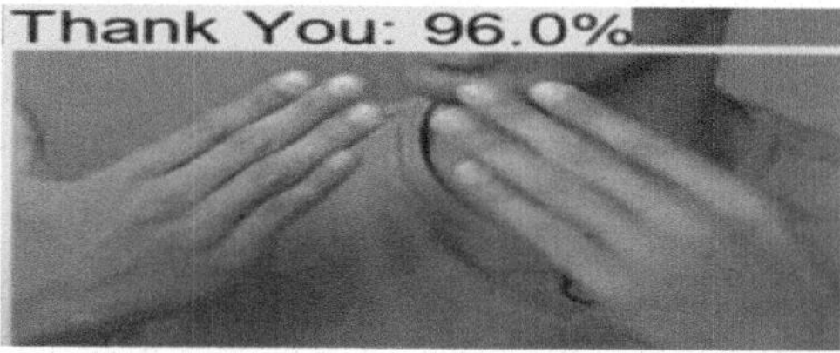

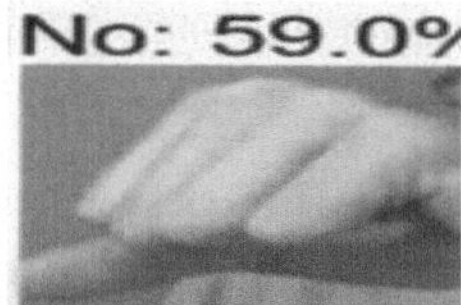


**Fig. 6.** Real-time detection and recognition

## 4 Results and Discussions

The model works well and accurately identifies the gesture. Each motion can then be given a label and translated into the native tongue for easier understanding. A total of 26 photos are gathered for a specific gesture, stored in a separate file, labeled with the name of the gesture, and the gesture is then recognized in real time. When various photos of each gesture are gathered and the model is trained, the losses may be calculated and minimized.

## 5 Conclusion and Future Works

Processing sign language involves a critical component called hand recognition. Machine learning and computer vision techniques have enhanced hand recognition, which is now widely employed in numerous sign language processing applications. The ultimate goal of this discipline is to create systems that can recognize sign language motions with high accuracy in real-time, allowing persons with hearing and speech disabilities to interact with others in meaningful ways. Hand recognition for sign language processing faces a number of difficulties, including occlusion, poor illumination, and variations in hand forms and postures.

Future research should therefore concentrate on creating stronger, more precise hand identification algorithms that can handle these difficulties. Hand recognition for sign language processing faces a number of difficulties, including occlusion, poor illumination, and variations in hand forms and postures. Future research should therefore concentrate on creating stronger, more precise hand identification algorithms that can handle these difficulties.

## References

1. Pathak, A.R., Pandey, M., Rautaray, S.: Application of Deep Learning for ObjectDetection: A School of Computer Engineering. Kalinga Institute of Industrial Technology (KIIT) University, Bhubaneswar, India (2018)
2. Thakur, P., et al.: Real- Time Hand Gesture Recognition Using TensorFlow and OpenCV. MES College Of Engineering, Pune (2014)
3. Mujahid, A., et al.: Real-Time Hand Gesture Recognition Based on Deep Learning YOLOv3 Model: Department of Computer Science, p. 54770. University of Management and Technology, Lahore (2021)
4. Srivastava, S., Gangwar, A., Mishra, R., Singh, S.: Sign Language Recognition System using TensorFlow Object Detection API: Department of ECE. University of Allahabad, Prayagraj, India (2021)
5. Real Time Hand Gesture Recognition using Tensorflow and OpenCV: link: https://techvidvan.com/tutorials/hand-gesture-recognition-tensorflow-opencv/
6. Soni, A.N.: Application and Analysis of Transfer Learning Survey: Volume 1 Issue 2, Department of Information Technology. Campbellsville University, USA (2018)
7. Wu, S., Yang, J., Wang, X., Li, X.: IOU- Balanced loss functions for single stage object detection (2020)
8. Oudah, M., Al-Naji, A., Chahl, J.: Hand Gesture Recognition Based on Computer Vision: A Review of Techniques. J. Imagine

# A Review of an Automated Model for Sexist Language Detection and Replacement of Sexist Terms

M. S. Shriram(✉), S. Sushmitha, and Shravanthi Murugesan

B.Tech Information Technology, SSN College of Engineering, Chennai, India
{shriram2010160,shravanthi2010346}@ssn.edu.in

**Abstract.** The perception of anonymity provided by the digital environment has contributed to the rise in harassment and hate speech in recent years on social media platforms. Particularly sexism and gender-based harassment have risen to frightening heights, increasing the number of people worldwide who experience or witness online abuse. Such offensive content frequently contains extremely vulgar and disparaging language, which causes great injury and distress to its targets. In this research paper we propose a way to automatically identify and replace insults directed at people based on their gender on social networks in order to solve this urgent problem and ultimately create a more secure and civil online community. Various machine learning and deep learning models are used to detect sexually explicit information, enabling thorough investigation using a range of performance measures to pinpoint the most efficient models. This research also includes the application of a word replacement algorithm to guarantee that abusive words completely lose their destructive context. The research seeks to reduce the negative effects of abusive information and promote a more positive digital environment by routinely substituting harsh language with neutral or constructive substitutes.

**Keywords:** Machine Learning · NLP · Deep Learning · Sexism

## 1 Introduction

A statement we can all agree with is that social media has become an essential part of our lives. Life before social media has become unimaginable. It has helped us stay connected with others and keep up with current events. The impact of social media on the climate of public opinion cannot be stressed enough. Social networks provide a platform for people from all over the world to share their opinion. However, social media has engendered increased hate and harassment towards some communities. Sexism can take several forms on social media including toxic comments, hate speech, violent threats generally containing offensive words or phrases. Women are subjected to gender-based discrimination in multiple ways on social media such as objectification, criticism for their choice of clothing as it may be considered too revealing by some, sexual harassment and death threats, lack of equal treatment, and respect. Harassment can severely affect


R. Geetha et al. (Eds.): AAIMB 2023, CCIS 2203, pp. 45–58, 2025.
https://doi.org/10.1007/978-3-031-73068-9_5

the mental health of victims. Imposing restrictions would destroy the notion of free speech. Therefore, it is necessary to come up with a solution to make social media a safer and more inclusive space. In this paper, we present a method to automatically detect and replace sexist language. The method is a pre-trained model that can automatically detect and replace sexist language. The model was trained on a large corpus of social media data. A dictionary containing technical terms for every alphabet was created for the replacement purpose. The remainder of this paper is organized as follows. First, we review the existing literature on detecting hate speech on social networks. Next, we describe the data and methodology used. Finally, we discuss the results, performance of our model.

## 2 Literature Review

The challenge of identifying and rebuking hate speech on social media has drawn a lot of attention recently. The prevalence and severity of hate speech on social media have increased along with the popularity of these platforms. Researchers have come up with a number of methods to identify and address hate speech in response. In this review of the literature, we give a general overview of the most well-known techniques and strategies applied in the subject, with a particular emphasis on the automatic identification of sexist and hateful language. The author of a study by Z. Mansur et al. [1] analyzed and emphasized the different strategies and tactics employed to combat the problem of hate speech on Twitter. It concentrates on studies released between 2015 and 2022. The study highlighted the difficulties in automatically detecting hate speech and featured techniques and practices of several investigations, along with their accuracy and performance. In a different study, Chiril, P., Pamungkas, E.W., Benamara, et al. [2] used multi-class classification to identify targets of racism, xenophobia, misogyny, sexism, and hate speech beyond generic binary classification. They constructed numerous models for this investigation. A linear support vector classifier model, which served as their baseline model, outperformed other models. With one dataset where ELMo performed best, Bidirectional Encoder Representations from Transformers (BERT) model outperformed model outperformed it in the majority of categories. Typically, only acts of violence or hatred directed at women are referred to as sexism or misogyny.

In their research on the automatic classification of sexism in social networks, authors F. Rodrguez-Sánchez et al. [3] also attempted to include more nuanced expression in online dialogues for classification. From Twitter, they have gathered a corpus of sexist remarks in Spanish. Models like logistic regression, Bidirectional Long Short-Term Memory (BiLSTM) with multiple word embeddings, BERT classifier, Support Vector Machines (SVM), and RF have been integrated.

The best accuracy score, 0.74, was provided by the models BERT, BiLSTM, and WEBSB. In a different study, the authors Harika Abburi, Shradha Sehgal, Himanshu Maheshwari, and Vasudeva Varma suggested classification models on a corpus of twitter and gab comments in their study of Knowledge-based Neural Framework for Sexism Detection and Classification [4]. The best model for classifying sexism exceeds all other models with an accuracy score of 0.716 and combines language characteristics with RoBERTa. Performance of deep learning models is superior to that of classical machine learning models. A new method for identifying gender stereotypes in tweets and improving sexism classification through stereotype injection was proposed in the work by authors Chiril, P., Dinu, G., and Morin, E., titled "Gender Stereotype Detection and Improvement of Sexism Classification in Tweets" [5]. The authors present a unique data augmentation strategy based on sentence similarity employing multilingual external resources using a French language dataset of 9,200 annotated tweets. For detecting gender stereotypes, many BERT structures have been developed. The authors also compare their findings to cutting-edge methods for detecting binary stereotypes against immigrants and classifying binary gender bias in job ads. The objective of another study by authors Dylan Grosz and Patricia CondeCespedes, Automatic Detection of Sexist Statements Commonly Used at the Workplace [6], was to create a cutting-edge model for detecting workplace sexism that can be used by sentiment analysis, AI chatbots, and human resources departments. The authors manually assembled the dataset. Convolutional Neural Networks (CNNs), Long Short-Term Memory (LSTM) networks, attention mechanisms, and pre-trained Word Embeddings GloVe were used to create seven alternative model versions for the study. A recall of 0.87 and a precision of 0.91 were achieved by the model with the highest performance, giving it an F1 score of 0.89.

In their paper, CyberPolice: Classification of Cyber Sexual Harassment [7], authors Basu, P., Roy, T. S., Tiwari, & Mehta sought to examine various machine learning and deep learning models and compare their effectiveness in identifying sexually abusive utterances. The study also sought to shed light on the occurrence, nature, and effects of online sexual harassment on its targets. SVM, Logistic Regression, Naive Bayes, Random Forest, and other machine learning models were employed in this work. CNN, LSTM, and BERT are some of the deep learning models used in this study. The model demonstrated the highest model accuracy at 83.56% with a loss of 0.727. Prior methods of detecting hate speech had drawbacks, such as high rates of false positives or an inability to distinguish between hate speech and other offensive words, according to the authors Davidson, T., Warmsley, D., Macy, & Weber. They create a more precise system for detecting hate speech on social media by separating it from other types of offensive language in their paper, Automated Hate Speech Detection and the Problem of Offensive Language [8]. The authors begin by reducing the dimensionality of the data using a logistic regression with L1 regularization. The models under test include logistic regression, naive Bayes, decision trees, random forests, and linear SVMs, all of which have been employed in earlier research.

They discovered that the Logistic Regression and Linear SVM models frequently outperformed other models by a significant margin. The model that performed the best overall had an F1 score of 0.90, a precision of 0.91, and a recall of 0.90. When does a Compliment Become Sexist?, the study's title, was written by Akshita Jha and Radhika Mamidi. Analysis and Classification of Ambivalent Sexism Using Data from Twitter [9] has investigated ambivalent sexism using data from Twitter and developed computational models to categorize tweets into three categories: "Benevolent," "Hostile," or "Others." The experiments used two classifier models: Sequence to Sequence (Seq2Seq) and Support Vector Machines (SVM). With recall values of 77% and 65%, respectively, the Seq2Seq model using RNN and LSTM beat the SVM model for the "Benevolent" and "Hostile" classes. Authors Harika Abburi, Pulkit Parikh, Niyati Chhaya, and Vasudeva

Varma propose a number of techniques, including self-training-based techniques, data augmentation methods, a neural model combining biLSTM and attention, and a multi-level training approach, to classify accounts of sexism in their study on fine-grained multi-label sexism classification using a semi-supervised multilevel neural approach [10]. On a recently available dataset for multi-label sexism categorization across multiple standard measures, the offered approaches outperform various baselines. Instance-based F-score, Fmac, and SA were all higher for the best combined approach than they were for the best baseline, which had F-scores of 0.714, 0.546, and 0.242.

## 3 Methodology

In this research paper, our primary goal was to develop and train eight distinct models to detect sexism in Twitter comments. To achieve this, we utilized the sexism dataset curated by Fatma Elsafoury [11], consisting of 14,881 records with five columns: index, id, Text, Annotation, and oh label. The 'Text' column contained the raw Twitter comments, while 'Annotation' and 'oh label' columns served as binary indicators to classify whether a tweet was considered sexist or non-sexist. Recognizing the complexities of analyzing unstructured text data, we employed crucial natural language processing (NLP) pre-processing methods to prepare the dataset for model training. Tokenization served as a fundamental step, breaking down the text into individual tokens or words, enabling the models to effectively understand the linguistic context and identify patterns related to sexism. Additionally, we opted for lowercase conversion to ensure uniformity in word representations, mitigating the impact of letter casing variations and promoting better generalization of the models.

Removing common and less informative words through stopword removal reduced noise in the data, allowing the models to focus on more relevant terms crucial for detecting sexism in Twitter comments. Furthermore, we conducted rigorous experimentation with lemmatization and stemming techniques, effectively reducing words to their base or root forms. This linguistic normalization enhanced the models' capacity to grasp semantic relationships between words, leading to improved accuracy in predicting sexism. Special attention was dedicated to handling the presence of special characters and URLs in Twitter comments, ensuring data integrity throughout the preprocessing phase. By effectively managing these elements, we safeguarded the quality of the dataset and prevented potential disruptions in model training and analysis. Furthermore, recognizing the significance of emojis and emoticons in expressing sentiments, we implemented specialized techniques to interpret and manage their impact on sentiment analysis. This ensured that the models could capture the emotional nuances present in the comments, contributing to more contextually-aware and accurate predictions. With our NLP pre-processing procedures meticulously applied, the dataset was transformed into a clean and optimized format, ready to be fed into the developed models.

The combined efforts of tokenization, lowercase conversion, stopword removal, lemmatization/stemming, and handling special characters, URLs, emojis, and emoticons enhanced the models' ability to grasp the linguistic context of the Twitter comments, leading to more informed and discriminating predictions of sexism. As part of our data preparation process, we also employed two essential techniques: one-hot encoding and

TF-IDF (Term Frequency-Inverse Document Frequency). One-hot encoding was utilized to convert categorical features, such as the 'Annotation' and 'oh label' columns, into binary vectors, enabling the models to process these labels effectively during training. On the other hand, TF-IDF was applied to transform the textual data into numerical feature vectors, accounting for the importance of words in each comment relative to the entire dataset. TF-IDF played a crucial role in assigning higher weights to more discriminative words and reducing the impact of common terms, further refining the model's ability to identify sexism in the Twitter comments. By leveraging one-hot encoding and TF-IDF, the dataset was successfully transformed into a format that could be effectively processed and fed into the developed models. These preparatory steps set the stage for a robust and informed analysis of sexism detection in Twitter comments, laying the foundation for our research's successful outcomes. After undergoing NLP preprocessing, the cleaned dataset was efficiently fed into both machine learning and deep learning models, enabling them to learn and detect sexism in Twitter comments with enhanced accuracy and contextual understanding.

a. **Machine learning models**

**Logistic Regression.** Logistic regression is a statistical technique employed to solve binary classification problems, in which the objective is to estimate the probability that an input belongs to one of two possible classes. It is classified as a supervised learning algorithm, indicating that it learns to make predictions by training on a dataset that has labeled outcomes. Throughout the training phase, the logistic regression model finds the best values for the weight coefficients by minimizing a loss function, like the binary cross-entropy loss.

**AdaBoost Classifier.** AdaBoost is an ensemble learning technique that merges multiple weak learners to form a robust and powerful classifier. In the training phase, AdaBoost assigns weights to every input data point and trains a new weak learner on the weighted data. The algorithm then uses a weighted voting scheme to combine the predictions of the weak learners. The end result is a strong classifier capable of making predictions on new data with high accuracy. While AdaBoost is a potent and adaptable technique that performs well on numerous problems, it can be susceptible to noisy data and outliers, and requires thorough tuning of hyperparameters to achieve optimal performance.

**XGBoost Classifier.** XGBoost is an ensemble learning technique that leverages gradient boosting to merge numerous weak learners. The algorithm incorporates several features, including regularization, early stopping, and estimation of feature importance, to improve its performance. XGBoost employs a specialized loss function that enables it to optimize for specific performance metrics, and it can handle both binary and multiclass classification and regression problems. With its remarkable performance on many benchmark datasets, XGBoost is considered a state-of-the-art technique. However, it can be sensitive to noisy data and necessitates careful hyperparameter tuning to achieve optimal performance.

**Decision Tree.** Decision Tree is a supervised learning technique that constructs a tree-based model to make predictions on new data. It is highly adaptable and suitable for both classification and regression problems. Decision Tree is straightforward to interpret and visualize, making it a valuable tool for comprehending the underlying patterns in the data. Nevertheless, it may be susceptible to noisy data and overfitting and necessitates meticulous hyperparameter tuning for optimal performance. Decision Tree is a potent and extensively employed algorithm, especially when the relationships in the data are nonlinear and intricate.

**Random Forest Classifier.** Random Forest is an ensemble learning technique that fuses several decision trees to form a robust and powerful classifier. The algorithm trains each decision tree by randomly selecting subsets of features and data points, which curbs overfitting and enhances generalization. Random Forest can handle both classification and regression problems and estimate the importance of features. It finds extensive applications across different domains, but it can be susceptible to noisy data and needs meticulous parameter tuning to achieve optimal performance.

**KNN.** K-Nearest Neighbors (KNN) is a supervised learning technique that assigns class labels to new data based on the labels of its nearest neighbors in the training data. It can be applied to both classification and regression problems and is a versatile approach. KNN is widely used because of its ease of comprehension and implementation and is often utilized in real-world scenarios. It can capture non-linear relationships in the data, but the choice of distance metric and the value of K can affect its performance. KNN is a potent and prevalent algorithm, particularly when the decision boundary is non-linear, and the amount of training data is relatively low.

**Gaussian Naive Bayes.** Gaussian Naive Bayes (NB) is a supervised learning technique that leverages Bayes' theorem under the assumption that the features are conditionally independent, given the class label. The algorithm is fast and straightforward, and can handle high dimensional feature spaces with limited training sets. Gaussian NB is also resistant to irrelevant features, which makes it a favored approach for various real-world applications. Gaussian Naive Bayes is a potent and extensively employed algorithm, primarily when the independence assumption is valid, and the data is well-differentiated by class.

**A Deep Learning Models.**

**Neural Network** Neural networks are a machine learning technique that takes inspiration from the human brain's structure and function. They comprise interconnected nodes organized into layers that process input data to produce output predictions. The training of neural networks involves modifying the nodes' weights and biases to minimize the difference between predicted and actual values. Different types of neural networks exist, each designed to address specific problems. One of their strengths is their ability to learn and extract relevant features from raw data, but they can be computationally intensive and may necessitate substantial amounts of data to achieve good performance. Neural networks are a potent and versatile class of machine learning models that have made a significant impact in many fields.

**LSTM.** LSTM is a type of recurrent neural network specially designed to capture long-term dependencies in sequential data by utilizing memory cells with specialized gating mechanisms that can selectively retain and update information. Due to their effectiveness in natural language processing and speech recognition, LSTMs have been widely used and have set state-of-the-art performance benchmarks in numerous related tasks. As a result, LSTM and its variants have become a fundamental tool in the field of deep learning and are frequently employed in various applications that involve sequential data.

**BiLSTM.** The Bidirectional Long Short-Term Memory model is an enhanced version of the LSTM model that can process input sequences in both forward and backward directions. This feature makes BiLSTMs particularly suitable for natural language processing tasks, including named entity recognition, sentiment analysis, and part-of-speech tagging. The ability to capture information from past and future words in the sequence by processing in both directions allows BiLSTMs to achieve better performance than traditional LSTM models. BiLSTMs are widely used and have become a popular choice for various applications that involve sequential data.

a **Evaluation Metrics**

**Accuracy score.** Accuracy is an evaluation metric that measures how well a model predicts the correct label for a given input. It is calculated by dividing the number of correctly predicted instances by the total number of instances in the dataset. The resulting score is typically expressed as a percentage. Accuracy = (number of correctly predicted instances) / (total number of instances) This can be given in terms of true positive, true negative, false positive and false negative.

$$Accuracy = (TP + TN)/(TP + TN + FP + FN) \quad (1)$$

**Precision.** The proportion of true positive (TP) predictions among all the positive predictions the model makes is measured by the accuracy metric. In other words, precision assesses the model's ability to make accurate positive predictions. In mathematics, precision is represented as:

$$\Pr ecision = TP/(TP + FP) \quad (2)$$

**Recall.** Recall assesses the model's ability to accurately separate out every positive case from the overall number of real positive instances in the data.

$$\mathrm{Re}call = TP/(TP + FN) \quad (3)$$

**F1 score.** A classification model's precision and recall are balanced by the F1 score measure. Since it is the harmonic mean of precision and recall, it may be expressed as a single number that accurately conveys both precision and recall. The mathematical definition of the F1 score is

$$F1Score = 2 * (\Pr ecision * \mathrm{Re}call)/(\Pr ecision + \mathrm{Re}call) \quad (4)$$

**AUC.** The true positive rate (TPR) vs the false positive rate (FPR) for various classification thresholds is plotted on the ROC (Receiver Operating Characteristic) curve, and the area under the curve (AUC) is the area under the curve. The ROC curve can be mathematically integrated to determine the AUC:

$$AUC = \int_0^1 (TPR(x) \quad FPR'(x)\partial x \quad (5)$$

## 4 Experimentation

**A Preprocessing**

14,881 records total in the Twitter sexism dataset, and each record's text content underwent preprocessing before analysis. The content was preprocessed by taking out URLs, hashtags, HTML elements, emoticons, and punctuation. Then, all of the text's characters were changed to lowercase. The stop words were also eliminated from the text, and the TF-IDF approach was used to tokenize and vectorize the remaining words. A numerical representation of the text that can be utilized for additional analysis or modeling was produced as a result of this operation.

**B Classification.**

We must determine if the text is sexist or not based on the information provided. Therefore, this is best handled as a binary classification issue. Predicting one of the two potential outcomes is the challenge at hand in a binary classification problem. Building a model that can precisely categorize fresh, undiscovered occurrences into one of the two categories based on a collection of features is the ultimate aim of binary classification. In our example, binary classification was carried out using the models indicated above. The logistic regression approach was used in the creation of the initial model. In order for this approach to function, training data are fitted with a logistic function that converts input variables into probabilities. We received an accuracy rating of 84.30% from this model. The Ada boost classifier was the second model to be created. The ensemble approach known as Ada boost (short for adaptive boosting) combines several weak classifiers to create a strong one. It functions by iteratively training a number of weak classifiers on the training data and then summing their weights. It concentrates on the training examples that were incorrectly categorized in the previous round and modifies their weights to give them more weight in the subsequent round. We received an accuracy rating of 83.87% with this model. The xgboost classifier was the next model to be created. Extreme gradient boost, commonly known as Xgboost, is a technique that combines a number of weak classifiers to produce a powerful one. However, it also uses a regularized objective function to prevent overfitting, enables parallel processing, and utilizes distributed computing to hasten the training process. Our accuracy score from this model was 83.89%. The decision tree classifier was the second model to be created. The most discriminating feature is chosen by the decision tree algorithm, which then makes use of it to separate the data into subsets according to the feature values. Recursively repeating this process until a halting requirement is satisfied. The decision tree creates a model that can forecast the class label of incoming instances based on their features by breaking the data into more homogeneous groups. We received an accuracy of 83.87% using this model. A random forest classifier was the next model to be created. This algorithm likewise uses ensemble learning and integrates different decision trees. The main goal of this technique is to improve generalization by reducing overfitting and adding unpredictability to the training process. Random forest is able to capture various parts of the data and lessen model variation by training each tree on a distinct subset of the data and characteristics. The accuracy rating this model provided was 83.04%. The K nearest neighbors model was the following one created. The K data points in the training set that are most similar to a new input instance are found by this technique. This is accomplished by employing a distance metric, such as the Euclidean or Manhattan distance, to calculate the separation between each instance in the training set and the input instance. The class label or regression value of the input instance is then predicted using the class label or regression value of the K nearest neighbors, which are chosen based on the shortest distances. The accuracy rating was 81.89%. The Gaussian Naive Bayes model is the following model created. Based on the training data, this method calculates the mean and standard deviation of each characteristic for each class. The conditional probability of a new input instance falling into each class is then determined for that instance using the Gaussian probability density function. The highest probability class is then chosen as the projected class for the input instance. A 38.84 percent accuracy rating was provided. Additionally, we included the LSTM, BiLSTM, and Neural Networks as three Deep Learning models. Artificial Neural Networks are the first model that was created. The final output layer in this model has a sigmoid activation function, and there are five hidden layers with Relu activation functions. In order for the model to produce the desired categorical output, the functions are used. Multiple layers of interconnected nodes or neurons make up a neural network, and each neuron in a neural network processes and transmits information to other neurons. The input layer's neurons take in the input data, while the output layer generates the final output or prediction. Multiple hidden layers may exist between them, where the neurons perform non-linear changes on the input data to help the model learn intricate connections and representations of the data. Additionally, we included the LSTM, BiLSTM, and neural networks as three Deep Learning models. Artificial Neural Networks are the first model that was created. The final output layer in this model has a sigmoid activation function, and there are five hidden layers with Relu activation functions. In order for the model to produce the desired categorical output, the functions are used. Multiple layers of interconnected nodes or neurons make up a neural network, and each neuron in a neural network processes and transmits information to other neurons. The input layer's neurons take in the input data, while the output layer generates the final output or prediction. Multiple hidden layers may exist between them, where the neurons perform non-linear changes on the input data to help the model learn intricate connections and representations of the data. The input gate controls how much fresh data is entered into the cell, the forget gate decides how much data should be removed from the cell, and the output gate manages how much of the data from the cell is used in the prediction. In the training phase, this model's accuracy score was 99.04 percent, while in the testing phase, it was 87.23 percent. The BiLSTM (Bidirectional LSTM's) was the next deep learning model to be created. A particular kind of LSTM called a bi-LSTM processes sequences both forward and backward. Because it can capture both past and future context, it is useful for tasks like sentiment analysis and speech recognition. Two LSTM layers make up the bi-LSTM, which reads the sequence in both ways and concatenates its outputs to get the final result. To reduce the loss, Bi-LSTM is trained utilizing backpropagation over time. In the training phase, this model's accuracy score was 99.67%, while in the testing phase, it was 86.39%.

a. **Replacement**

In the replacement phase, we are identifying the sexist terms from the texts that were earlier classified to be sexist and replacing them with alternate words that is taken from a dictionary we built containing computer science based terms. This is done in an intention to remove the sexist terms and at the same time give the sentence turns out meaningless and funny completely cancelling out an abuser's intentions. Post classification we're maintaining a threshold of 0.5. If the prediction result is below 0.5, this implies the input text is not sexist and will return the text as it is. If the prediction is above 0.5, we proceed with replacement. Based on the accuracy results, we chose to pre-train the LSTM and BiLSTM models on the large corpus of twitter sexism dataset. This pretrained model will be used to perform the prediction to the given textual input. Once the user gives the textual input, the input is padded to make the input sequence of the same length. This padded sequence is used an input for our model prediction and based on the prediction the appropriate replacement is performed.

## 5 Results and Discussion

Using the above mentioned models, the comparison was performed by taking accuracy, precision, recall, F1 score and AUC into account. The results of these metrics are presented in Table 1.

**Table 1.** Comparison of models based on evaluation metrics

| Models | Accuracy | Precision | Recall | F1 score | AUC |
|---|---|---|---|---|---|
| LR | 0.8429 | 0.8120 | 0.4007 | 0.5366 | 0.6867 |
| AdaBoost | 0.8429 | 0.8948 | 0.3277 | 0.4797 | 0.6582 |
| XGBoost | 0.8389 | 0.8951 | 0.3287 | 0.4808 | 0.6587 |
| DT | 0.8387 | 0.8948 | 0.3277 | 0.4797 | 0.6582 |
| RF | 0.8304 | 0.9571 | 0.2645 | 0.4145 | 0.6305 |
| KNN | 0.8189 | 0.7303 | 0.3208 | 0.4458 | 0.6430 |
| GNB | 0.3884 | 0.2614 | 0.9289 | 0.4080 | 0.5793 |
| NN | 0.8434 | 0.8434 | 0.8434 | 0.8434 | 0.7061 |
| LSTM | 0.8723 | 0.7446 | 0.6651 | 0.7026 | 0.7991 |
| BiLSTM | 0.8639 | 0.6991 | 0.7022 | 0.7006 | 0.8067 |

It is interesting to note that accuracy and AUC score captures different aspects of model performance and can vary independently, leading to cases where models with the same accuracy have different AUC values. The reason why models with the same accuracy can have different AUC values is because accuracy only considers the overall correctness of predictions, while AUC takes into account the prediction probabilities or scores produced by the model. Two models with the same accuracy may have different AUC values due to variations in their ability to rank positive and negative samples correctly in terms of the predicted probabilities or scores. For example, the Logistic Regression model and the AdaBoost model both have an accuracy of 84.29%, meaning they make correct predictions 84.29% of the time. However, the Logistic Regression model seems to produce more distinct and well-separated prediction scores for positive- and negative samples, resulting in a higher AUC. The AdaBoost model, on the other hand, might produce prediction scores that are more overlapping, leading to a lower AUC even though the accuracy is the same. Taking into account the training accuracies as well, we are taking LSTM and BiLSTM to perform desired replacements. The graphs for training accuracy vs validation accuracy for the same are shown in Figs. 1 and 2 respectively.

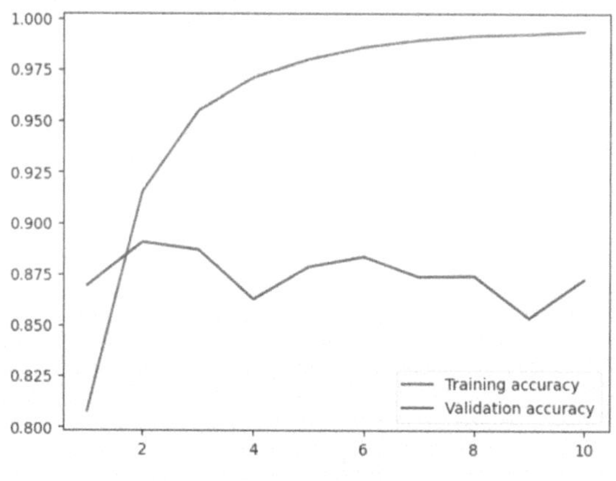


x-axis:Epochs y-axis:Accuracy

**Fig. 1.** Training and validation accuracy LSTM

Figures 1 and 2 represents training and validation accuracy for LSTM and BiLSTM model with 10 epochs. As we can see here the training accuracy increases with number of epochs and goes for a maximum value whereas the validation accuracy plateaus at a particular validation accuracy. The pre-trained model uses the dictionary shown in Table 2 and replaces sexist language identified by it.

A sample identification of a sexist comment and replacement of slur words is shown in Table 3.

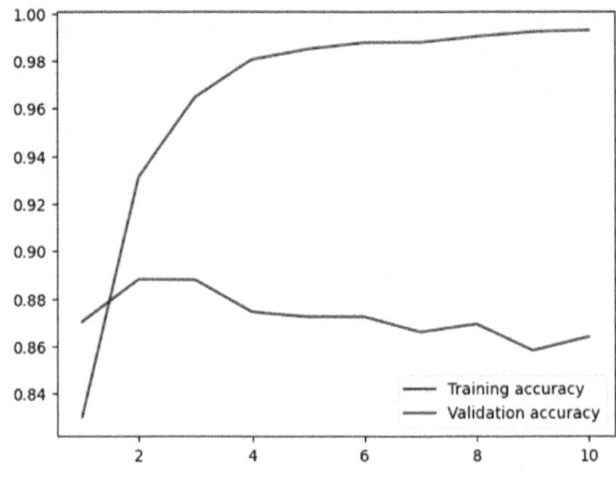


x-axis:Epochs y-axis:Accuracy

**Fig. 2.** Training and validation accuracy BiLSTM

**Table 2.** Dictionary used for replacement

| Possible slur words | Alternative words Fuck Function |
|---|---|
| Awful | Array |
| Bitch | Bert |
| Sexist | Linkedlist |
| Shit | Git |
| Slut | Struct |
| Sexism | Lexism |
| Whore | Core |
| Kitchen | Knn |
| Sex | Sets |
| Hoe | Hacker |
| Devil | Dictionary |
| Hate | Hash |

**Table 3.** Sample replacement

| Sample input | Sample output this bitch is a slut this bert is a struct |
|---|---|

## 6 Conclusion

In this study, we delved into a comprehensive exploration of different techniques for detecting sexism in texts, including Logistic Regression, AdaBoost, XGBoost, KNN, Decision Tree, Random Forest, Gaussian Naive Bayes, Neural Networks, LSTM, and BiLSTM. While these models exhibited promising results in identifying sexist content, there remains a need for further research to enhance their performance and accuracy. A potential direction for future investigations involves the implementation of hybrid models that amalgamate the strengths of diverse algorithms. By combining ML models with DL models or exploring ensembles of decision trees with deep neural networks, we can potentially achieve improved detection accuracy. Leveraging successful hybrid approaches from various NLP tasks could foster even more precise sexism detection. Moreover, expanding the focus of this research to encompass other offensive speech components, such as homophobic, transphobic, and racist language, would significantly contribute to creating a safer and more inclusive online environment. The development of models capable of identifying and replacing such slurs and abusive language would be instrumental in fostering respectful digital communities. Furthermore, we can extend and adapt the current models to recognize and address different forms of hate speech and cyberbullying. Utilizing the knowledge gained from this research, we can implement measures to prevent psychological harm to individuals. In conclusion, the detection of sexism in texts is a crucial endeavor to safeguard people from psychological harm in online environments. Future research in this area has the potential to advance the development of more secure and welcoming online communities. By investigating hybrid models, broadening the focus to include other types of hate speech, utilizing sophisticated pretraining techniques, enhancing replacement generation, and increasing dataset size and diversity, we can contribute to a more respectful and empathetic digital space. The application of our findings can extend not only to researchers in the same field but also to researchers in other domains and the general public, promoting a safer and more inclusive online environment for all.

## References


1. Mansur, Z., Omar, N., Tiun, S.: Twitter hate speech detection: a systematic review of methods, taxonomy analysis, challenges, and opportunities. IEEE Access **11**, 16226–16249 (2023). https://doi.org/10.1109/ACCESS.2023.3239375
2. Chiril, P., Pamungkas, E.W., Benamara, F., et al.: Emotionally informed hate speech detection: a multi-target perspective. Cogn. Comput. **14**, 322–352 (2022). https://doi.org/10.1007/s12559-021-09862-5
3. Rodríguez-Sánchez, F., Carrillo-de-Albornoz, J., Plaza, L.: Automatic classification of sexism in social networks: an empirical study on twitter data. IEEE Access **8**, 219563–219576 (2020). https://doi.org/10.1109/ACCESS.2020.3042604
4. Abburi, H., Sehgal, S., Maheshwari, H., Varma, V.: Knowledge-based Neural Framework for Sexism Detection and Classification. IberLEF@SEPLN (2021)
5. Chiril, P., Benamara, F., Moriceau, V.: Be nice to your wife! The restaurants are closed: Can Gender Stereotype Detection Improve Sexism Classification?. In: Findings of the Association for Computational Linguistics: EMNLP 2021, pp. 2833–2844, Punta Cana, Dominican Republic. Association for Computational Linguistics (2021)

6. Grosz, D., Conde-Cespedes, P.: Automatic detection of sexist statements commonly used at the workplace. In: Lu, W., Zhu, K.Q. (eds.) Trends and Applications in Knowledge Discovery and Data Mining. PAKDD 2020. Lecture Notes in Computer Science(), vol 12237. Springer, Cham (2020). https://doi.org/10.1007/978-3-030-60470-7_11
7. Basu, P., Singha Roy, T., Tiwari, S., Mehta, S.: CyberPolice: classification of cyber sexual harassment. In: Marreiros, G., Melo, F.S., Lau, N., Lopes Cardoso, H., Reis, L.P. (eds.) Progress in Artificial Intelligence. EPIA 2021. Lecture Notes in Computer Science(), vol 12981. Springer, Cham (2021). https://doi.org/10.1007/978-3-030-86230-5_55
8. Davidson, T., Warmsley, D., Macy, M., Weber, I.: Automated hate speech detection and the problem of offensive language. Proc. Int. AAAI Conf. Web and Social Media **11**(1), 512–515 (2017). https://doi.org/10.1609/icwsm.v11i1.14955
9. Jha, A., Mamidi, R.: When does a compliment become sexist? Analysis and classification of ambivalent sexism using twitter data. In: Proceedings of the Second Workshop on NLP and Computational Social Science, pp. 7–16. Vancouver, Canada. Association for Computational Linguistics (2017)
10. Abburi, H., Parikh, P., Chhaya, N., et al.: Fine-grained multi-label sexism classification using a semi-supervised multi-level neural approach. Data Sci. Eng. **6**, 359–379 (2021). https://doi.org/10.1007/s41019-021-00168-y
11. Elsafoury, F.: Cyberbullying datasets. Mendeley Data **1** (2020). https://doi.org/10.17632/jf4pzyvnpj

# Machine Learning

# Detection of DoS Attack Using Machine Learning in Software Defined Network

M. A. Gunavathie[1], S. Kousalya[2](✉), J. Celina Rekha[3], S. Rasikaranjani[3], and D. Mahalakshmi[3]

[1] Department of Information Technology, Easwari Engineering College, Chennai, India
[2] Department of CSE, Panimalar Engineering College, Chennai, India
kousikousi3006@gmail.com
[3] Departmant of IT, Panimalar Engineering College, Chennai, India

**Abstract.** Nowadays, Software-Defined Networking (SDN) offers benefits in the areas of automation, elasticity, and resource consumption. However, evidence is there that SDN controllers may undergo certain defeat for the network structure, particularly as they are targeted by attacks like Denial of Service (DoS) attacks. Due to this network traffic has increased tremendously and attacked the server severely. To handle this issue, we used the RYU controller and Mininet tool to identify and alleviate the DoS attack by the Machine learning algorithm. Since machine learning (ML) is deemed as the main method for detecting peculiarities, the detection of DoS attacks was done through Machine learning-based classification. In this paper, several machine learning techniques were used to identify the DoS attack, and the traffic which is causing the attack has been dropped immediately to avoid congestion. The proposed work has been simulated in Mininet and the results show that the proposed work detects DoS attacks well and achieves good accuracy.

**Keywords:** Software Defined Network (SDN) · Ryu Controller · Mininet

## 1 Introduction

A DoS attack is a malicious assault on a network or its resources, with the primary objective of rendering critical services and resources inaccessible to authenticated users. By exhausting the target system's capacity with an excessive amount of traffic, DoS attacks aim to disrupt normal network operations and cause service outages. As traditional network security mechanisms struggle to keep pace with the sophistication and scale of such attacks, there is an increasing need for innovative and intelligent approaches to detect and mitigate them effectively. Machine Learning (ML), an emerging field in the realm of artificial intelligence, has shown remarkable promise in transforming various domains, and network security is no exception. Leveraging the power of ML techniques, Software Defined Networks (SDNs) can be fortified with intelligent and proactive defense mechanisms that autonomously analyze network traffic patterns, enabling early detection and response to potential DoS attacks. The application of ML in DoS attack detection in


R. Geetha et al. (Eds.): AAIMB 2023, CCIS 2203, pp. 61–71, 2025.
https://doi.org/10.1007/978-3-031-73068-9_6

SDNs offers several distinct advantages. Firstly, it allows the network to learn from historical traffic data and recognize patterns associated with normal network behavior, thus enabling it to distinguish between legitimate and malicious activities. Secondly, ML algorithms can adapt and evolve over time, continuously improving their detection accuracy as they encounter new attack patterns.

The core principle behind ML-based DoS attack detection involves the training of algorithms on labeled datasets containing examples of normal network behavior and known DoS attack instances. These algorithms then generalize from the training data to detect anomalous patterns in real-time network traffic. By employing features such as traffic volume, packet rate, and network flow characteristics, ML models can discern subtle deviations that indicate potential DoS attacks. In this study, we aim to discover the potential of ML techniques in detecting DoS attacks within Software Defined Networks comprehensively. By investigating various ML algorithms, including decision trees, support vector machines, deep learning, and ensemble methods, we endeavor to identify the most effective models for enhancing the security posture of SDNs. This paper shows the proposed work which detects the Dos attacks in SDN using the RYU controller and Mininet tool.

## 2 Literature Survey

Meng Yue et al. [1], suggested a method for DoS attack recognition using features in the flow table and then designed a neural network to distinguish DoS attacks from Flash crowds at the same time. Shang Gao et al. [2], introduced Defender, to prevent SDN networks from network-directed DoS that leverage new structures in attack recognition, as well as three new mitigation methods. J. E. Varghese et al. [3], by incorporating data layer intelligence in the SDN data plane architecture, authors suggested a new context that addresses both the effectiveness and design issues of Intrusion Detection Systems related to DDoS attacks. J. A. Pérez-Díaz et al. [4], presented an adaptable, flexible architecture that lets Low-Rate DDoS attacks to be detected and alleviated in SDN settings. Mohammed Anbar et al. [5], detected DDoS attacks on SDN controllers. A hybrid approach could also be developed to combine the advantages of existing approaches to detect DDoS attacks more comprehensively.

Dan Tang et al. [6], proposed a lightweight method for identifying and lessening LDoS attacks in a SDN network. L. Tao Wang et al. [7] suggested a mitigation strategy for the DoS attacks against an SDN controller which mainly comprised four mechanisms: Detecting the DoS attack component, predicting engine, priority manager, and scheduler. Noe Marcelo Yungaicela-Naula et al. [8], demonstrated a way to identify various types of transport layer DDoS attacks, and the application layer DDos attacks using artificial intelligence. Tan et al. [9], suggested a method for the identification and defense of DDoS attacks in the SDN by using the methods that combine control plane and data plane cooperative identifying methods to increase detection accuracy and performance. S. Dong et al. [10], suggested two ways to categorize DDoS attacks in Softwaredefined networks. The first method for identifying DDoS attacks is to take into account the degree of attack. A second method employs Machine Learning to determine the DDoS attack's neighbors based on K- Nearest Neighbors (KNN). Zakaria Abou El Houda et al.

[11], dealt with DNS amplification attacks while preserving the security of the network (i.e., protecting the data plane resources). This work will also be extended to introduce inter-domain reducing strategies based on Blockchain. W. Zhijun et al. [12], described a method for defense using dynamic flow rule deletion. Lingfeng Yang et al. [13], suggested a framework that uses machine learning to identify and protect against DDoS attacks on the institute campus network. Abimbola O. Sangodoyin et al. [14], use several popular machine learning (ML) algorithms to identify and categorize DDoS floods on SDNs for identification and categorization. K. S. Sahoo et al. [15] presented a method for recognizing DDoS attacks and mitigating them for SDN systems. Specifically, this method employs a Support Vector Machine for detection purposes.

Kandoi et al. [16], examined DoS attacks directing SDN network OpenFlow. They illustrated how a flow rule's timeout value and bandwidth of control plane affect the switch's performance. Peng Zhang et al. [17], discuss several existing solutions to defend SDN environments against denial-of-service attacks. R. Durner et al. [18], optimized a statistical detection method and a lightweight countermeasure. The Simulation and analysis were used to evaluate the detection speed and adaptability. Lohit Barki et al [19], analyzed the DDoS attacks based on the traffic traces. The traffic was categorized as normal or abnormal using a combination of algorithms such as Naive Bayes, K-Nearest Neighbor, K-means, and K medoids. Signature IDS uses an algorithm with superior accuracy, and then Advanced IDS processes its results to detect anomalous behavior based on open connections.

Obaid Rahman et al. [20] developed an SDN framework that detects and protects controllers and switches from DDoS attacks. Santos et al. [21], used Mininet simulated environment for SDN and recommended four machine learning techniques to categorize DDoS attacks. Mykhailo klymash et al. [22], suggested a log analysis method for identifying DDoS attacks in SDN networks. W. Sun et al. [23] have established a method of identifying DDoS attacks. BiLSTM-RNN is applied to the DDoS system in an SDN environment in order to calculate Renyi entropy. Babatunde Hafis et al. [24] presented a real-time security solution with the ability to find and decrease attacks on the SDN. They provided a real-time detection method that uses sFlow mitigation technology to find and lessen DDoS flood attacks on the SDN. M. Ruaro et al. [25], presented a complete and protected SDN framework, which covers the procedures required to support SDN. In addition, the paper describes how the hardware, operating system, and user interface interact.

A.B. Feroz Khan et al [26] present a significant contribution to the domain of IoT security. Their study focuses on mitigating DDoS attacks through the introduction of a trust-based protocol, specifically designed for enhancing the security of IoT devices and networks. Gunavathie et al [27] presented an innovative approach to enhance the efficiency and performance of Software Defined Networks. This proactive re-routing mechanism ensures that critical and time-sensitive traffic receives optimal routing paths, ultimately improving overall network performance. Authors in [28] proposed the BBR algorithm, designed to leverage real-time bandwidth information for making routing decisions. Authors in [29] provides a comprehensive and insightful exploration of the existing techniques and methodologies in the domain of traffic prediction and classification within SDN environments.

## 3 Proposed System

In the developed networking architecture, the threat to the safety of the network system is a splendid task. The major threat is found to be the DDoS attack. The proposed work is to identify the DDoS attack in the SDN using the techniques of Machine Learning. Also, to mitigate the attack by specifically identifying the port which causes the attack and blocking the port. The machine learning algorithm is trained with the data collected in training mode to detect and differentiate between the attack and normal traffic. In training mode one of the ports is made to ping one of the servers and by generating normal traffic and the collected data is saved in a CSV file format. The attack traffic is made by creating a syn attack on the server by pinging from any of the ports and the collected data is stored in the same CSV file. The stored file is given as input to the ML algorithm and the Ryu controller is run in detection mode. The ML algorithm classifies normal and attack traffic and displays the result as either 0 or 1 meaning normal or attack respectively. Also, this port that attacks the network is identified and blocked by dropping the packets from that port for a specified time interval. Once the time interval expires the port is set free to ping over again in normal conditions. This method not only identifies the attack but also reduces the attack and ensures the security of the network. The network topology mentioned in Fig. 1, consists of three switches and a pair of servers for sending and receiving requests and responses. The entire network topology is connected with a Ryu controller, a type of SDN controller which helps in efficient management and control of the network traffic.

The network is of single-layered topology with four ports in which each port is run using some transport layer protocols. The used protocols are ICMP, TCP/IP, and UDP. Each of the port run in all the protocols and the network traffic is formed through one of the ports and the data in testing mode is collected. The attack traffic is generated via the xterm terminal on one of the servers from one of the hosts. The type of attack caused is a syn-attack which is a type of DoS attack which makes use of the internet communication protocol TCP/IP and bombards the server with multiple requests to make a large queue for requests of services and thus making the host unresponsive due to the heavily loaded traffic

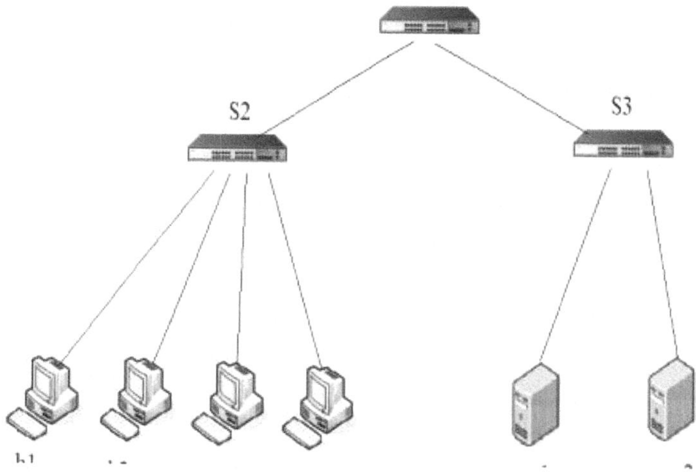


**Fig. 1.** Network Topology

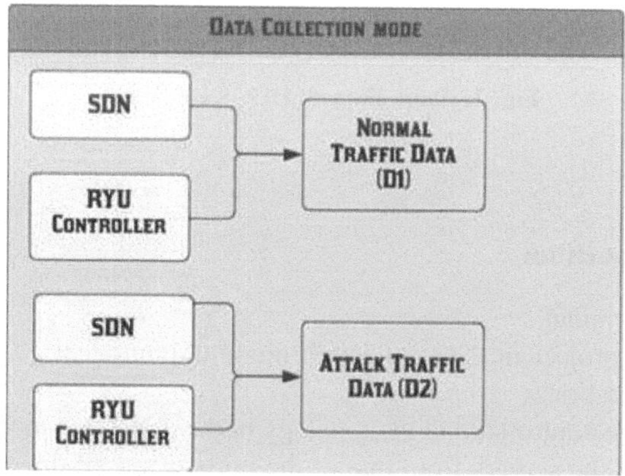


**Fig. 2.** Work Flow of Data Collection

The machine learning algorithm is trained with the data collected in training mode to detect and differentiate between traffic that could be either normal or attack. The workflow of Data collection is shown in Fig. 2. In training mode one of the ports is made to ping one of the servers and by generating normal traffic and the collected data is stored in a CSV file format. The attack traffic is made by creating a syn attack on the server by pinging from any of the ports and the collected data is stored in the same CSV file. The data collected in normal traffic mode D1 and attack traffic mode D2 is trained using machine learning algorithms and it is depicted in Fig. 3.

The flow of the project starts with the data collection for training the system with the normal traffic and attack traffic, and the detection of the attack is made by using the ML classifier to differentiate between attack and normal, and then the attack is mitigated which is referred to in Fig. 4.

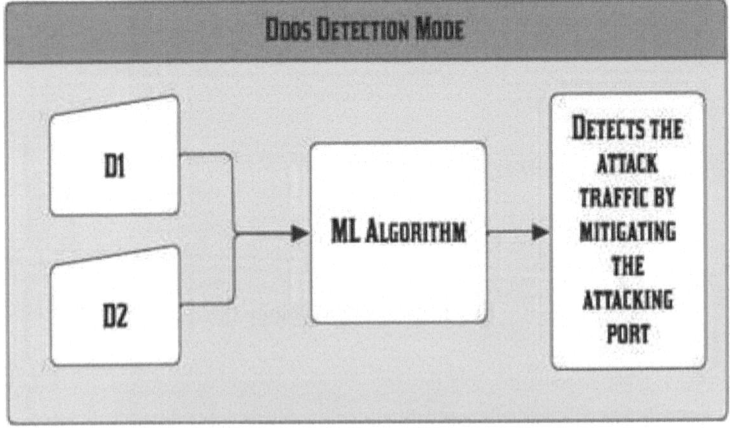


**Fig. 3.** Work Flow of DDOS Detection

### 3.1 Proposed Algorithm

Step1: Open the terminal.
Step2: Set up the project in data collection mode (Appmode=0, Trafficmode=0) for normal traffic generation.
Step3: Start the Ryu controller and the topology in the collection mode.
Step4: Ping any of the servers from one of the ports using all three protocols. Step5: Save the collected data in a .csv file.
Step6: Stop the controller and topology after the data collection in normal traffic.
Step7: Again start the Ryu controller and the topology in attack traffic mode (App mode=0, Traffic mode=1) Step8: Ping any of the servers from one of the ports with the syn attack traffic and store the collected data in the same CSV file created in Step 5.
Step9: Repeat step 6 and open a new terminal.
Step10: Edit the algorithm to DoS detection mode (App mode=1) and repeat step 1 for dos attack detection mode.
Step11: Testing the attack by generating normal traffic and attack traffic by pinging the server from the ports. Step12: While an attack is detected it will be mitigated simultaneously and the flow can be viewed in the dump flow of the packets.
Step13: Once the attack is identified, the particular port will be suspended for a specific time and will be again active and eligible to send the packets again.
Step 14: Thus the Dos attack is successfully detected and mitigated in the SDN architecture

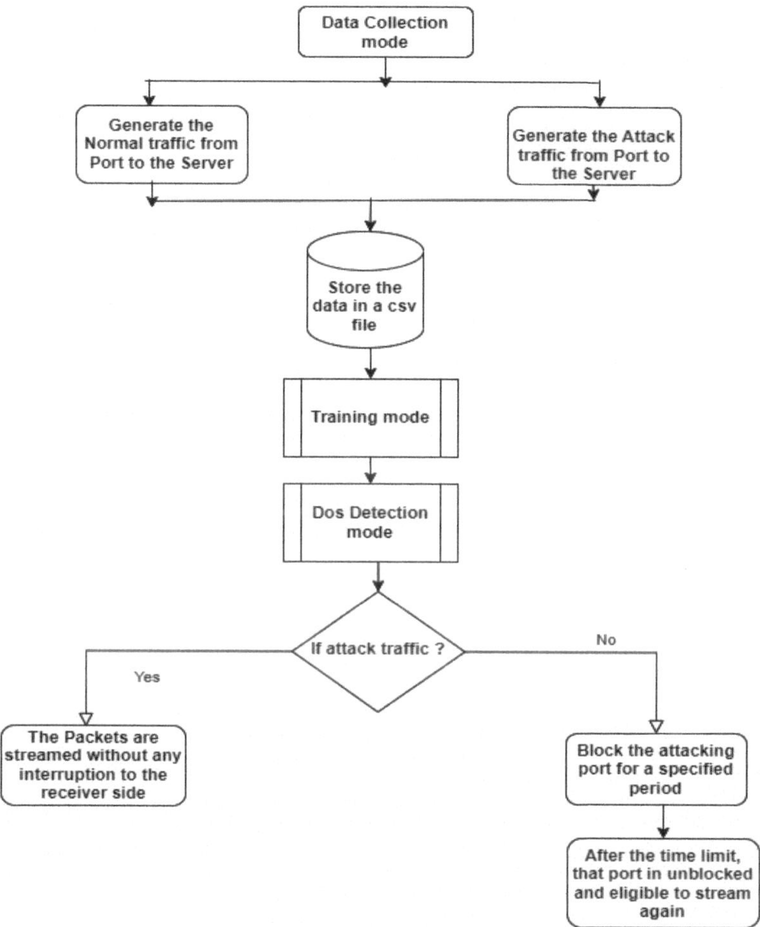


**Fig. 4.** Flow of proposed work

## 4 Experimental Results and Discussion

The experiments are especially conducted on Mininet, which has been the testing environment for many experimenters for the last few years. Researchers conduct experiments using hardware testbeds with the result of experimental platforms. Here we estimated our method's performance in this section by comparing it to other methods and proving its effectiveness through test-bed experiments.

The testbed contains three switches, four hosts, and two servers in it. The network metrics are collected as training data and then we train the Machine Learning Algorithm with the training data. In live Traffic, the attack is made in the network and using the ML algorithm the traffic is categorized based on the collected training data as an attack or normal traffic. Machine Learning Algorithm then classified as ATTACK traffic or NORMAL traffic. Based on the response generated, if it is an attack, SDN Controller performs the prevention (Block the attacked port). We then compare and scrutinize the performance metric values such as Accuracy score, Detection rate, Confusion matrix, Precision, recall, and F-1 score.

Subsequently, we trained and tested the six classifiers (SVM, Decision tree, Gaussian Naïve Bayes, Random Forest, Extra tree classifier, and Neural network classifier). Furthermore, we compared variable values for the identical classifier parameter across different classifiers to optimize each classifier. From the result fetched we have classified the obtained results and tabulated them below. The comparisons are illustrated in graphs between the metrics used for classification.

### 4.1 Metrics Used

ACCURACY SCORE:

The accuracy is determined based on the number of data sets in which the model analysts are accurate, which is calculated as:

$$Accuracy = \frac{TN + TP}{FN + TN + TP + FP} \tag{1}$$

PRECISION:

A precision (P) measure shows how many packets got attacked actually in the data packet considered as attack type by the model, and is calculated as follows:

$$\Pr ecision(P) = \frac{TP}{TP + FP(2)} \tag{2}$$

RECALL:

The Recall rate (R) represents a percentage of all packets the model identifies as attack types, which is calculated as follows:

$$\mathrm{Rec}all(R) = \frac{TP}{TP + FN} \tag{3}$$

F-1 SCORE:

Using the F1 score (F1) makes it possible to determine the accuracy of model performance by using the harmonic average of precision and recall

$$F^1 Score = 2 * \frac{PR}{P + R} \tag{4}$$

Table 1 displayed above offers a comparison of the performance outcomes achieved by diverse machine learning algorithms employed in the proposed DoS attack detection system. The algorithms evaluated encompass SVM, Decision Tree, Gaussian Naïve Bayes, Forests of Random Trees, and Extra Tree Classifier. The Accuracy Score indicates the percentage of correctly classified instances. The Decision Tree, Forests of Random Trees, and Extra Tree Classifier achieved perfect accuracy scores of 100%, indicating that they correctly classified all instances. Gaussian Naïve Bayes achieved a relatively high accuracy score of 96.05%, while SVM performed slightly lower with an accuracy score of 94.35%.

**Table 1.** Comparison of ML algorithms in the detection of DoS

| Algorithm | SVM | Decision Tre | Gaussian Naïve Bayes | Forests of Random Tre | Extra Tree eClassifier |
|---|---|---|---|---|---|
| Accuracy sco | 94.35 | 100.0 | 96.05 | 100.0 | 100.0 |
| Cross Validation Score | 0.96 | 0.99 | 0.97 | 0.99 | 1 |
| Detection Ra | 1.0 | 1.0 | 1.0 | 1.0 | 1.0 |
| False Alarm rate | 0.65 | 0.0 | 0.42 | 0.0 | 0.0 |
| Precision | 0.941 | 1 | 0.961 | 1 | 1 |
| Recall | 1 | 1 | 0.996 | 1 | 1 |
| F1- Score | 0.969 | 1 | 0.96 | 1 | 1 |

The Cross Validation Score represents the generalization performance of the models on unseen data. Extra Tree Classifier achieved a perfect crossvalidation score of 1, indicating excellent generalization capabilities. The Decision Tree and Forests of Random Trees performed well with crossvalidation scores of 0.99, while Gaussian Naïve Bayes and SVM achieved scores of 0.97 and 0.96, respectively. The Detection Rate, also called as True Positive Rate, measures the ability of the algorithms to correctly identify DoS attacks. All algorithms, except Gaussian Naïve Bayes, attained a detection rate of 1, indicating that they successfully detected all DoS attacks. Gaussian Naïve Bayes achieved a slightly lower detection rate of 0.99.

The False Alarm Rate, also known as False Positive Rate, represents the rate at which the algorithms falsely classify normal instances as DoS attacks. Decision Tree, Forests of Random Trees, and Extra Tree Classifier had zero false alarm rates, indicating that they did not produce any false positives. Gaussian Naïve Bayes had a false alarm rate of 0.42, and SVM had the highest false alarm rate of 0.65. Precision, Recall, and F1-Score serve as metrics to assess the model's ability to accurately identify positive instances. All the algorithms demonstrated impressive performance with high Precision, Recall, and F1-Score values, approaching 1, indicating their effectiveness in correctly classifying DoS attacks.

In conclusion, the Extra Tree Classifier outdone all other algorithms in accuracy, generalization performance, detection rate etc. It achieved perfect scores in most of the metrics, demonstrating its superiority in accurately detecting DoS attacks. The Decision Tree and Forests of Random Trees also performed exceptionally well, while Gaussian Naïve Bayes and SVM showed slightly lower performance but still exhibited good classification capabilities. Overall, the results prove the efficiency of these ML algorithms in detecting DoS attacks, with the ensemble-based Extra Tree Classifier standing out as the top-performing model.

## 5 Conclusion

In this work, we have examined the summary of SDN and the outcomes of DoS and DDoS attacks in SDN. The centralization of network management functionality at the controller makes DoS attacks more likely in SDN, where they are vital in traditional networks. With the refined technique of SDN, the protection of SDN has evolved into one of the most important parts of network technology. With SDN, programmable networking has also gradually developed. We have proposed a solution using ML techniques to predict DDoS attacks against SDN networks and we have accomplished this using the tool Mininet. The proposed work has revealed high precision and efficiency under two types of attacks. we have tabulated a summary of ML methods for recognizing DDoS attacks in SDN. In the future, we will focus on security challenges in DDoS attack identification and will also provide a reduction in the SDN domain. Behind that, we will examine the attack and scan the mechanisms that are currently in place in present networks and concern whether they can be deployed in environments.

## References

1. Yue, M.: Detecting DoS Attacks Based on Multi-Features in SDN. IEEE Access **8**, 104688–104700 (2020)
2. Gao, S. et al.: Detection and Mitigation of DoS Attacks in Software Defined Networks. IEEE/ACM Transactions on Networking **28**(3), 14191433 (2020)
3. Varghese, J.E., Muniyal, B.: An Efficient IDS Framework for DDoS Attacks in SDN Environment. IEEE Access **9**, 6968069699 (2021)
4. Pérez-Díaz, J.A., Valdovinos, I.A., Choo, K.-K.R., Zhu, D.: Flexible SDN-Based Architecture for Identifying and Mitigating LowRate DDoS Attacks Using Machine Learning. IEEE Access **8**, 155859–155872 (2020)
5. Aladaileh, M.A., et al.: Detection techniques of distributed denial of service attacks on software-defined networking controller–a review. IEEE Access **8**, 143985143995 (2020)
6. Tang, D., Yan, Y., Zhang, S., Chen, J., Qin, Z.: Performance and Features: Mitigating the LowRate TCP-Targeted DoS Attack via SDN. IEEE J. Sel. Areas Commun. **40**(1), 428–444 (2022)
7. Wang, T., Guo, Z., Chen, H., Liu, W.: BW manager: mitigating denial of service attacks in software-defined networks through bandwidth prediction. IEEE Trans. Netw. Serv. Manage. **15**(4), 1235–1248 (2018)
8. Prediction: IEEE Transactions on Network and Service Management **15**(4), 1235–1248 (2018)
9. Yungaicela-Naula, N.M., Vargas-Rosales, C., Perez-Diaz, J.A.: SDN-based architecture for transport and application layer DDoS attack detection by using machine and deep learning. IEEE Access **9**, 108495–108512 (2021)
10. Tan, L., Pan, Y., Wu, J., Zhou, J., Jiang, H., Deng, Y.: A New Framework for DDoS Attack Detection and Defense in SDN Environment. IEEE Access **8**, 161908–161919 (2020)
11. Dong, S., Sarem, M.: DDoS attack detection method based on improved KNN with the degree of DDoS attack in software-defined networks. IEEE Access **8**, 5039–5048 (2020)
12. Abou El Houda, Z., Khoukhi, L., Senhaji, A.: Hafid: Bringing intelligence to software defined networks: mitigating DDoS attacks. IEEE Trans. Netw. Serv. Manage. **17**(4), 2523–2535 (2020)

13. Zhijun, W., Qing, X., Jingjie, W., Meng, Y., Liang, L.: Low-Rate DDoS attack detection based on factorization machine in software defined network. IEEE Access **8**, 17404–17418 (2020)
14. Yang, L., Zhao, H.: DDoS attack identification and defense using SDN based on machine learning 14. In: Sangodoyin, A.O., Akinsolu, M.O., Pillai, P., Grout, V. (eds.) Detection and Classification of DDoS Flooding Attack on Software-Defined Networks: A Case Study for the Application of Machine Learning. in IEEE Access, vol. 9, p. 122495122508 (2021)
15. Sahoo, K.S., et al.: An Evolutionary SVM Model for DDOS attack detection in software defined networks. IEEE Access **8**, 132502132513 (2020)
16. Kandoi, R., Antikainen, M.: Denial-of service attacks in OpenFlow SDN networks. In: IFIP/IEEE International Symposium on Integrated Network Management (IM), p. 13221326. IEEE (2015)
17. Zhang, P., Wang, H., Hu, C., Lin, C.: On denial of service attacks in software defined networks. IEEE Network **30**(6), 28–33 (2016)
18. Durner, R., Lorenz, C., Wiedemann, M., Kellerer, W.: Detecting and mitigating denial of service attacks against the data plane in software defined networks. In: IEEE Conference on Network Softwarization (NetSoft), pp. 1–6. IEEE (2017)
19. Barki, L., Shidling, A., Meti, N., Narayan, D.G, Moin Mulla, M.: Detection of distributed denial of service attacks in software defined networks. In: International Conference on Advances in Computing, Communications and Informatics (ICACCI), pp. 2576–2581. IEEE (2016)
20. Rahman, O., Gauhar Quraishi, M.A., Lung, C.-H.: DDoS attacks detection and mitigation in SDN using machine learning. IEEE world congress on services (SERVICES), vol. 2642, p. 184189. IEEE (2019)
21. Santos, R., Souza, D., Santo, W., Ribeiro, A., Moreno, E.: Machine learning algorithms to detect DDoS attacks in SDN. Concurrency and Computation: Practice and Experience **32**(16) (2020)
22. Klymash, M., Shpur, O., Peleh, N., Maksysko, O.: Concept of Intelligent Detection of DDoS Attacks in SDN Networks Using Machine Learning. In: IEEE International Conference on Problems of Info communications. Science and Technology (PIC S&T), pp. 609–612 (2020)
23. Sun, W., Li, Y., Guan, S.: An improved method of DDoS attack detection for controller of SDN. In: IEEE 2nd International Conference on Computer and Communication Engineering Technology (CCET), pp. 249–253 (2019)
24. Lawal, B.H., Nuray, A.T.: Real-time detection and mitigation of distributed denial of service (DDoS) attacks in software defined networking (SDN). In: 26th Signal Processing and Communications Applications Conference (SIU), pp. 1–4 (2018)
25. Ruaro, M., Caimi, L.L., Moraes, F.G.: A systemic and secure SDN framework for NoCBased ManyCores. IEEE Access **8**, 105997–106008 (2020)
26. Feroz Khan, A.B., Hussain, M.M., Kalpana Devi, S., Gunavathie, M.A.: DDoS attack modeling and resistance using trustbased protocol for the security of Internet of Things. J. Eng. Res. **11**(2), 100058 (2023). ISSN 2307-1877, https://doi.org/10.1016/j.jer.2023.100058
27. Gunavathie, M.A., Umamaheswari, S.: MLPRS: a machine learning-based proactive re-routing scheme for flow classification and priority assignment. J. Eng. Res.
28. Gunavathie, M.A, Umamaheswari, S.: BBR - A Novel Bandwidth Based Routing Algorithm in Software Defined Networking. In: 1st International Conference on Computational Science and Technology, ICCST 2022 - Proceedings, pp. 1038–1041 (2022)
29. Gunavathie, M.A., Umamaheswari, S.: A survey on traffic prediction and classification in SDN. Adv. Parallel Comput. **37**, 367–370 (2021)

# Regression Model Approach Towards Concrete Compressive Strength Prediction and Evaluation

Vijayalakshmi G. V. Mahesh(✉), CP Achyutha Gowda, Alla Vamsi Krishna, and Leti Manish Kumar

Department of Electronics and Communication Engineering, BMS Institute of Technology and Management, Bengaluru, India
vijayalakshmi@bmsit.in

**Abstract.** The quality and durability of any structure depends upon the mechanical property of the concrete that is concrete compressive strength. The concrete strength relates to its influencing factors such as the quality and proportions of the binding materials, aggregates, water and admixtures. This paper proposes to use machine learning aided approach to predict concrete compressive strength which is required for construction applications determined on the factors/attributes such as cement, blast surface slag, superplasticizer, fly ash, water, age, coarse and fine aggregate. For this study, 1030 samples were collected and utilized with varying proportions of 8 attributes and used both bagging and boosting regression ensemble machine learning algorithms. The regression models: Bagging Regressor, Extra Tree Regressor, Adaboost regressor, Random Forest Regressor, Gradient boosting and Histogram gradient boosting regressor developed were created and tested to find compressive strength of the concrete and also to identify the significant attribute effecting the concrete compressive strength. The performance of the models was quantitatively evaluated and compared. The experimental results and their analysis using the metrics, Explained Variance Score, Mean Absolute Error, $R^2$ and Mean Square Error showed a better performance from histogram gradient boosting regressor and indicated age as a significant influencing factor that plays key role towards the concrete compressive strength.

**Keywords:** Concrete compressive strength · mechanical properties · superplasticizer · fly ash · aggregates · construction · ensemble methods · performance metrics · regression

## 1 Introduction

Developing nations are always in need of evolving or modern construction techniques. Civil engineering plays a major role in design, implementation and managing constructions. The key component for construction is concrete, a composite material created by combining binding materials, aggregates with water and admixture in certain proportions. The binding material can be either cement or lime while the aggregates could be fine (sand) or coarse (stone, chips, gravel, brick chips, recycled glass etc.,). The quality and strength of the concrete are dependent upon the proportions of the mixture.


R. Geetha et al. (Eds.): AAIMB 2023, CCIS 2203, pp. 72–83, 2025.
https://doi.org/10.1007/978-3-031-73068-9_7

During concrete creation, water and the binding material (usually cement) are combined to create a paste that coats the aggregate, binds it and hardens to form a hard structure. Depending on the variations in concrete material, proportions considered to form a concrete mixture and applications (floors, walls, beams, pillars, slabs, sidewalks), concrete can fall into any of the following types [1] that include (a) Asphalt concrete, (b) Cement concrete, (c) Glass concrete, (d) High performance concrete, (e) High strength concrete, (f) Lime concrete, (g) Limecrete concrete, (h) Light transmitting concrete, (i) polymer concrete, (j) Reinforced cement concrete, (k) Rapid strength concrete, (l) Regular concrete, (m) Roller compacted concrete, (n) Stamped concrete, (o) Self-consolidating concrete and (p) Vacuum concrete.

Once the concrete is ready, it is imperative to know the strength of the concrete, a mechanical property as it provides the characteristics/qualities and performance of concrete that is suitable for various applications. Different structures constructed have their definite acceptable concrete strength that has to be created with quality materials and proportions. Stronger concrete is more durable and last longer. The strength is represented as Megapascals (MPa). To put to use the prepared concrete in construction it is important to know its strength (an important metric). Thus different tests are to be conducted on concrete to know its (i) Concrete compressive strength (CCS), (ii) Tensile strength (TS) (iii) Flexural strength (FS) and (iv) Shear strength (SS). As compared to other strengths, it is imperative to find concrete compressive strength (CCS) because it relates to how well the concrete can support heavy structures for a longer duration. Normally the CS ranges from 20 -60 Mpa and tensile strength is about 10–15% of the CS.

The factors that influence the concrete strength are [2] quality of the materials (cement, aggregates, water), water to cement ratio, coarse and fine aggregation ratio, aggregate to cement ratio, concrete age, concrete compactness, temperature, curing and relative humidity. These factors decide the acceptable concrete strength required for the civil structures. Variations could produce a strength that is unsuitable for the structure for which it is intended to be used for. Thus maintaining the factors for concrete preparation, measuring the strength and putting it in use for a particular application is crucial. This paper focuses on predicting the CCS based on the above mentioned parameters.

Traditionally, the concrete strength is measured at the application (construction) site by engineers. In the process, cylinders or cubes are prepared with concrete, hardened and cured for 28 days, later the compressive strength test is conducted on the cured material to find the strength. The literature and survey finds the availability of various methods [3, 4] for testing the compressive strength of the concrete. The methods include: Rebound hammer test, ultrasonic pulse velocity, penetration resistance test, pull out test, cast in place cylinders, drilled and usage of wireless maturity sensors. To set the seal on the quality of the structure, sometimes combination of tests is also conducted. The benefits of these tests include (i) Few of the methods are easy to use and can be performed in the construction site, (ii) provide good accuracy and (iii) deployment of wireless sensors provides real time concrete strength. There are also few shortcomings related to these methods such as (i) there is requirement of pre calibration for accurate concrete strength measurements, (ii) as some of the methods are destructive, it involves damaging or crushing of the concrete, (iii) few methods require more number of test samples acquired at different positions for providing accurate results.

The results regarding computing CCS can be enhanced with the evolution in technology. With the advancements in Machine learning (ML), a subsection of artificial intelligence (AI) it has become possible to provide solutions to civil engineering tasks. ML with its abilities in solving complex problems and availability of various algorithms to carry-out classification, clustering and regression tasks has attracted the research community to explore its abilities and strengths in civil engineering domain. ML has been applied and has found success in the areas of infrastructure design, safety checks, construction, maintenance and project management [5, 6].

This has motivated domain experts to use ML algorithms to develop models and predict concrete compressive strength based on its influencing factors (attributes). Thus the paper proposes to create regression model to predict concrete strength. The objectives of the work include (i)analysis of the concrete data represented quantitatively and apply regression methods to predict the CCS, (ii) investigation of the contribution or significantly influencing attributes towards CCS prediction, (iii)quantitative performance evaluation of the regression models and comparison. Subsequent part of the paper is organized into four sections, where Sect. 2 presents a survey on the related ML based methodologies or frameworks available. Section 3 provides the details about the methodology. The results obtained after simulation is presented and discussed in Sect. 4. Finally, the paper is concluded in Sect. 5.

## 2 Related Work

The literature indicates several works and studies on the application of advanced ML techniques on concrete data to investigate and predict the concrete strength. This section provides an overview of the methods. [7] has presented and discussed about the part of ML algorithms in concrete, applications, challenges and best procedures/practices. An investigation into the application of gene expression programming (GEP), artificial neural network (ANN) and decision tree (DT) algorithms in predicting the CCS was presented in [8]. The work considered 98 samples with seven parameters: cement, superplasticizer, coarse aggregate, fine aggregate, age and water. The results were validated to find the best performance of the bagging algorithm indicating a high correlation coefficient R2 value of 0.95. [9] created three machine learning models using Catboost gradient boosting (CGB), support vector regression (SVR) algorithms and k nearest neighbours (kNN) to predict the CS of self-compacting concrete and further their performances were compared. The results indicated the best performance of kNN model with least errors of MAE, MSE, RMSE, MAPE with 1.97, 6.85,2.62,6.15 respectively and high R2 score of 0.99.

A method to forecast concrete compressive strength for high performance concrete with high volume substitution of ground granulated blast furnace slag was proposed by [10]. The work used boosting machine learning algorithms such as gradient boosting regressor (GBR), light gradient boosting machine (LGBM), extreme gradient boosting (XGBoost), CGB and adaboost (ADA) regressor. The performance evaluation of the results specified a high prediction accuracy and least error rates. Similarly, a study on high performance concrete and its prediction was done in [11, 12] and [13]. [11] created ML regressor models based on extreme gradient boosting(XGBoost), SVR and random forest (RF) algorithms using the data from 60 concrete samples. XGBoost proved to be best among other models in terms of the prediction accuracy and rates of error, while [12] worked with Adaboost, XGBoost and GBDT methods to find the compressive and also tensile strengths. The experimental works to find CS and TS were carried out with variable conditions such as (a) combination of silica fume and fly ash and (b) silica fume, fly ash and polypropylene fiber with triple bending. The analysis on results pointed towards the acceptable performance from GBDT model. The influence of variation in age on the concrete strength was studied by [13] using nonlinear regressor models. The CCS was predicted at each varying age in days of 3, 7, 14, 28 and 91 to find high correlation coefficient with different types of curing.

Artificial neural networks (ANN) are also successful in concrete science [14, 15] for predicting the concrete strength. [16] considered the data taken from cylindrical concrete parts with various characteristics and provided to ANN algorithm. The algorithm was optimized using genetic algorithm (GA) to create an optimized model for. The method proved to be capable, flexible and accurate in concrete prediction. Selforganizing fuzzy neural networks devised based on extreme learning machine optimization and clustering was used in [17]. The method with optimized fuzzy rules turned out to be valid in concrete strength prediction. Further [18] contemplated the usage of ANN for predicting the strength of the concrete that was created using waste paper ash as a partial replacement for cement as binding material. To be utilized as a binding material in this work, the percentage of waste paper ash was varied from 0 to 20 percent in incremental steps of 5 for replacement of cement. The simulation results indicated an effective performance of ANN in predicting CCS.

Support vector machine (SVM) and ANN were used in [19] to examine the impact of adding natural silica to the concrete at different levels in regard to its compressive strength. The work utilized 90 data samples for the study. The performance of the models was compared to find better correlation coefficient provided by ANN and the method also highlighted the importance of adding the silica to the concrete.

The survey found a work [20] that acquired the data from the prepared 100 concrete samples by conducting various tests such as pulse velocity, Windsor probe penetration and Schmidt hammer. Later the data was utilized to model and test ML models under three categories: Lazy learners, tree based and function based. Together 12 different models were created and four validation experiments were conducted to identify the promising performance from the regression models in predicting CCS. [21] created a machine learning model based on extreme learning machine to predict the compressive strength, splitting tensile of the rubber fiber based concrete and flexural strength. The work considered 114 sets of data as per the influencing factors: polypropylene fiber, rubber content, particle size of the rubber. The results emphasized the application of extreme learning machine in providing high prediction accuracy and generalizing ability as compared to other conventional methods.

Recent years are also witnessing the application deep learning models [22, 23] in civil engineering tasks. [24] worked on developing the deep learning model using long short term memory(LSTM) algorithm utilizing the factors that include cement, fly ash, water, blast furnace slag, superplasticizer, coarse and fine aggregates to predict CCS. The model performed well providing high correlation coefficients and low statistical errors in comparison to other ML methods. Additionally, it was discovered that LSTM is reliable for finding high performance concrete strength prediction [25] tried analyzing the effects of utilizing IoTs and diatomite on the concrete and developed the model using convolutional neural network (CNN) architecture for predicting the concrete properties. Further a method utilizing convolutional neural networks (CNN) with softmax regression was studied by [26] to predict the CCS of recycled concrete. This method outperformed ANN model with improved generalization ability and efficiency.

From the survey, it is clear that ML and deep learning models have a greater impact on concrete science in analyzing the mechanical attributes of the concrete by forecasting the concrete strength. So, the idea in this paper is to extended the prediction task making use of ensemble ML algorithms with both boosting and bagging methods. Also the work tries to examine the effect of the influencing factors in determining the CCS.

## 3 Methodology Formulation

The process flow based on ensemble Machine Learning algorithms to predict the CCS is exhibited in Fig. 1. The concrete data set to conduct the experiment was acquired from [27]. The dataset has 1030 samples with 8 input attributes and concrete strength measured in Megapascals (MPa). Here, in this work the relation between the input attributes and the concrete strength has to be established using a machine learning algorithm to create a model, then the model is further used to predict the CCS. The details of the eight attributes, also known as the CCS influencing factors are provided in Table 1.

The influencing factors or the attributes or the materials are mixed in a definite proportion to form a concrete that is to be used in a particular project or application. The strength of the concrete created depends upon the proportions considered.

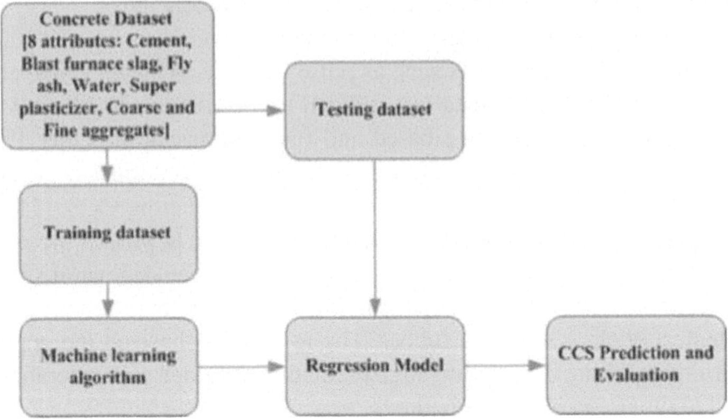


**Fig. 1.** The process flow of the methodology

**Table 1.** The details of the eight attributes

| Attribute | Measurement Unit | Symbol | Range | Average value |
|---|---|---|---|---|
| Cement | kg in a $m^3$ mixture | $x_1$ | (102, 540) | 281.16 |
| Blast furnace slag | kg in a $m^3$ mixture | $x_2$ | (0, 359.4) | 73.8958 |
| Fly ash | kg in a $m^3$ mixture | $x_3$ | (0, 200.1) | 54.1883 |
| Water | kg in a $m^3$ mixture | $x_4$ | (121.8, 247) | 181.5673 |
| Superplasticizer | kg in a $m^3$ mixture | $x_5$ | (0, 32.2) | 6.2047 |
| Coarse aggregate | kg in a $m^3$ mixture | $x_6$ | (801,1145) | 972.9189 |
| Fine aggregate | kg in a $m^3$ mixture | x7 | (594, 992.6) | 773.5805 |
| Age | Days | $x_8$ | (1,365) | 45.6621 |

The attributes with the concrete strengths as a complete dataset is fragmented into two sets for the experiment in the ratio of 70:30, where 70% forms the training dataset and the remaining 30% includes the testing set. The attributes from the training dataset is input to the ensemble machine learning algorithms with both bagging and boosting methods. During the training process, various trainable parameters are tuned to create optimized regression models. These regressions models were validated with the samples from the testing test and the performance was assessed based on the quantitative measures: Mean Squared Error(MSE), Explained Variance Score(EVS), Mean Absolute Error(MAE) and $R^2$ error or Coefficient of determination.

## 4 Results and Discussion

The methodology for the proposed work was formulated to work on three objectives. One of the objectives was to predict the compressive strength of the concrete and other two are to investigate the impact of the influencing factors or attributes on the CCS

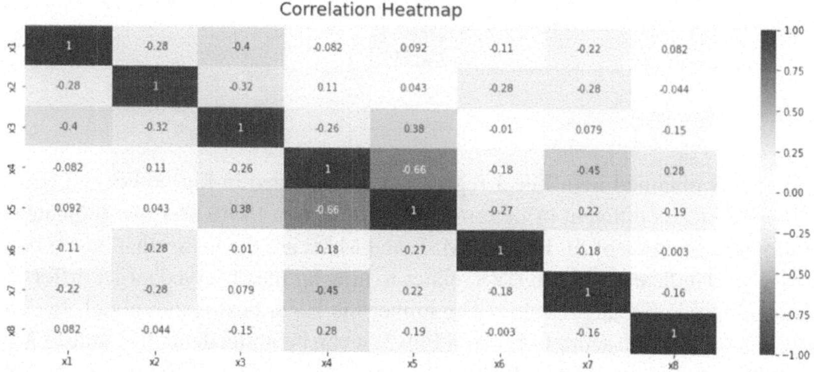


**Fig. 2.** Correlation map of attributes. Here the labels x1 till x8 indicate the input attributes from concrete to age. The heat map bar on the right hand side indicates the degree of correlation. (0 till 1) indicates positive correlation with 1 pointing towards high correlation. This can be observed on the principal diagonal. While (0 till −1) indicates negative correlation

and quantitative performance evaluation of the regression models and comparison. Thus considering all the objectives, as a preliminary step, the correlation between the attributes was computed. The correlation is displayed in Fig. 2 as a heat map.

Based on the heat map, from the original dataset (indicated as X) two datasets are derived. It is observed that there is a high correlation of 0.38 between $x_3$ and $x_5$ that is fly ash and superplasticizer, thus one of the attribute - $x_5$ is dropped from X to form X1. Further there is correlation of 0.28 between $x_4$ and $x_8$, so $x_8$ is also dropped along with $x_5$ from X to form X2. Thus there are three datasets available now to work on (i)X = ['$x_1$', '$x_2$', '$x_3$', '$x_4$', '$x_5$', '$x_6$', '$x_7$', '$x_8$'], (ii)X1 = ['$x_1$', '$x_2$', '$x_3$', '$x_4$', '$x_6$', '$x_7$', '$x_8$'] and (iii)X2 = ['$x_1$', '$x_2$', '$x_3$', '$x_4$', '$x_6$', '$x_7$']. Now the experiments of CCS prediction (predicted value is termed as y here) is conducted on all the three datasets.

The experiment is started with linear regression(LR) where all the three datasets were considered. The three linear regression models were created, tested and the performance was evaluated with the measures [28] EVS, MAE, MSE and $R^2$. The metrics are computed using the equations from Eqs. (1) till (4),

$$\mathrm{EVS} = 1 - \frac{Var(y - \overline{y})}{Var(y)} \tag{1}$$

$$\mathrm{MAE} = \frac{1}{N}\sum_{j=0}^{N-1} |y_j - \overline{y}_j| \tag{2}$$

Here, N is the number of samples, y is the actual strength and $\overline{y}$ is the predicted strength. This is also termed as L1 norm loss

$$\mathrm{MSE} = \frac{1}{N}\sum_{j=0}^{N-1} (y_j - \overline{y}_j)^2 \tag{3}$$

This corresponds to quadratic loss.

$$\mathrm{R}^2 = 1 - \frac{\sum_{j=0}^{N-1} (y_j - \overline{y}_j)^2}{\sum_{j=0}^{N-1} (y_j - \hat{y}_j)^2} \tag{4}$$

where, $\hat{y} = \frac{1}{N}\sum_{j=0}^{N-1} y_j$

The results obtained with linear regression is displayed in Fig. 3. For the best predictive model, R2(coefficient of determination) has to be 100% or 1 that indicates best correlation or goodness of fit. Further, MSE and MAE are the losses that has to be less, i.e., as low as 0 indicates best fit. EVS relates to how far the observed target differs from the averaged predicted target value. The higher the EVS, best is the model. In case of EVS the best evaluation score is 1. From Fig. 3, it can be noticed that the dataset X1 has provided best result with R2 of 0.6099, MSE of 10.419, MAE, 0.6105 and EVS, 8.2463 as compared to datasets X and X2.

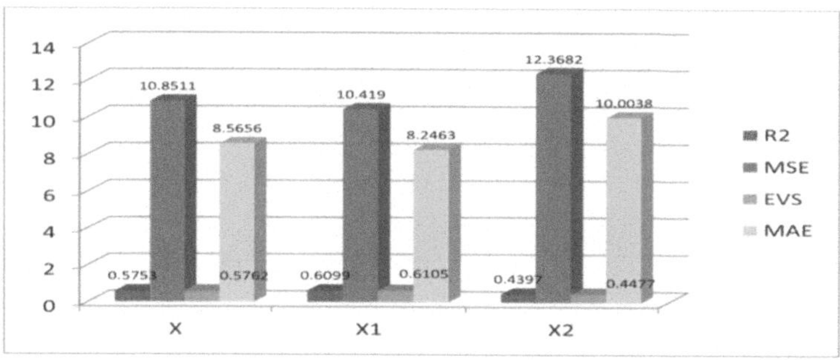


**Fig. 3.** The performance metrics obtained with linear regressor

Next, the experiment was conducted with ML ensemble bagging methods: Random forest Regressor(RFR), variant of RFR, Bagging Regressor(BR), Extra Tree regressor(ETR). In this method, the total number of estimators was positioned to 100 and error condition was varied considering Absolute error, Squared error, Poisson and Friedman_mse. With each variation, the model was validated, assessed using the performance metrics and the metrics are presented in Fig. 4. Though the performance is undesirable with BR method, but comparatively X1 has provided better results in comparison to X and X2. From Fig. 4(b) and (c), it is further noted that RFR and ETR has performed well with criterion = Absolute error and dataset X.

The Final experiment was carried out with ML ensemble boosting methods: Gradient Boosting Regressor(GBR),Adaboost Regressor(ABR) and Histogram Boosting Regressor(HBR). In this method, the learning rate was adjusted to 0.1, number of estimators to 100 and error criterion = Friedman_mse and the loss function was varied with: squared error, absolute error, Hubel and quantile for GBR and squared error, absolute error, poisson and least square for HBR. The validation results are plotted in Fig. 5.

From the experimental results it is observed that ABR has performed well with square loss function providing a $R^2$ of 0.7992, MAE, 5.9678, EVS, 0.7997 and MSE, 7.3222. Moving towards GBR and HGR, it is noticed that GBR performed well with squared error and HGR with least square error for X1 datasets. A highest $R^2$ of 0.9, MSE, 5.4239, EVS, 0.8918 and MAE, 3.9736 was obtained with HGR. As compared to LR, ensemble bagging methods and boosting methods have performed well in terms of prediction accuracy. Further to add on to that, it has been noted that all the methods have produced meaningful outcomes with X and X1 datasets and poor performance with X2 dataset.

With this it can be decided that, X1 dataset formed by removing $x_5$(super plasticizer) which was highly correlated to $x_3$ has predominantly provided good performance with boosting methods, indicating that there can be choice in selection between fly ash and super plastizer to create concrete for acceptable compressive concrete strength. But dropping x8(age) from X was a bad choice as the effectiveness of the models came down, indicating that the attribute age is a significant influencing factor that plays key role towards the CCS.

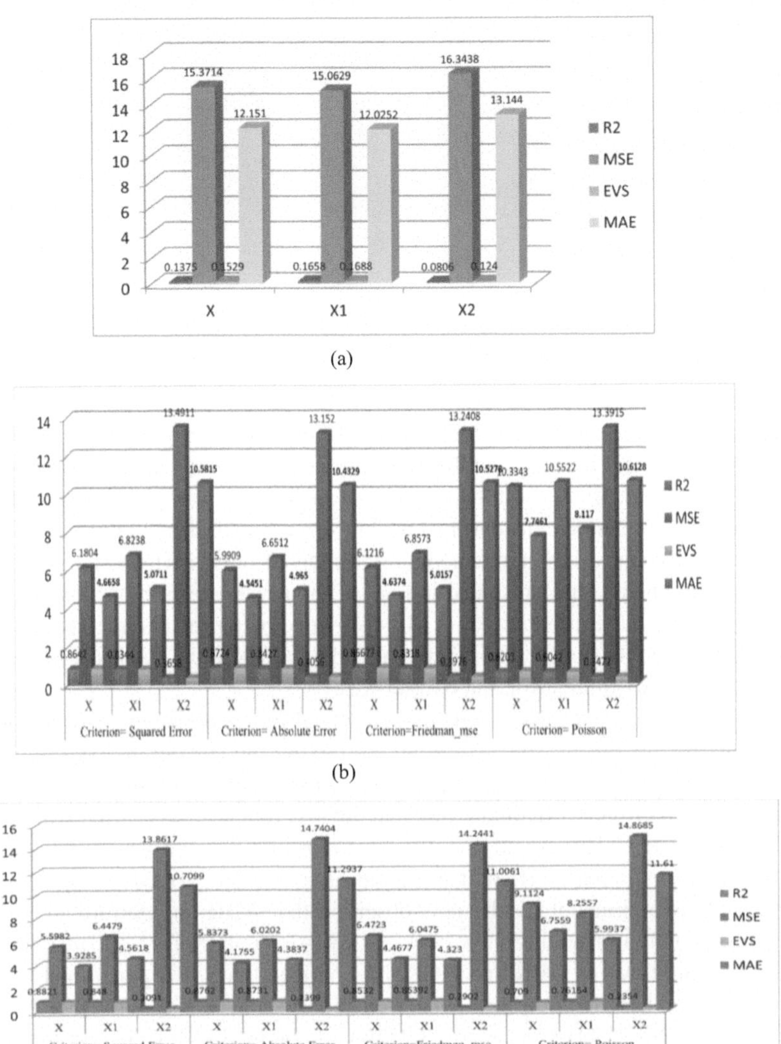


(c)

**Fig. 4.** The performance metrics obtained with Bagging methods (a)Bagging Regressor (b)Random Forest Regressor and (c) Extra Tree Regressor

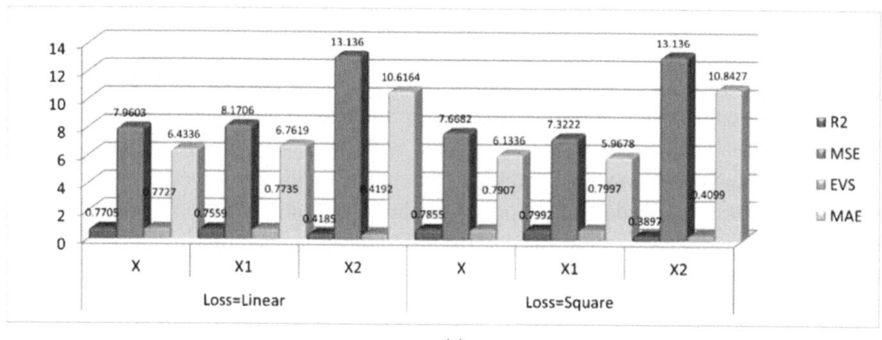


(a)

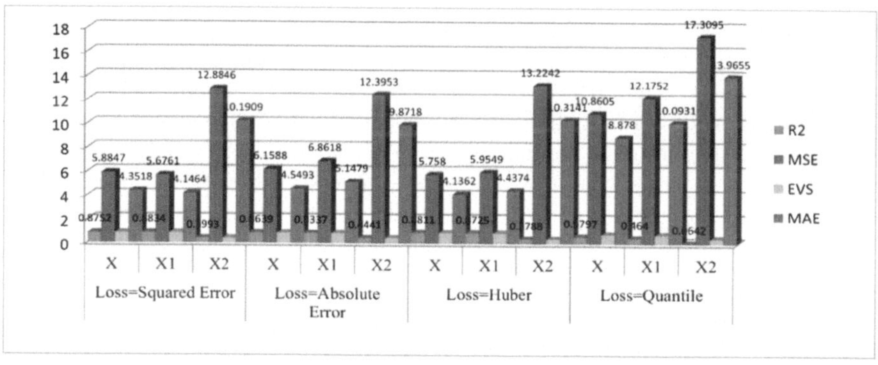


(b)

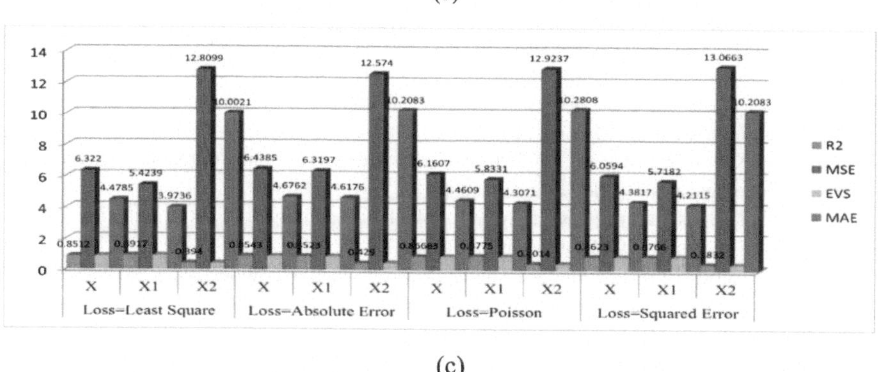


(c)

**Fig. 5.** The performance metrics obtained with Boosting methods (a)Adaboost Regressor (b)Gradient Boosting Regressor and (c) Histogram Gradient Regressor

## 5 Conclusion

The paper presented an ensemble machine learning based methodology to predict the compressive strength of the concrete based on the attributes/factors: cement, blast surface slag, water, fly ash, superplasticizer, age, coarse and fine aggregates. Different construction structures have their own acceptable concrete strengths that depend on the proportions of the materials used. The proposed methodology utilized a concrete dataset of 1030 samples with 8 above mentioned attributes to develop ensemble regression models for prediction task. The study considered both bagging and boosting ensemble methods that include Bagging Regressor, Random Forest Regressor, Extra Tree Regressor, Adaboost regressor, Gradient boosting and Histogram gradient boosting regressor for experimentation. The simulations were conducted by tuning all the required parameters of the algorithms and also by dropping few attributes from the dataset to understand their significance towards the concrete strength. The results were assessed using the quantitative measures: EVS, $R^2$, MSE and MAE to find that the boosting ensemble method HGR has performed well providing high coefficient of determination and lower errors. The results also demonstrated the significance of the attribute 'age' as a key influencing factor that plays vital role towards the concrete compressive strength. Subsequently, the proposed methodology based on ML can be applied to determine additional significant mechanical properties such as flexural strength, tensile strength and shear strength of the concrete.

## References

1. Mehta, P.K.: Concrete. Structure, properties and materials (1986)
2. Ozyildirim, C., Carino, N.J.: Concrete strength testing. In Significance of tests and properties of concrete and concrete-making Materials. ASTM International (2006)
3. Carino, N.J., Lew, H.S.: The maturity method: from theory to application. In: Structures 2001: A structural engineering odyssey, pp. 1–19 (2001)
4. Trtnik, G., Kavčič, F., Turk, G.: Prediction of concrete strength using ultrasonic pulse velocity and artificial neural networks. Ultrasonics **49**(1), 53–60 (2009)
5. Melhem, H.G., Nagaraja, S.R.I.N.A.T.H.: Machine learning and its application to civil engineering systems. Civil Engineering Systems **13**(4), 259–279 (1996)
6. Deka, P.C.: A primer on machine learning applications in civil engineering. CRC Press (2019)
7. Li, Z., et al.: Machine learning in concrete science: applications, challenges, and best practices. npj Computational Materials **8**(1), 127 (2022)
8. Song, H., et al.: Predicting the compressive strength of concrete with fly ash admixture using machine learning algorithms. Constr. Build. Mater. **308**, 125021 (2021)
9. Beskopylny, A.N., et al.: Concrete Strength Prediction Using Machine Learning Methods CatBoost, k-Nearest Neighbors, Support Vector Regression. Applied Sciences **12**(21), 10864 (2022)
10. Rathakrishnan, V., Bt. Beddu, S., Ahmed, A.N.: Predicting compressive strength of high-performance concrete with high volume ground granulated blast-furnace slag replacement using boosting machine learning algorithms. Scientific Reports **12**(1), 9539 (2022)
11. Liu, Y.: High-performance concrete strength prediction based on machine learning. Computational Intelligence and Neuroscience (2022)
12. Li, Q.F., Song, Z.M.: High-performance concrete strength prediction based on ensemble learning. Constr. Build. Mater. **324**, 126694 (2022)

13. Zain, F.M.M., Abd, M.S.: Multiple regression model for compressive strength prediction of high performance concrete. Journal of applied Sciences **9**(1), 155–160 (2009)
14. Tam, V.W., Butera, A., Le, K.N., Da Silva, L.C., Evangelista, A.C.: A prediction model for compressive strength of CO2 concrete using regression analysis and artificial neural networks. Constr. Build. Mater. **324**, 126689 (2022)
15. Pan, L., Wang, Y., Li, K., Guo, X.: Predicting compressive strength of green concrete using hybrid artificial neural network with genetic algorithm. Struct. Concr. **24**(2), 1980–1996 (2023)
16. Nikoo, M., Torabian Moghadam, F., Sadowski, Ł.: Prediction of concrete compressive strength by evolutionary artificial neural networks. Advances in Materials Science and Engineering (2015)
17. Zhang, X., Wang, H., Wang, D., Li, C.: Prediction of concrete strength based on self-organizing fuzzy neural network. In: Proceeding of the 11th World Congress on Intelligent Control and Automation, pp. 5631–5634. IEEE. (2014)
18. Azmi, N.A., Salim, W.S.W.: Compressive strength prediction model of concrete by utilizing wastepaper ash as partial replacement of cement using artificial neural network (ANN). In: AIP Conference Proceedings, Vol. 2532, No. 1, p. 040011. AIP Publishing LLC (2022)
19. Shaqadan, A.: Prediction of concrete strength using support vector machines algorithm. In: Materials Science Forum, Vol. 986, pp. 9–17. Trans Tech Publications Ltd. (2020)
20. Erdal, H., Erdal, M., Simsek, O., Erdal, H.I.: Prediction of concrete compressive strength using non-destructive test results. Comput. Concr. **21**(4), 407–417 (2018)
21. Zhang, J., Xu, J., Liu, C., Zheng, J.: Prediction of rubber fiber concrete strength using extreme learning machine. Frontiers in Materials **7**, 582635 (2021)
22. Jacobsen, E.L., Teizer, J.: Deep learning in construction: Review of applications and potential avenues. J. Comput. Civ. Eng. **36**(2), 03121001 (2022)
23. Ocak, A., Nigdeli, S.M., Bekdaş, G., Işıkdağ, Ü.: Artificial intelligence and deep learning in civil engineering. In: Hybrid Metaheuristics in Structural Engineering: Including Machine Learning Applications, pp. 265–288. Springer Nature Switzerland, Cham (2023)
24. Latif, S.D.: Concrete compressive strength prediction modeling utilizing deep learning long short-term memory algorithm for a sustainable environment. Environ. Sci. Pollut. Res. **28**(23), 30294–30302 (2021)
25. Lv, Z., Jiang, A., Liang, B.: Development of eco-efficiency concrete containing diatomite and iron ore tailings: Mechanical properties and strength prediction using deep learning. Constr. Build. Mater. **327**, 126930 (2022)
26. Deng, F., He, Y., Zhou, S., Yu, Y., Cheng, H., Wu, X.: Compressive strength prediction of recycled concrete based on deep learning. Constr. Build. Mater. **175**, 562–569 (2018)
27. Yeh, I.-C..: Modeling of strength of high performance concrete using artificial neural networks. Cement and Concrete Research **28**(12), 1797–1808 (1998)
28. Plevris, V., Solorzano, G., Bakas, N.P., Ben Seghier, M.E.A.: Investigation of performance metrics in regression analysis and machine learning-based prediction models. In: 8th European Congress on Computational Methods in Applied Sciences and Engineering (ECCOMAS Congress 2022). European Community on Computational Methods in Applied Sciences (2022)

# An Enhanced Artificial Neural Network Mode for Type 2 Diabetes Classification Using SMOTE and SMOTE-Tomek with Effective Feature Selection Methods

E. Sabitha and M. Durgadevi(✉)

Department of Computer and Science Engineering, SRM Institute of Science and Technology, Vadapalani Campus, Chennai 600026, India
se7648@srmist.edu.in

**Abstract.** The advancement in technology and changes in lifestyle is a root cause for the significant decrease in physical activities. A decrease in physical activity produces problems that are linked to an individual's day-to-day life choices. If an individual's daily habits are poor, it might lead to chronic noncommunicable diseases. Diabetes is a chronic noncommunicable disease that exposes people at risk. So, early diabetes prediction model can help to reduce the risk level of diabetes. The study proposes an effective Artificial Neural Network (ANN) model with SMOTE and SMOTE-Tomek that can achieve better classification accuracy for diabetes prediction. The study consisted of two fundamental parts. Firstly, the study explored handling the imbalanced data by performing both oversampling and under sampling using SMOTE and SMOTE-Tomek. Secondly, ANN model was employed with and without dropout layer for diabetes prediction. SMOTE and SMOTE-Tomek are two different data augmentation method which performs Oversampling and Under sampling on the dataset to improve the performance of the model by overcoming the imbalance dataset problem. ANN is utilised as a classifier in this study to make accurate predictions on the PIMA Indian Diabetes dataset. Important features are selected by implementing Filter, Wrapper and Embedded methods. The selected features are ranked by their priority. The performance of the ANN model is evaluated with and without dropout layer. Based on the experimental result ANN -SMOTE with dropout layer performed well by scoring accuracy 95.1 and ROC curve 90.4%.

**Keywords:** Artificial Neural Network · SMOTE · SMOTE-Tomek · Feature selection · diabetes Mellitus

## 1 Introduction

The growth of the social economy, population ageing, and other behavioural risk factors including smoking, drinking, and inactivity are all contributing to an increase in the prevalence of non-communicable diseases, also referred to as chronic diseases. The World Health Organisation (WHO) estimates that diabetes caused 1.5 million fatalities,


R. Geetha et al. (Eds.): AAIMB 2023, CCIS 2203, pp. 84–104, 2025.
https://doi.org/10.1007/978-3-031-73068-9_8

with 48% of these deaths occurring in those under the age of 70. According to the International Diabetes Federation (IDF), diabetes will be responsible for 6.7 million fatalities in 2021 [1]. Figure 1 displays graphically the most recent IDF data. 537 million people, or over 10% of all people in the 20 to 79 age group worldwide, have diabetes as of 2021. Additionally, 240 million people worldwide are believed to have diabetes but are still waiting for a diagnosis.

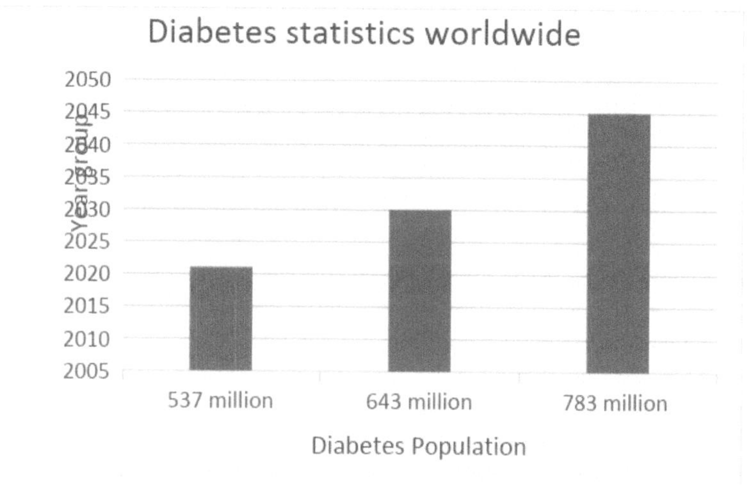


**Fig. 1.** Diabetes statistics chart

The inability of the body to produce or use enough insulin causes diabetes, a chronic illness. Blood sugar levels increase and bodily cells begin to starve in the absence of insulin. Hyperglycemia, as this disease is also known, can harm the body's functions if left untreated. Thankfully, many of these issues can be avoided by managing diabetes properly and getting regular check-ups. However, if it is not adequately managed, diabetes can be a serious, even fatal, condition [3]. The two most common types of diabetes are type 1 diabetes (T1D) and type 2 diabetes (T2D) [4]. Strokes, heart conditions, chronic renal failure, diabetic foot syndrome, and a host of other conditions are among the life-threatening complications of both kinds of the disease [5]. The development of computer vision technology over the past few years has been greatly aided by Deep Learning (DL), a current research area in the field of machine learning. The neural network was developed initially using computer perception. The neural network differs from computer perception because it has multiple hidden layers that connect it [6]. Diabetes can be quickly and effectively diagnosed with accuracy, sensitivity, and specificity utilising DL-based ANN models. For a more precise diagnosis, which has a significant impact on treatment and preventative methods, using modern technologies in diabetes prediction is crucial. The following are the key goals of the paper:

To apply both oversampling and under sampling techniques on the PIMA dataset and overcome the imbalance data problem and investigate the performance of the proposed model.

To select the important features by using filter, wrapper and embedded feature selection techniques and best features are selected by ranking.

To propose an effective ANN model with dropout layer using SMOTE and SMOTE-Tomek for diabetes prediction.

Finally, the performance of ANN model with SMOTE and SMOTE-Tomek is evaluated with Accuracy and ROC curve performance metrics.

The final paragraph is organised as follows. The related works are discussed in Sect. 2. Section 3 describes the suggested work flow. The experimental findings are explained in Sect. 4. The work is concluded in Sect. 5 and provided with future scope.

## 2 Related Works

The classification of DM dataset using computational intelligence as the primary strategy—either using a single methodology or a hybrid approach—will be the focus of the discussion and sharing in this section.

Using deep learning structures, Swapna, G. et al. offer a method for identifying diabetes and healthy HRV signals. This long short-term memory (LSTM) use CNN and both of them to extract complex temporal dynamic components from the input HRV data. The characteristics are classified by the Support Vector Machine (SVM). Having a 91.7% accuracy rate, the output was quite accurate. [7]

A model created by D. Sisodia et al. can predict the diabetes in people with the highest degree of accuracy. In this research, diabetes is detected early on using the machine learning classification methods Decision tree, SVM, and Naive Bayes. The best performance is achieved by Nave Bayes, whose accuracy is 76.30%.[8]

In order to predict type 2 diabetes mellitus, Wu, H. et al. have developed a novel data mining-based algorithm. The upgraded K-means algorithm and the logistic regression technique make up the two components of the model. In comparison to other studies, the model's prediction accuracy was 3.04% greater. [9]

Zou, Q., et al. employed neural networks, decision trees, and random forests in order to predict diabetes mellitus. The data were obtained from physical examinations carried out at Luzhou, China, hospitals. To perform better for conducting independent test experiments, the model incorporates five-fold cross validation. Minimal reliability maximum relevance (mRMR) and PCA (principal component analysis) are both used to decrease the dimensionality issue. The model with the highest accuracy, the random forest model, has an ACC of 0.8084 [10].

Perveen, S., et al., the author, created the J48 in order decision tree and Nave Bayes approaches to forecast the potential start of diabetes utilising significant risk factors discovered by logistic regression analysis, employing balanced and unbalanced datasets. The outcomes showed that randomised under-sampling, over-sampling, and no sampling were all outperformed by Naive Bayes utilising the K-medoids under-sampling technique. The enhanced true positive rate allows for a median receiver's operational performance to be as high as 79% [11].

The goal of the study, conducted by Choi, B.G., et al., was to construct a Type 2 diabetes predictive model utilising EMRs (electronic medical records) and machine learning and to compare the model's performance to conventional statistical techniques. Models for the logistical regression (LR), linear discriminant analysis (LDA), quadratic discriminant analysis (QDA), and K-nearest neighbour (KNN) algorithms were constructed in order to assess the cross-validation test in accordance with the prediction model. The LR learning model provided the best prediction performance overall, with an ACC of 0.78 [12].

Deep learning methods were used by Garcá-Ordás, M. T., et al. to create a pipeline for predicting who would get diabetes. Variational autoencoders (VAE), sparse autoencoders (SAE), and convolutional neural networks are used, respectively, for data augmentation and feature augmentation. The CNN classifier achieved a final accuracy rating of 92.3% after being trained using a balanced dataset [13].

P. Nuankaew, S. et al. offer the Average Weighted Objective Distance (AWOD), a unique prediction approach based on the presumption that the individual has a variety of health issues brought on by distinct individual attributes, in order to construct an effective prediction model. By utilising information acquisition, AWOD modifies Weighted Objective Distance (WOD) by disclosing large and minor individual components with different priority, which are represented by variable weights. Accuracy ratings of 93.22% were obtained using the suggested method [14].

Naz, H., and Ahuja, S. describe a strategy for diabetes prediction utilising a variety of machine learning techniques using the PIMA dataset. Artificial Neural Networks (ANN), Naive Bayes (NB), Decision Trees (DT), and Deep Learning (DT) are functional classifiers that have an accuracy of 90–95 percent. With a 92.07% accuracy rate on the PIMA dataset, DL outperforms the other three in terms of diabetes onset prediction [15].

## 3 Proposed Workflow

With the help of data-driven applications, particularly in the healthcare industry, ANN have revolutionised the study of various medical disciplines. It spans a broad spectrum of fields, including disease diagnosis, decision support systems (DSS), and medical image processing. Artificial intelligence and other cutting-edge technologies, particularly in the healthcare industry, depend on a variety of contemporary ANN models and genetic algorithms. ANN is a perfect instrument in identifying and analysing thrombo-embolic stroke illness, bone densitometry, hepatitis B, and breast cancer [19]. The ANN prediction model's suggested structure, which methodically validates and tests the network with the purpose of boosting significant certainty and self-reliance, is illustrated in the accompanying image. Comparing the operations and parameters produced by ANN training to the actual values obtained during ANN optimisation. In this study we built an ANN model with two different aspects, ANN with SMOTE augmentation along with dropout layer and ANN with SMOTE-Tomek along with dropout layer. The study concentrates on the impact of oversampling and under sampling on the dataset using the ANN model. To emphasise the importance of the data augmentation and ANN model in disease prediction, the work process in two different approaches as follows (Fig. 2):

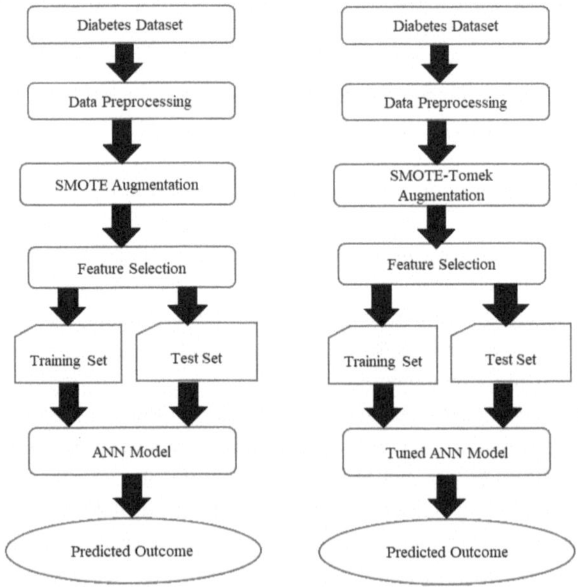


**Fig. 2.** ANN with two different approaches

In accordance with the Fig. 3, the network compiles data on the specific characteristics of female diabetes patients with the intention of identifying the condition. Before doing an evaluation, the training data needs to be pre-processed. This is necessary because some features could cause the outcome to be misinterpreted because of noise or null data. Data augmentation is done to enlarge the class instance after data pre-processing. The minority class instance, or the class with fewer results, is used for data augmentation. After the augmentation is finished, the best attributes that are highly connected with diabetes prediction must be identified. The main features are selected using a number of different feature selection strategies. Next, the selected features are added to a fresh dataset and split into the train and test sets using the coordinates (x, y). To create and fit the ANN model, the train dataset is utilised. The test set is used to evaluate the model and quantify performance.

### 3.1 Data Set

All of the patients in this investigation are females at least 21 years old with PIMA Indian ancestry ("Dataset 1"), as shown in Fig. 3. The datasets utilised in the study were collected from the National Institute of Diabetes and Digestive and Kidney Diseases [20]. The sample contains a total of 9 characteristics, or 768 in all. Among the features are Pregnancies, Glucose, Blood pressure, Insulin, Age, BMI, Diabetes pedigree Function, Skin Thickness, and Outcome. The data is divided between cases with and without diabetes (Fig. 4).

| Pregnanci | Glucose | BloodPres | SkinThickn | Insulin | BMI | DiabetesP | Age | Outcome |
|---|---|---|---|---|---|---|---|---|
| 2 | 138 | 62 | 35 | 0 | 33.6 | 0.127 | 47 | 1 |
| 0 | 135 | 68 | 42 | 250 | 42.3 | 0.365 | 24 | 1 |
| 0 | 173 | 78 | 32 | 265 | 46.5 | 1.159 | 58 | 0 |
| 4 | 99 | 72 | 17 | 0 | 25.6 | 0.294 | 28 | 0 |
| 8 | 194 | 80 | 0 | 0 | 26.1 | 0.551 | 67 | 0 |
| 2 | 83 | 65 | 28 | 66 | 36.8 | 0.629 | 24 | 0 |
| 2 | 89 | 90 | 30 | 0 | 33.5 | 0.292 | 42 | 0 |
| 4 | 99 | 68 | 38 | 0 | 32.8 | 0.145 | 33 | 0 |
| 3 | 80 | 0 | 0 | 0 | 0 | 0.174 | 22 | 0 |
| 6 | 166 | 74 | 0 | 0 | 26.6 | 0.304 | 66 | 0 |
| 5 | 110 | 68 | 0 | 0 | 26 | 0.292 | 30 | 0 |
| 2 | 81 | 72 | 15 | 76 | 30.1 | 0.547 | 25 | 0 |
| 7 | 195 | 70 | 33 | 145 | 25.1 | 0.163 | 55 | 1 |
| 6 | 154 | 74 | 32 | 193 | 29.3 | 0.839 | 39 | 0 |
| 2 | 117 | 90 | 19 | 71 | 25.2 | 0.313 | 21 | 0 |
| 3 | 84 | 72 | 32 | 0 | 37.2 | 0.267 | 28 | 0 |
| 6 | 0 | 68 | 41 | 0 | 39 | 0.727 | 41 | 1 |
| 7 | 94 | 64 | 25 | 79 | 33.3 | 0.738 | 41 | 0 |
| 3 | 96 | 78 | 39 | 0 | 37.3 | 0.238 | 40 | 0 |
| 10 | 75 | 82 | 0 | 0 | 33.3 | 0.263 | 38 | 0 |
| 0 | 180 | 90 | 26 | 90 | 36.5 | 0.314 | 35 | 1 |
| 1 | 130 | 60 | 23 | 170 | 28.6 | 0.692 | 21 | 0 |
| 12 | 84 | 72 | 31 | 0 | 29.7 | 0.297 | 46 | 1 |
| 0 | 139 | 62 | 17 | 210 | 22.1 | 0.207 | 21 | 0 |
| 9 | 91 | 68 | 0 | 0 | 24.2 | 0.2 | 58 | 0 |

**Fig. 3.** PIMA Dataset

### 3.2 Data Preprocessing

Data preprocessing is a necessary step for preparing the data for deep learning and machine learning models, which also improves a model's accuracy and effectiveness. Pre-processing data removes missing or inconsistent data values resulting from human or computer error, which can improve the accuracy and quality of a dataset, making it more reliable. It makes data consistent. In PIMA dataset, there is no missing values, but there are inconsistent values like zero for Glucose, BMI, DPF, ST, BP, Insulin. These values can reduce the performance of the model, and also degrades the quality of the database. Not only zero values, Outliers can also be figured in the PIMA dataset i.e. out of range values can identified by fixing the threshold values for each instance in the attribute. Figure 5: represents the plot histogram of each numerical variable in each attribute.

From the above plots the number of zero's present in each attribute can identified easily. Generally, as stated early few, zero value as meaning in few attributes like pregnancies, Diabetes pedigree function, Age. But in attributes like Glucose, Blood pressure

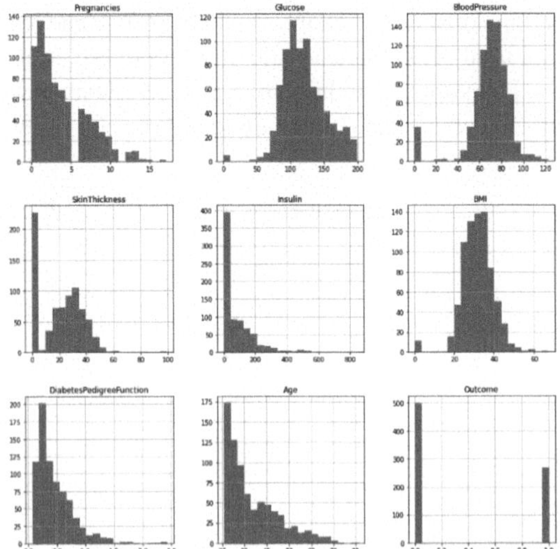


**Fig. 4.** Plot histograms for numerical variables

Skin thickness, Insulin, BMI it senses meaning less and also sounds bad. Such meaning less values is called as Anomalies or Noisy data in dataset. These anomalies or noisy data can be removed by performing data pre-processing. In this paper, these anomalies are removed by replacing by zero values with mean-median calculation. Based on the above mention figure, the table represents the numerical values of zero in each attribute.

```
Pregnancies                   0
Glucose                       5
BloodPressure                35
SkinThickness               227
Insulin                     374
BMI                          11
DiabetesPedigreeFunction      0
Age                           0
Outcome                       0
dtype: int64
```

```
Pregnancies                 0
Glucose                     0
BloodPressure               0
SkinThickness               0
Insulin                     0
BMI                         0
DiabetesPedigreeFunction    0
Age                         0
Outcome                     0
dtype: int64
```

**Fig. 5.** Before Pre-processing and After Pre-processing

### 3.3 Imbalance Classification Problem

When the class distribution in training data has a significant skew, the imbalanced classification problem arises. Although the skew may not be very significant, it is nonetheless thought to be an imbalanced classification issue that affects the mode's performance. Imbalance classification problem can be overcome by performing data augmentation. Data augmentation is a technique of artificially increasing the training set by creating modified copies of a dataset using existing data. It includes making minor changes to the dataset or using deep learning to generate new data points. The augmentation techniques used in deep learning applications depends on the type of data, are popular. These techniques are generally used to address the class imbalance problem in classification tasks (Fig. 6).

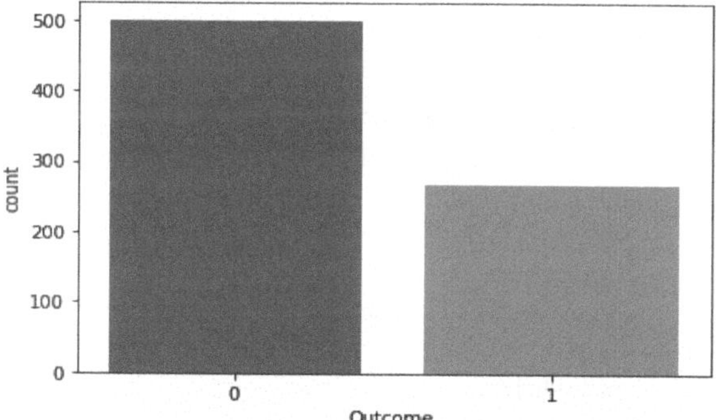


**Fig. 6.** Diabetes outcome

PIMA dataset consists of 768 females of aged 21 or above who took part in a national health check program aimed at diagnosing diabetes. There are 500 non-diabetes and 268 diabetes. Based on the outcome, we can find there is no balanced data between two classes i.e. non-diabetes (0)- (500) and diabetes (1)- (268). The imbalance classes can lower the performance of the model. The reason is model will be trained more to non-diabetes class rather than diabetes class. The proposed work is focused to overcome the class imbalance issues by performing both oversampling and under sampling which improves the performance of the model (Fig. 7).

**SMOTE (Synthetic Minority Oversampling Technique)**
Unlike random oversampling, which just copies a few random instances from the minority class, SMOTE is one of the most widely used oversampling strategies. SMOTE create components for the minority class. The method creates a new point along a line that is drawn between samples in the feature space that are close to one another. The samples created by SMOTE differ from the original minority class because they are based on the distance between each data point using Euclidean distance and the minority class's nearest neighbours.

The process to generate the synthetic samples are as follows:

1. Choose random data from q1 q2, from the minority class.

2. Calculate the Euclidean distance between the random data and its k-nearest neighbors.

   i.e., d(q1, k1) $=\sqrt{\sum_{i=1}^{n}(q1i - k1i)}\ ^{2}$
   q1, k = random sample, k1 neighbor
   q1i, ki = vectors starting from the origin of the space (initial point).
   n = n-space.

3. Using random number multiply the difference between (0 and 1) then add the synthetic sample as a result to the minority class.
4. Until desired proportion of minority class is met, repeat the procedure.

**Algorithm**

```
Inputs :X1minor, Npresent ,K1
Output: X1smote
Function SMOTE(X1minor, Npresent ,K1)
X1smote □ {}
for i□ 1 to len (X1minor) do
    nn□ K1 Nearest neighbors (X1i, X1minor, K)
     p□ [Npercent/100]
     while p!=0 do
     X1neighbor □ select random (nn)
      X1smote□ X1i + rand(0,1) * |X1neighbor – X1i |
      p□p-1
      end while
 end for
return□X1smote
```

Since the number of diabetes patients in the PIMA dataset is lower than that of non-diabetic. More negative than positive data will be used to train the model. Whatever diabetes prediction model is chosen, it is more crucial to take care to ensure that the model is trained equally with both examples, i.e., both positive and negative outcome. SMOTE approach is employed in this case to generate synthetic data for the positive cases, making it practicable. Duplication is prevented, and overfitting data in the results is eliminated, when the data is supplemented. The results of the SMOTE technique are shown in the following Table 1.

**Table 1.** SMOTE Augmentation

| Before SMOTE Augmentation | | After SMOTE Augmentation | |
|---|---|---|---|
| Outcome | | Outcome | |
| 0 | 1 | 0 | 1 |
| 500 | 268 | 500 | 500 |

```
<class 'pandas.core.frame.DataFrame'>
RangeIndex: 1000 entries, 0 to 999
Data columns (total 9 columns):
 #   Column                    Non-Null Count  Dtype
---  ------                    --------------  -----
 0   Pregnancies               1000 non-null   int64
 1   Glucose                   1000 non-null   float64
 2   BloodPressure             1000 non-null   float64
 3   SkinThickness             1000 non-null   float64
 4   Insulin                   1000 non-null   float64
 5   BMI                       1000 non-null   float64
 6   DiabetesPedigreeFunction  1000 non-null   float64
 7   Age                       1000 non-null   int64
 8   Outcome                   1000 non-null   int64
dtypes: float64(6), int64(3)
memory usage: 70.4 KB
```

**Fig. 7.** Outcome after SMOTE Augmentation

**SMOTE-Tomek Oversampling and Under Sampling**

The number of the majority class is significantly larger than the minority class in a real-world application, making it difficult for the model to effectively learn from the minority class. When data from the minority class is more essential than data from the majority class, such as in disease prediction datasets, churn datasets, and fraud detection datasets, a serious issue arises. Either oversample the minority class or under sample the majority class is one way to address the imbalance dataset problem. When the original dataset is oversampled, too many duplicates are added; conversely, when the original dataset is under sampled, some of the original data's content is lost. Combining the oversampling and under sampling methods, or combining SMOTE and Tomek Links, is one way to address the problem.

**Combination of SMOTE and Tomek Links**

A combination of over-sampling the minority class and under sampling the majority class can achieve better classifier performance than only under-sampling the majority class. The process of SMOTE -Tomek Links is as follows:

1. Start of SMOTE: choose random data ( R) from the minority class (M).
2. Calculate the distance (D) between the random data ( R) and its k nearest neighbors.

$$D = \sqrt{(R - k)^2}$$

3. Multiply the difference with a random number between 0 and 1, then add the result to the minority class as a synthetic sample.
4. Repeat step number 2–3 until the desired proportion of minority class is met.(End of SMOTE).
5. Start of Tomek Links: choose random data from the minority class.
6. If the random data nearest neighbor is the data from the minority class, then remove the Tomek Link.

```
SMOTE with Tomek Algorithm
Inputs :X2minor, Npresent , D-Dataset to resample,TH-Threshold ,K2
Output: X2smote, TL-Tomek Link instances
Function SMOTE(X2minor, Npresent ,K2)
X2smote □ {}
for i□ 1 to len (X2minor) do
   nn□ K Nearest neighbors (X2i, X2minor, K2)
    p□ [Npercent/100]
   while p!=0 do
   X2neighbor □ select random (nn)
    X2smote□ Xi + rand(0,1) * |X2neighbor – X2i |
    p□p-1
    end while
 end for
 return□X2smote
TL□new emptylist of instances
CheckedSamples □new empty list of instances
for each samples in D do
   if (samples in checkedSamples) then
     continue
  end if
  NN□ Nearest Neighbor(sample)
 checkedSamples□ checkedSamples U sample
 dist□AdjustedHammingDist(sample,NN)
 if(dist >= TH)then
   TL□TL U sample
 end if
 end for
 return TL
```

The minority data are enhanced in SMOTE, in contrast. As result 1 in the PIMA dataset only has minimal value that is only increased through oversampling, it is regarded as minority data. Oversampling and under sampling are both done in SMOTE-Tomek, though. The distinction between SMOTE and Tomek is that SMOTE is an oversampling method that creates new credible examples in the minority class. A technique known as Tomek linkages is used to find nearest neighbours in a dataset whose classifications differ from one another. The SMOTE-Tomek technique's results are displayed in the Table 2 below:

**Table 2.** SMOTE-Tomek Augmentation

| Before SMOTE-Tomek Augmentation | | After SMOTE-Tomek Augmentation | |
|---|---|---|---|
| Outcome | | Outome | |
| 0 | 1 | 0 | 1 |
| 500 | 268 | 481 | 481 |

## 3.4 Feature Selection

**Feature Selection/Reduction**

The most important factors are typically not related to one another in the accurate diagnosis of diabetes. Medical professionals carefully consider the characteristics and determine whether each is necessary in order to identify the correct diagnosis. These characteristics, which ANN refers to as features, can be used to identify diabetic symptoms and other pertinent details that help predict diabetes. The method of feature selection, which involves choosing the most pertinent features to extract from the entire set, improves the likelihood that diabetes patients will be correctly diagnosed. The final dataset chosen contains one responder output variable and eight explanatory input variables. Each element is crucial in determining the specific type of diabetes. In this study, the most efficient feature selection technique is selected by evaluating some features selections methods and finally identify the most relevant features that fits best for diabetic prediction. Filter, Wrapper and Embedded feature selection techniques are used in this study.

**Table 3.** Multiple Feature Selection

| S. no. | Feature Selection Methods | | Selected Features |
|---|---|---|---|
| 1 | Filter | Information Gain | Glucose, Age, BMI, Insulin |
| | | Chi-square | Pregnancies, Glucose, Insulin, Age |
| | | Fisher Score | Skinthickness, Pregnancies, Age, BMI |
| | | Correlation Features | Glucose, BMI, Age, Pregnancies |
| 2 | Wrapper | Forward Feature Selection | Pregnancies, Glucose, BMI, DPF |
| | | Exhaustic Feature Selection | Pregnancies, Glucose, BP, Skinthickness |
| | | Recursive Feature Elimination with Cross Validation | Pregnancies, Glucose, BMI, DPF |
| 3 | Embedded | Recursive Feature Selection with Random Forest | Glucose, BMI, DPF, Age |

Table 3 describes the Filter, Wrapper and Embedded feature selection methods and selected features based on their importance for the diabetes prediction. First method, Filter is a univariate statistics approach that selects the features easily and quickly rather than cross-validation. Filter method is computationally less expensive. The method consists of Information gain, Chi-square, Fisher score and Correlation features selection techniques. Second method, Wrapper is also known as greedy algorithm, that trains the algorithm iteratively using subset of features. The fundamental benefit of wrapper approaches over filter methods is that they offer an ideal collection of features for training the model, producing improved accuracy compared to filter methods but at the expense of high computational costs. Some strategies include: Forward Selection, Exhaustic selection, Recursive Feature Elimination method. Third method is Embedded, Embedded method combines the qualities of filter and wrapper methods. The algorithm is implemented by its own built in feature selection methods. By using the above-mentioned techniques few features are selected based on the importance in diabetes prediction. From the Table 3, the selected features are manually ranked based on the number times selected in overall features. The selected features are ranked manually and represented in Table 4. From the Table 4, Glucose, Pregnancies, Body mass index and Age is considered to be more important in diabetic prediction based on number of times selected feature selection methods. Figure 9 represents the graphically representation of the feature importance. For this study, based on the feature importance Glucose, Pregnancies, Body mass index and Age features are used i.e., from eight attributes only four attributes are used for the diabetic prediction (Fig. 8).

**Table 4.** Feature selection

| S. no. | Features Selected | Feature Importance |
|---|---|---|
| 1 | Glucose | 7 |
| 2 | Pregnancies | 6 |
| 3 | Body mass index | 6 |
| 4 | Age | 5 |
| 5 | Insulin | 2 |
| 6 | Skin thickness | 2 |
| 7 | Blood Pressure | 2 |
| 8 | Diabetes Pedigree function | 2 |

**Train and Test Split**

Generally, the PIMA dataset consists of 768 instances. In 768, 500 is non-diabetic and 268 is diabetic patients. The outcome is not equally balanced, by data augmentation the outcome is balanced by performing both oversampling and under sampling. Based on outcome the dataset is augmented, after data augmentation, the dataset set increased to 1000 instances and 962 instances. Based on the augmentation and feature selection, the dataset is splitted into train and test. Generally, the train and test set will be splitted on ratio like 90:10, 80:20, 70:30, 60:40 and 50: 50. In this study the dataset is splitted in all categories to check the performance of the model and also analysis on which split-up the model is finely evaluated.

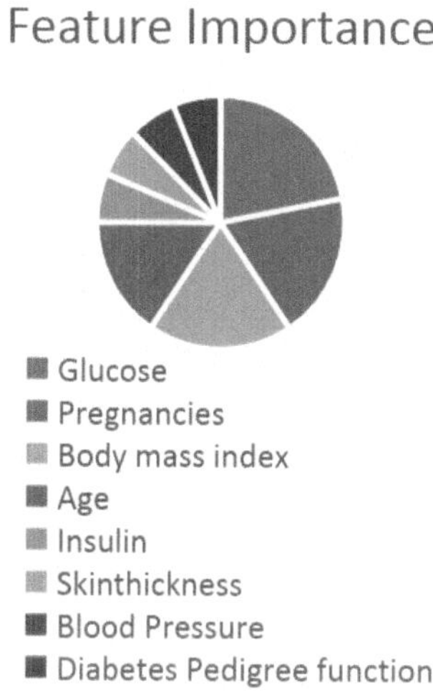


**Fig. 8.** Feature selection important

## 4 Artificial Neural Networks (ANN)

In this study two different ANN model is built with three hidden layers. In the first ANN, no dropout layers is added. Each hidden layers consist of 100 to 150 neurons. Four selected attributes units (Glucose, Age, BMI, Pregnancies) are forwarded to the first hidden layer consists of 100 neurons. These input units computes with weight (w1) and bias (b0) and then forwarded to the second hidden layers with 150 neurons likely after computes with weight (w2) and bias (b1), then forwarded to the third hidden layer. The third hidden layer consists of 200 neurons and then finally the units forwarded to the output layer. In these model, the ANN computes with general computation process. Dropout layers and other tuning functionalities are applied to the model. ANN model works on two different data size. PIMA dataset is improvised by augmenting with SMOTE and SMOTE-Tomek. When augmented with SMOTE the dataset size improved from 768 to 1000 instances likely by using SMOTE-Tomek the dataset size increased from 768 to 950. The augmentation process is mentioned in Tables 1 and 2.

In second approach ANN model is built with input layer, three hidden layers (each hidden layer with different units) and output layer with one unit. The main objective of the paper is to show the performance of the ANN model with dropout layer. In order to prevent the overfitting issue, dropout layer is crucial in ANN models. A training dataset will most likely quickly get overfitted by deep learning models. Using the same dataset to fit multiple separate neural networks and averaging the outcomes from each model is one method to reduce overfitting. An approximation can be made using a few different models since this is not practical in real life. A normalization technique called dropout comes close to parallel neural network training for an extensive amount of distinct neural network topologies. Dropout has the consequence of generating noise during the training

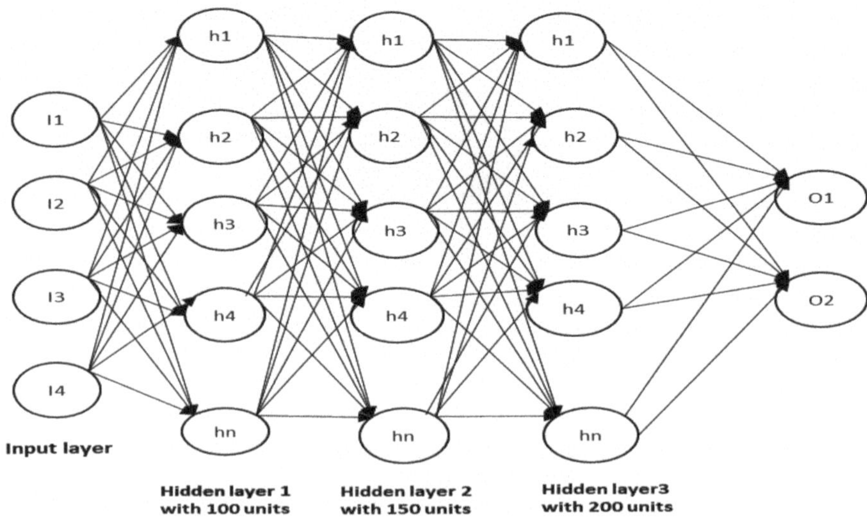


**Fig. 9.** ANN without dropout layer

process, prompting nodes in a layer to statistically assume a greater or lesser the liability of the input units (Figs. 10, 11 and 12).

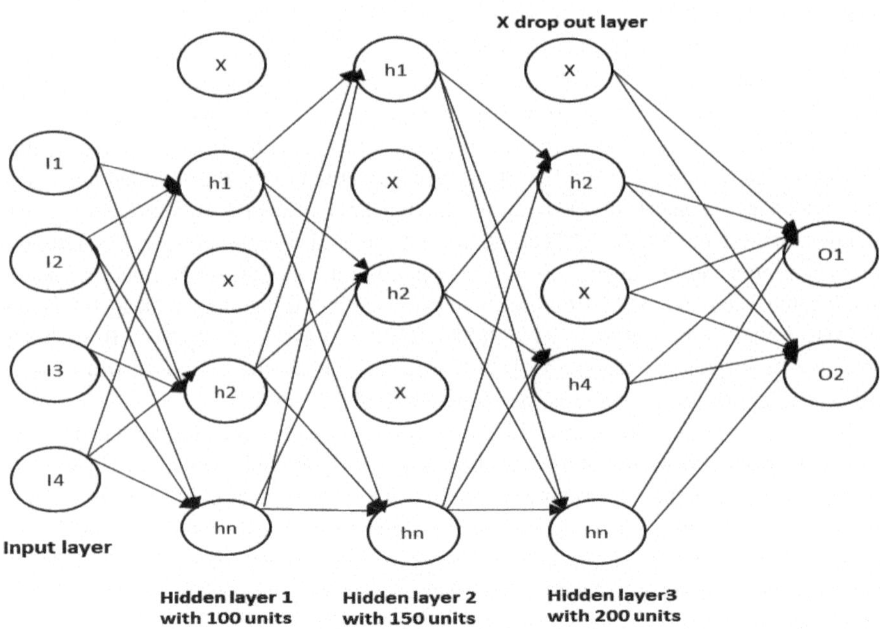


**Fig. 10.** ANN with drop out layer

## 5 Implementation and Result Analysis

In this research work, outcomes were achieved by applying ANN model with SMOTE and SMOTE-Tomek to achieve the maximum accuracy for diabetes predictions. The work also involved in comparing the performance of the ANN model with and without dropout layer. The impact of the dropout layer shows much difference in the performance of the model. Table 5: shows the outcome of the ANN model with SMOTE and SMOTE-Tomek. In our proposed system we use the PIMA dataset, we pre-processed the data and apply it on an ANN model. From the nine features, four features are selected based on the effective feature selection techniques. From these techniques, the selected features are tuned and finest features are applied are as the input to the model for the prediction. The feature selection techniques, selected features ranking is represented in Tables 3 and 4. In this work two different ANN model is built, First ANN without drop out layer, three hidden layers where each layer consists of different units starts from 100 to 200.Second ANN model with dropout layer, three hidden layers consists of same unit 500. Both models fit with data which augmented by both SMOTE and SMOTE-Tomek. The main objective of the study is to present the most promising features that needed to predict the patient having diabetes at an early stage. This proposed methodology consists of two main parts, first how accuracy is obtained using oversampling the data and by performing both oversampling & under sampling on the data using ANN classification models and second is how dropout layers improve the performance of the ANN model. Figure 13 graphically represents the result of an ANN model with both SMOTE and SMOTE-Tomek with and without dropout layer. ANN with dropout layer comparatively provide promising accuracy (95.1%) which can be proven as a prominent tool for the prediction of diabetes at an early stage. The performance of the model is evaluated by using two metrics likely Accuracy and ROC curve. Figure 13: shows the performance of the ROC curve metrics. ROC curve shows the performance of a classification model at all classification thresholds. The curve plots two parameters: True Positive Rate and False Positive Rate.

**Table 5.** Outcome of ANN model with SMOTE and SMOTE-Tomek

| Model | Data Augmentation | Neurons size | Activation Function | Loss | batch size | Epochs | Optimizer | Performance Metrics |
|---|---|---|---|---|---|---|---|---|
| ANN-3 hidden layers without drop out layers | SMOTE Befor SMOTE:768 After SMOTE: 1000 | Input with 4 attributes 1st hidden layer = 100 2nd hidden layer = 150 3rd hidden layer = 200 Output with 2 nodes | Relu Sigmoid | Binary_cross entropy | 10 | 100 | Adam | **Accuracy:89.2 ROC Curve: 92.1** |
| | SMOTE-Tomek Befor SMOTE-Tomek:768 After SMOTE-Tomek: 950 | | Relu Sigmoid | Binary_cross entropy | 10 | 100 | Adam | **Accuracy:87.3 ROC Curve: 90** |
| ANN-3 hidden layers with dropout layer | SMOTE Befor SMOTE:768 After SMOTE: 1000 | Input with 4 attributes 1st hidden layer = 500 2nd hidden layer = 500 3rd hidden layer = 500 Output with 2 nodes | Relu Sigmoid | Binary_cross entropy | 20 | 500 | Adam | **Accuracy:95.1 ROC Curve:90.4** |
| | SMOTE-Tomek Befor SMOTE-Tomek:768 After SMOTE-Tomek: 950 | | Relu Sigmoid | Binary_cross entropy | 20 | 500 | Adam | **Accuracy:88 ROC Curve: 89** |

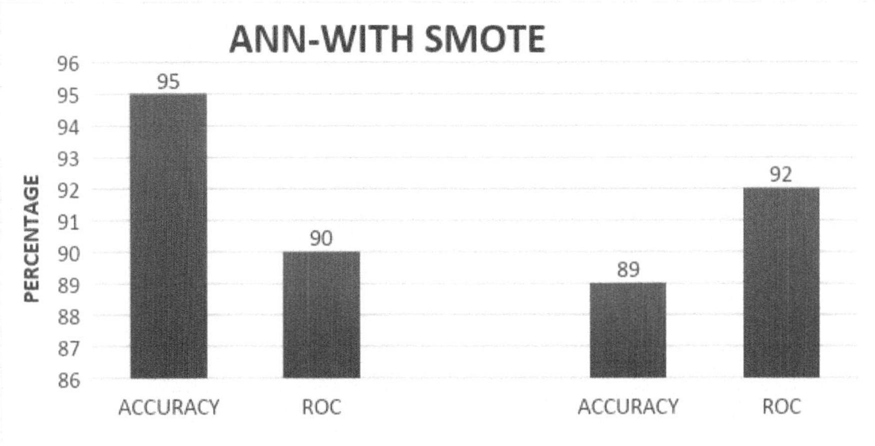


**Fig. 11.** ANN with SMOTE

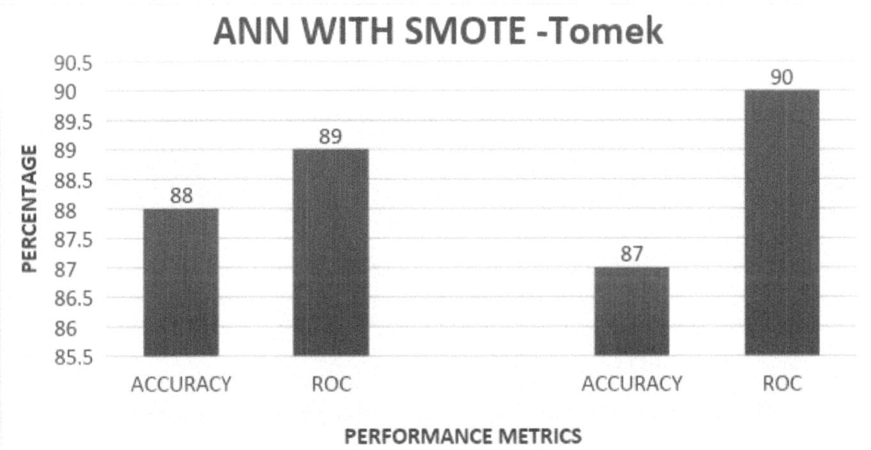


**Fig. 12.** ANN with SMOTE-Tomek

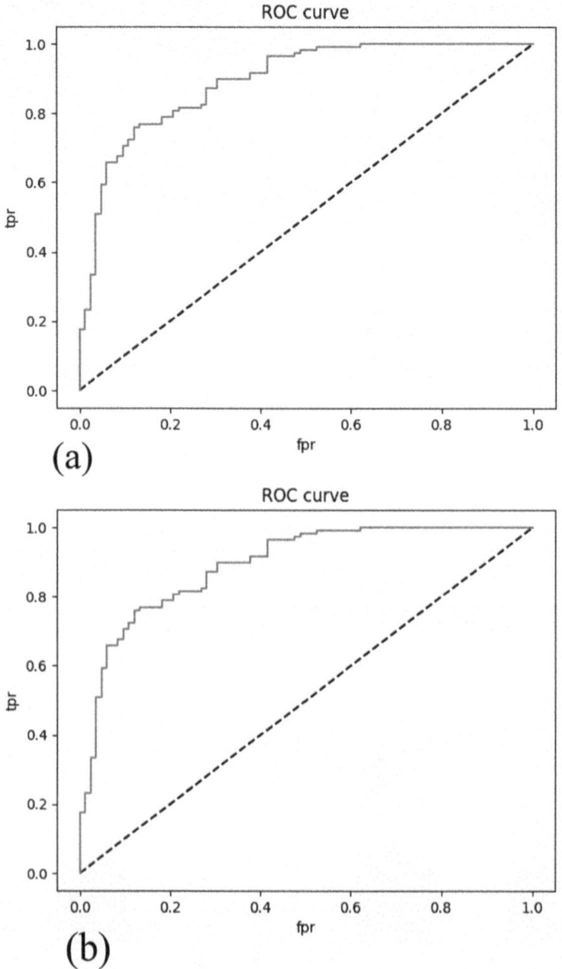


**Fig. 13. (a&b).** ROC Curve using ANN with and without drop out layer

## 6 Conclusion

In this study, we propose ANN with SMOTE and ANN with SMOTE-Tomek to figure out the performance in diabetes identification. ANN performance is also evaluated by adding the dropout layer. The motive of this study is to detect the abnormality and identify the diabetic patient quickly in a short time. We have shown how diabetes identified in two ways using SMOTE and SMOTE-Tomek, also the efficiency of the dropout layer. For the experimental analysis PIMA dataset is used. However, PIMA dataset consists of Outliers and Zero values in some features which is not suitable for the prediction. Dataset is pre-processed and important features are selected by using filter, Wrapper and Embedded method. Experimental results implied better accuracy 95.1% and ROC curve 90.4% using ANN-SMOTE with drop out layer, Where ANN-SMOTE Tomek with drop out layer obtained accuracy 88.1% and ROC curve 89.2%. In second Approach ANN-SMOTE without layer obtained accuracy 89.2% and ROC curve 92.1%, where ANN-SMOTE Tomek without dropout layer obtained accuracy 87.5% and ROC curve 89.7%. From result, ANN SMOTE with dropout layer performed better by achieving best accuracy and ROC curve comparative to other approach. Due to the small sample size in our study, further generalization to other populations or contexts is required. In the future, the work will be expanded by adding more samples and attributes, which could give detailed insights into the elements that help our model more accurately identify diabetes patients.

## References

1. International Diabetes Federation: IDF Diabetes Atlas, 8th edn (2017)
2. https://worldpopulationreview.com/country-rankings/diabetes-rates-by-country
3. American Diabetes Association: Classification and diagnosis of diabetes: standards of medical care in diabetes-2018. Diabetes Care **41**, 13–27 (2018)
4. https://cdn11.bigcommerce.com/s-a69c6/product_images/uploaded_images/type-1-vs-type-2.jpg
5. Zhou, H., Myrzashova, R., Zheng, R.: Diabetes prediction model based on an enhanced deep neural network. J. Wireless Com. Netw. **2020**, 148 (2020). https://doi.org/10.1186/s13638-020-01765-7
6. Xu, X., He, C., Xu, Z., Qi, L., Wan, S., Bhuiyan, S.: Joint optimization of offloading utility and privacy for edge computing enabled IoT. IEEE Internet Things J. (2019). https://doi.org/10.1109/JIOT.2019.2944007
7. Swapna, G., Vinayakumar, R., Soman, K.P.: Diabetes detection using deep learning algorithms. ICT Express **4**(4), 243–246 (2018)
8. Sisodia, D., Sisodia, D.S.: Prediction of diabetes using classification algorithms. Procedia Comput. Sci. **132**, 1578–1585 (2018)
9. Wu, H., Yang, S., Huang, Z., He, J., Wang, X.: Type 2 diabetes mellitus prediction model based on data mining. Inform. Med. Unlocked **10**, 100–107 (2018)
10. Zou, Q., Qu, K., Luo, Y., Yin, D., Ju, Y., Tang, H.: Predicting diabetes mellitus with machine learning techniques. Front. Genet. **9**, 515 (2018). https://doi.org/10.3389/fgene.2018.00515
11. Perveen, S., Shahbaz, M., Keshavjee, K., Guergachi, A.: Metabolic syndrome and development of diabetes mellitus: predictive modeling based on machine learning techniques. IEEE Access **7**, 1365–1375 (2019)
12. Choi, B.G., Rha, S.W., Kim, S.W., Kang, J.H., Park, J.Y., Noh, Y.K.: Machine learning for the prediction of new-onset diabetes mellitus during 5-year follow-up in non-diabetic patients with cardiovascular risks. Yonsei Med. J. **60**(2), 191–199 (2019)
13. García-Ordás, M.T., Benavides, C., Benítez-Andrades, J.A., Alaiz-Moretón, H., García-Rodríguez, I.: Diabetes detection using deep learning techniques with oversampling and feature augmentation. Comput. Methods Programs Biomed. **202**, 105968 (2021)
14. Nuankaew, P., Chaising, S., Temdee, P.: Average weighted objective distance-based method for type 2 diabetes prediction. IEEE Access **9**, 137015–137028 (2021). https://doi.org/10.1109/ACCESS.2021.3117269
15. Naz, H., Ahuja, S.: Deep learning approach for diabetes prediction using PIMA Indian dataset. J. Diabetes Metab. Disord. **19**(1), 391–403 (2020)
16. Khan, F.A., Zeb, K., Al-Rakhami, M., Derhab, A., Bukhari, S.A.C.: Detection and prediction of diabetes using data mining: a comprehensive review. IEEE Access **9**, 43711–43735 (2021)

17. Sowah, R.A., Bampoe-Addo, A.A., Armoo, S.K., Saalia, F.K., Gatsi, F., Sarkodie-Mensah, B.: Design and development of diabetes management system using machine learning. Int. J. Telemed. Appl. **2020**, 1–17 (2020). https://doi.org/10.1155/2020/8870141
18. Reddy, D.J., et al.: Predictive machine learning model for early detection and analysis of diabetes. Mater. Today Proc. (2020)
19. Sapon, M.A., Ismail, K., Zainudin, S.: Prediction of diabetes by using artificial neural network. In: Proceedings of the 2011 International Conference on Circuits, System and Simulation, vol. 2829. Singapore, Singapore (2011)
20. https://www.kaggle.com/datasets/uciml/pima-indians-diabetes-database

# Machine Learning Based Traffic Congestion and Accident Prevention Analysis

A. Sathya Sofia[1(✉)], C. P. Thamil Selvi[2], S. Suganya[3], P. Francis Antony Selvi[4], and M. Shanthalakshmi[5]

[1] PSNA College of Engineering and Technology, Dindigul, India
sathyasofia@psnacet.edu.in
[2] Rathinam Technical Campus, Coimbatore, India
[3] SSM Institute of Engineering and Technology, Dindigul, India
[4] RVS College of Engineering and Technology, Dindigul, India
[5] Sri Venkateshwara College of Engineering, Sriperumbudur, India

**Abstract.** To enhance decision-making and encourage urban development in VANET, a subclass of mobile ad hoc network, the suggested approach elucidates how to enhance interconnection in VANET in order to prevent and identify traffic congestion. The solution is constructed on an algorithm grounded on machine learning. The recommended algorithm is called YOLO (You Only Look Once) algorithm. This proposed method improves the connectivity of the GPS and several VANET network devices. Reducing the effects of traffic congestion besides traffic accidents considering those two as external aspects of transportation, is one of the main goals of transportation policy makers. Road traffic accidents have also had a substantial economic impact on people, property, and society at large. It is crucial to lessen the effects of accidents and traffic congestion because both burden societies. Augmented travel time caused through congestion of traffic enforces unrestricted value towards road users which leads to the reduced quality of life and mobility. Owing to adjacent spacing of vehicles and overheating of vehicles, congestion in traffic increases the risk of traffic accidents. YOLO (You only look once) implemented technology is used in this proposed system to predict automobile traffic congestion. Machine learning models for predicting traffic congestion is developed. In order to anticipate future traffic congestion, a machine learning system can be trained using the data. Decisions concerning routes, speeds, and other traffic control tactics can then be made using the trained model. Machine learning can also be incorporated to find data patterns which can be processed further to increase the precision of traffic predictions.

**Keywords:** VANET (Vehicular Adhoc Network · Traffic Congestion · Accident Prevention · Machine learning · YOLO Algorithm

## 1 Introduction

Human mobility suffers greatly as a result of urbanization since city dwellers tend to increase traffic flow. Traffic congestion occurs in densely populated urban centers because city road infrastructure does not expand at a rate that is greater than the number


R. Geetha et al. (Eds.): AAIMB 2023, CCIS 2203, pp. 105–115, 2025.
https://doi.org/10.1007/978-3-031-73068-9_9

of cars. We are attempting to decrease traffic accidents by means of better connectivity happening around networking field by enhancing connectivity among automobile to automobile while they remain connected near one another using VANET [1]. This is done to augment the functionality of network applications for connecting vehicles on highways and in metropolitan areas. Road accidents and traffic congestion [2] are two unusual transportation issues, and minimizing their effects is frequently one of the main goals of transportation policy makers.

The primary need of the hour in highly populated countries like India is to prevent accidents in denser cities and to increase the network system's vehicle connection. It is possible to use an algorithm based on machine learning [3] to analyses the system's traffic congestion. Vehicle speed can be recognized, and if it exceeds the maximum speed restriction or not, the user will instantly be informed. In populous countries, the incidence of accidents can be decreased if the driver's sleepiness and fatigue are identified early on and if they can be informed simultaneously. The usage of Vehicular Ad Hoc Networks (VANETs) for driving safety support systems is becoming more popular in recent years.

Vehicles have the ability to set up the network and interact with one another using VANETs without the assistance of infrastructure. It is possible to increase road traffic safety and transit effectiveness by using the aforementioned vehicle-to-vehicle interaction successfully. For instance, in the Cooperative Collision Warning (CCW) implementation [4], the scheduling component uses periodic information transmission among vehicles, such as their location and speed, to estimate the likelihood of accidents and alert drivers to them in advance. In a manner similar to this, vehicles can notify other vehicles as well as roadside units regarding road congestion employing Congested Road Notification (CRN) technologies. This enables route in addition trip planning and guarantees a smooth drive.

Vehicle representation is a useful tool for encouraging safe driving because it can significantly offset the driver's limited field of vision by showing hidden automobiles on navigational maps or in-car displays, which can reduce accidents that could otherwise occur. Furthermore, it can be tough to identify other vehicles when utilizing simply on-vehicle sensors and cameras at a congested intersection or a narrow road with a hazy route.

## 2 Background and Related Work

In recent years, traffic accidents have become a major global issue and the leading cause of death worldwide. India now has a significant problem with vehicle accidents because of the staggering number that occur each year. It is completely immoral and distressing that a country would permit its citizens to die in traffic accidents. The authors in [5] afford insights into traffic controlling and produced outcomes of data imputation in detecting accidents. The technology for predicting traffic is essential to an intelligent transportation system. Vehicle Ad-hoc Networks (VANET) willpower as a key component of the future smart transport system. They provide the drivers with excellent service on the roadways. The recommended method by Singh and Verma [6] makes use of Random Forest scheduled using VANET to assess route remodeling to acquire a road without congestion, accident exposure as well as prevention with delivering alternative facilities for passengers as well as drivers at the same QOS.

The characterized method by Beak, & Chung [7] builds data sets by combining organized and unorganized modality data on an underlying basis of statistical plus evaluative fusion. This allows it to forecast the danger of an accident. It has the ability to address the issue of data scarcity and reliably get a variety of information. Using the equation of road risk plus accident risk, suggested technique first performs preprocessing before generating data sets containing event risk information. Then, using the regression analysis model as well as the association coefficient investigation model, the correlation factor besides regression factor for each variable are calculated.

Inattentive drivers, broken traffic regulations, inadequate road infrastructure, driving in terrible weather conditions, and other factors contributing to traffic accidents, road crashes, and injuries were explained in [8]. To more effectively manage traffic, the authors in [9] employed instantaneous information from sensors and connected devices. Estimating traffic metrics including speed, weather, currents, and accident likelihood is one of the main measures used in that work. More features, especially traffic, highway, and climate conditions have been extracted, which has increased the performance of congestion prediction. Collision avoidance systems take into account network as well as application-related issues and potential solutions in the context of 5G architecture. The investigation done by authors in [10] encompasses the various message types in the V2VRU interaction that relies safety application and includes standardization, investigations, and project initiatives linked to the use case. They look into how applications can utilize the resources of networks effectively to offer dependable facilities by adjusting the request settings to the network state and the devices existing as resources for traffic prediction. Utilizing the characteristics of the previous obstructions in accident zones, a warning signal is generated on requirement through a processing rapidity of 0.09 seconds each frame and live-distance amongst vehicles and additional objects is successfully estimated. A multivariate analysis is used to examine the real-time output data in [11]. The outcome demonstrates that multimodal premature accident alarming structure is quite efficient for car users who are distracted.

## 3 Methodology

The proposed methodology tries to identify and reduce the causes of traffic congestion and accidents using data and algorithms. Various techniques are used in the current system, including data collecting, analysis, modelling, and optimization. Along with GPS tracking devices, traffic cameras, and other sensors, information regarding traffic patterns is gathered from a variety of sources. To find patterns and trends in traffic congestion and accident frequency, this data is then evaluated numerically. The suggested system employs predictive modelling to forecast future traffic patterns and identify possible regions for congestion or accident risk. Scheduled on the basis of historical data, machine learning algorithms remain employed to analyses the data and generate predictions about future traffic patterns. Alternative key element is optimization. Algorithms for optimization can be used to develop methods to lower these risks by identifying places that are at high risk for congestion or accidents.

The current machine learning-based approach for analyzing traffic congestion and accident prevention is made to give transportation planners and officials the resources, they need to make well-informed decisions concerning traffic management and safety. The technology can classify potential problems before they become significant hiccups and develop solutions by examining vast volumes of data and employing sophisticated algorithms.

### 3.1 Usage of YOLO Algorithm

YOLO [12] is a method which offers instantaneous object identification employing neural networks. The efficiency and speed of this algorithm account for its popularity. Currently, it is used in many different contexts to identify people, animals, traffic signals as well as parking meters. This algorithm (which operates in real-time) can find and identify different items in images. The object categorization method known as YOLO, that is implemented as a regression problem, can be used to determine the classification probabilities of the seen pictures. Due to its ability to forecast objects happening real time, this algorithm improves the rapidity of revealing the output. YOLO is a predictive method which vintages accurate outcomes with insignificant errors. Because of the system's outstanding abilities for learning, it can acquire visualizations of objects and use those representations for object detection. It offers a more accurate generalization for new domains, which makes it great for applications relying on fast and robust object detection within a limited time.

### 3.2 YOLO Algorithm Techniques

Three methods are used by the YOLO algorithm [13] to work: the residual block method, bounding box regression with intersection over union. In residual block method, involving image is initially divided into numerous grids in leftover blocks. Here, every grid cell determines the items that cross the threshold into it. For example, particular cell is responsible of identifying an object when its center looks inside that cell. In the second method of bounding box regression, the prescribed box is an framework which pulls consideration to an entity in a image. Identified image takes the following attributes: bx – considered to be the x coordinates, and by – y coordinates which are the of the bounding box's midpoint in relation to the surrounding grid cell. The bounding box's height and width in relation to the surrounding grid cell are denoted by the letters bh, bw. Center of the bounding box is given as (bx, by).

YOLO employ just one bounding box regression to ascertain the width, height, center of a given object. The item recognition feature is referred to as intersecting over union (IOU) might be utilized for clarifying box overlapping. YOLO uses IOU to build a result container that correctly encloses the components. Each grid cell is accountable for the anticipated bounding boxes and their confidence scores. The IOU is set to 1 if the expected bounding box is 1.

i. **Combination of three techniques in YOLO**

If a traffic-based image contains numerous object classifications, all predictions are produced simultaneously using a single convolutional neural network. By using intersection over union (IOU), the expected bounding boxes are assured to be equal to the real containers of the items [14]. This phenomenon results in the elimination of bounding boxes whose height and width do not correspond to the object dimensions. Distinct bounding boxes designed specifically to fit the items will comprise the final detection. As an illustration, if a car is the first object, it can be represented by a pink bounding box, followed by a yellow bounding box for the truck, a blue bounding box for the lorry, and so on.

### 3.3 Usage of COCO Dataset for Traffic Analysis

The appellation COCO, that signifies Common Objects in Context, has been assigned to the image dataset with the goal of improving image recognition. The COCO collection features difficult, excellent visual computer vision datasets that are mostly employed by innovative neural networks. In this regard, COCO [23] has been extensively utilized to evaluate algorithm performance and assess object detection speed. The format of the COCO dataset is automatically interpreted by sophisticated neural network libraries. Microsoft made available the MS COCO dataset, a huge object recognition, image segmentation, and captioning dataset. This is used by machine learning programs for numerous performance investigates. This dataset's main objective is to understand visual situations, which entails identifying the things that are present, localizing them in 2D and 3D, figuring out what they are made of, and describing them and portraying the association amongst items.

## 4 System Design

The already existing system implicates the use of various methods, including data collection and analysis, analytical modelling, and optimization. Several sources of data are used to gather information about traffic patterns, together with GPS tracking devices,

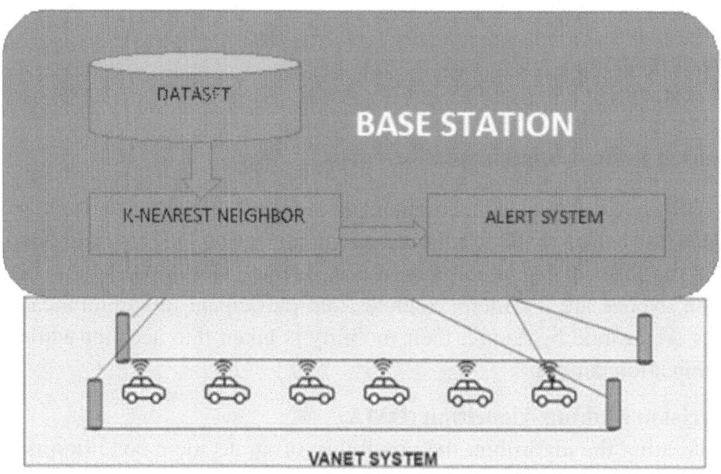


**Fig. 1.** System Design

traffic cameras, and other sensors. This data is then analyzed using numerical methods to identify patterns and trends in traffic congestion [15, 22] and accident amount as shown in Fig. 1. Predictive modelling is also used in the existing system to estimate future traffic patterns and recognize potential areas of congestion or accident risk.

\Machine learning procedures are practiced to analyses important data and make predictions about future traffic patterns based on ancient data. Optimization is alternative key component. Recognizing areas of congestion or accident risk, optimization algorithms can be used to grow strategies to reduce these risks. The existing system for traffic congestion and accident prevention analysis using machine learning is designed to afford transportation planners and officials with the tools they need to make informed results about traffic management and safety. By analyzing large amounts of data and using advanced algorithms, the system can classify potential issues before they become serious glitches and develop strategies to address them successfully.

### 4.1 Proposed System

The suggested method uses machine learning algorithms to analyze traffic congestion and prevent accidents. The system will combine information from a number of sources, including traffic sensors, security cameras, and social media sites. Machine learning algorithms will be applied to proceed to examine the facts to avoid traffic jams and locate probable accident hotspots. In order to prevent accidents and ease traffic congestion, the system will then generate alerts and suggestions for motorists and traffic authorities [16]. Using outdated data, machine learning models are used to forecast traffic density, speed, and flow. Models will be utilized to detect potential accident flashpoints depending on the density and speed of traffic, classifying traffic circumstances such as heavy traffic, prudent traffic, and light traffic.

**Advantages:**

- Cost-Effective
- High Performance
- High Accuracy

### 4.2 Proposed System Implementation Steps

By broadcasting messages using communications from vehicle to vehicle (V2V) besides vehicle to infrastructure (V2I), traffic congestion detection [17] is made possible. It is presumable that the car has a communication device. The impacted car initiates the conversation. Nodes are the motor vehicles that participate in communication. Since these nodes are mobile by nature, their mobility is taken into account while choosing the communication pattern.

**Step 1: Decision Making Algorithm (DMA)**

When creating the algorithm, the possibility of an accident condition occurring is taken into account. We want broadcasts to reflect any congestion that may result from one vehicle having an accident that blocks one lane of the road. In our system, the message is produced by the impacted car. According to our strategy, the impacted vehicle creates a warning message, or message for notifying an unintended event. In this case, the broadcasting was started by accident acting as an external trigger. The short-range vehicle makes a judgment and forwards the alert after receiving it, signaling the approach of congestion. The traffic flow is consequently directed through V2V and V2I.

**Step 2: Gathering of Information**

The collection of data from the environment provides the most fundamental information. Each vehicle must have one or more sensors that can collect information like the current location and speed. Vehicles must exchange the information. The ability to transmit and receive information to and from other cars must be built into every vehicle using VANET.

**Step 3: Propagation of Knowledge**

The final decision is disseminated so that vehicles away from the event zone, not capable to make this determination with the information received from vehicles around them, can benefit from it.

**Step 4: Making a decision**

The right judgment must be made in order to detect the congestion, as information from adjacent vehicles as well as information relayed from distant roadside units is received. All the adjacent vehicles receive the information sent by the impacted vehicle. The geo broadcast pattern is utilized for broadcasting the message since it has very low message latency to tell addressed cars as soon as possible. When information is received, the receiving vehicle makes a decision and changes the message. Geo broadcast messages are unidirectional and often delivered in response to external event, Geo broadcast messages are not propelled constantly, while messages with the same content may be repetitive from time to time.

## 5 Modules and Module Description

### 5.1 Data Set Collection

Traffic datasets [18] that are utilized for various trainings can be broadly classified into two classes:

Stationary as well as probing data. Sensor data and fixed cameras are two additional categories of stationary data. The studies of probe data consisted of GPS data collected from various types of automobiles. Every step of the technique is important, if approved out inadequately, could take an effect on the outcomes. Distribution of data, which occurs after data collection, is essential for creating the drill and testing datasets. For each investigation, a separate collection of test cases is used. The model is validated by comparison with other base models after it has been developed.

a. **Data base maintenance**

We have a variety of elements for car specifications in the database. An online tool called the VANET Data Base Traffic Congestion (VDBTC) [19] offers up-to-date data on the movement of traffic and congestion happening in larger cities all over the world. The aforementioned offers the most contemporary information on traffic conditions, including flow, closures of roads, accidents, and more. Additionally, the VDBTC offers historical information on the flow of traffic in a certain area.

### 5.2 Traffic Flow Rate Analysis

In order to pinpoint regions where traffic congestion is expected to occur and provide solutions, traffic flow rate analysis [20, 24] of VANETs is used. Additionally, it aids in identifying and avoiding any risky driving behaviors that might cause collisions and fatalities. Data about the number of cars in a specific area, their speed, the categories of routes where they are, and the travelling time in a particular day they are moving are often collected in order to do the research. The average vehicle flow rate through the area is then determined using this data. To determine wherever congestion on traffic is most probable to happen, the rate of traffic will then have compared to the movement proportions of other locations.

### 5.3 Traffic Congestion Prediction

To produce precise predictions, the machine learning algorithm [21] builds many decision trees and aggregates their data. Given enough training data, producing real outcomes is highly useful. This approach demonstrated an accuracy of 87.5% when used to solve the road cramming issue. In this instance, input variables are modelled using weather, time, specific road conditions, road quality, and vacations in order to detect traffic congestion.

## 6 Results and Discussion

### 6.1 Vehicle Detection

**Fig. 2.** Vehicle Detection Output Screen

In the Fig. 2, the illustration of using YOLO algorithm is presented here. The car is detected before the two lines and the calculation is done, according to the speed limit of the vehicle, before it reaches the signal, to avoid congestion on further distance.

In the Fig. 3, vehicle count detector using traffic controller is done by signal optimizer in which the direction of vehicles is categorized as south and north, and in the corresponding direction the entire amount of vehicles counted as well as name of each vehicle is displayed that is moving towards the particular direction.

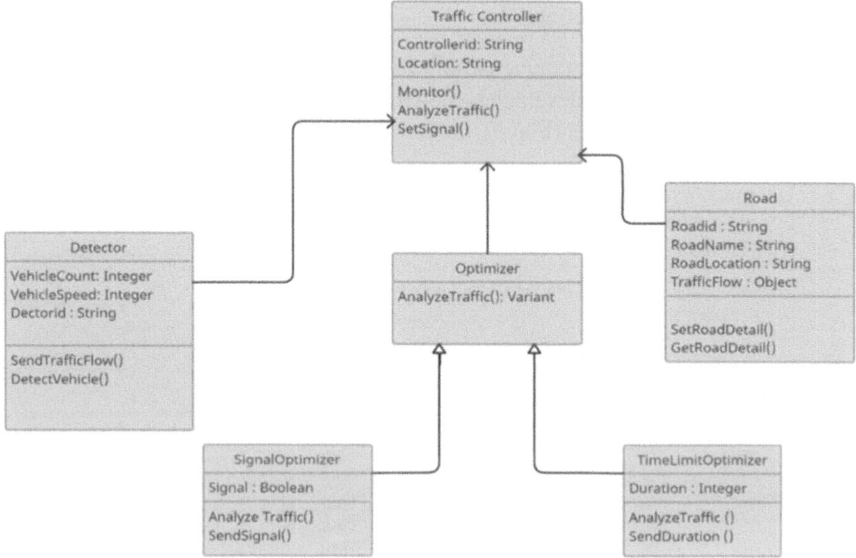


**Fig. 3.** Vehicle Count Detector using Traffic Controller

## 6.2 Speed Calculation

In the Fig. 4, the vehicle speed separately is calculated and the name of particular vehicle is displayed that is passing towards the signal and the count of incoming and outgoing traffic is calculated and displayed before it reaching the signal to avoid congestion.

**Fig. 4.** Speed Calculation Output Screen

### 6.3 Graph of Vehicle Congregation

The above graph portrays the congregation of vehicles at a transportation junction that demonstrates the unit time of vehicles showing the waiting besides queuing time of individual vehicle is displayed (Fig. 5).

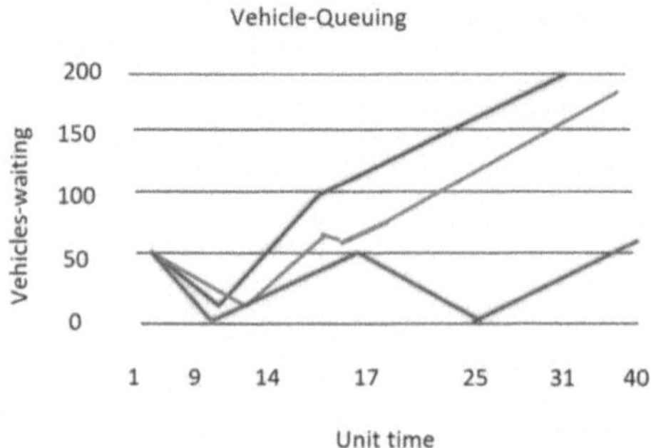


**Fig. 5.** Graph of congregation of vehicles

## 7 Conclusion

Machine learning algorithm has created as effective predictive model that can be used to enhance VANET traffic management. As a result, there will be less traffic congestion, more safety and greater effectiveness. The transportation and technology sectors must work together and form partnerships to address the issue of traffic safety. Aside from that, utilizing the potential of cutting-edge network technologies like ADAS, IOT, etc. can result in operational advancements, regional discoveries, and roadside lifesaving. Deep learning grounded algorithms can be incorporated in future in order to handle smart transportation system which can handle huge number of population. This type of work can save the life of the people who are travelling busy with their works.

## References

1. Xia, X., Li, X., Hou, W., Hua, S., Huang, Q.: AI-based efficient wireless technologies and infrastructure-based networks with VANET for smart transportation high performance. Soft Comput. 1–14 (2023)
2. Sehrawat, P., Chawla, M.: SDTMRP (software defined traffic management routing protocol) for efficient and reliable communication in vehicular networks. Wireless Pers. Commun. **131**(3), 1595–1629 (2023). https://doi.org/10.1007/s11277-023-10515-z
3. León, J.P.A., de la Cruz Llopis, L.J., Rico-Novella, F.J.: A machine learning based Distributed Congestion Control Protocol for multi-hop wireless networks. Comput. Netw. **231**, 109813 (2023). https://doi.org/10.1016/j.comnet.2023.109813
4. Li, W., Song, W., Lu, Q., Yue, C.: Reliable congestion control mechanism for safety applications in urban VANETs. Ad Hoc Netw. **98**, 102033 (2020)

5. Qi, H., Zhao, X., Yao, Y., Yang, H., Chai, S., Chen, X.: BGCP-based traffic data imputation and accident detection applications for the national trunk highway. Accid. Anal. Prev. **186**, 107051 (2023)
6. Singh, S., Verma, S.K.: Congestion and accident alerts using cloud load balancing random forest in VANET. Wireless Pers. Commun. **128**(1), 43–65 (2023)
7. Baek, J.W., Chung, K.: Accident risk prediction model based on attention-mechanism LSTM using modality convergence in multimodal. Pers. Ubiquitous Comput. 1–11. (2021).
8. Gururaj, H.L., Janhavi, V., Tanuja, U., Flamini, F., Soundarya, B.C., Ravi, V.K.: Predicting traffic accidents and their injury severities using machine learning techniques. Int. J. Transport Dev. Integr. **6**(4), 363–377 (2022)
9. Abdullah, S.M., et al.: Optimizing traffic flow in smart cities: soft GRU-based recurrent neural networks for enhanced congestion prediction using deep learning. Sustainability **15**(7), 5949 (2023)
10. Zoghlami, C., Kacimi, R., Dhaou, R.: 5G-enabled V2X communications for vulnerable road users safety applications: a review. Wireless Netw. **29**(3), 1237–1267 (2023)
11. Kabir, M.F., Roy, S.: Real-time vehicular accident prevention system using deep learning architecture. Expert Syst. Appl. **206**, 117837 (2022)
12. Jiang, P., Ergu, D., Liu, F., Cai, Y., Ma, B.: A Review of Yolo algorithm developments. Procedia Comput. Sci. **199**, 1066–1073 (2022)
13. Shinde, S.V., Lakshmi, S.V., Sabeenian, R.S., Bhavani, K.D., Murthy, K.V.S.R.: Traffic optimization algorithms in optical networks for real time traffic analysis. Optik 170418 (2022)
14. Jiang, C., et al.: Object detection from UAV thermal infrared images and videos using YOLO models. Int. J. Appl. Earth Obs. Geoinf. **112**, 102912 (2022)
15. Cherkaoui, B., Beni-Hssane, A., El Fissaoui, M., Erritali, M.: Road traffic congestion detection in VANET networks. Procedia Comput. Sci. **151**, 1158–1163 (2019)
16. Ning, H., et al.: Modeling and analysis of traffic warning message dissemination system in VANETs. Vehicular Commun. **39**, 100566 (2023)
17. Ali, M., Malik, A.W., Rahman, A.U.: Clustering-based re-routing framework for network traffic congestion avoidance on urban vehicular roads. J. Supercomput. **79**(18), 21144–21165 (2023). https://doi.org/10.1007/s11227-023-05455-1
18. Chu, J., Zhang, C., Yan, M., Zhang, H., Ge, T.: TRD-YOLO: a real-time, high-performance small traffic sign detection algorithm. Sensors **23**(8), 3871 (2023)
19. Vinoth kumar, S., Varadhaganapathy, S., Shanthakumari, R., Pradeev, S., Pragatheeswaran, S., Annamalai, K.S.: Traffic sign detection using hybrid network of yolo and Resnet. In: 2023 International Conference on Computer Communication and Informatics (ICCCI), pp. 1–7. IEEE (2023)
20. Majumder, M., Wilmot, C.: Automated vehicle counting from pre-recorded video using you only look once (YOLO) object detection model. J. Imag. **9**(7), 131 (2023)
21. Mittal, U., Chawla, P., Tiwari, R.: EnsembleNet: a hybrid approach for vehicle detection and estimation of traffic density based on faster R-CNN and YOLO models. Neural Comput. Appl. **35**(6), 4755–4774 (2023)
22. Vijay, K., Vijayakumar, R., Sivaranjani, P., Logeshwari, R.: Scratch detection in cars using mask region convolution neural networks. Adv. Parallel Comput. **37**, 575–581 (2020)
23. Singh, N., Saini, P., Shubham, O., Awasthi, R., Bharti, A., Kumar, N.: Improved YOLOv5l for vehicle detection: an application to estimating traffic density and identifying over speeding vehicles on highway scenes. Multimed. Tools Appl. **83**(2), 5277–5307 (2023). https://doi.org/10.1007/s11042-023-15520-9
24. Vijay, K., . Sabarish Abishek, W.R, Sabarish, V.U., Sanjeev Krishnan, R.: Private cloud storage using raspberry PI via virtual network computing – an analysis. In: 2023 International Conference on Computer Communication and Informatics (ICCCI), pp. 1–7. Coimbatore, India (2023). https://doi.org/10.1109/ICCCI56745.2023.10128489

# A Comparative Evaluation of Machine Learning Methods for Predicting Chronic Kidney Disease

K. Navaz[1], S. Yazhinian[1(✉)], N. Muthuvairavan Pillai[2], and N. Purushotham[3]

[1] Vel Tech Rangarajan Dr. Sagunthala R&D Institute of Science and Technology, Avadi, Chennai 600062, Tamil Nadu, India
navazit@gmail.com, yazhinian.s@gmail.com

[2] Department of CSE, R.M.D Engineering College, Kavaraipettai, India
muthuvikni@gmail.com

[3] Department of CSE, Mohan Babu University, A.Rangampeta, Tirupathi, India
nimmalanimmala10@gmail.com

**Abstract.** Chronic Kidney Disease (CKD) is a chronic renal failure condition that is one of the most serious and rapidly spreading diseases in today's world. Each kidney is composed of several filtrating units called Nephrons. It is difficult to predict renal failure at an early stage because there are no specific symptoms. This can have long-term consequences for the patient's health and even result in death. People who have type 2 diabetes or high blood pressure have the highest risk for renal failure. The objective of this paper is to analyze and evaluate the performance of a few existing predictive models for Chronic Kidney Disease (CKD) utilizing different Machine Learning Classification Algorithms. A powerful Machine Learning approach that is frequently applied to prediction is Classification. Different classifiers such as Random Forest, Decision Tree, K-Nearest Neighbour, Ada Boosting, Gradient Boosting, and Naive Bayes are used for the classification of CKD. Out of all the classifiers Random Forest and Gradient Boosting produced the best accuracy result when compared with other classifiers. A comparison of the accuracy score along with the confusion matrix to identify correct and incorrect features is presented to detect the optimal model.

**Keywords:** CKD · Machine Learning · KNN · Random Forest · Decision Tree · Ada Boosting · Gradient Boosting · Naive Bayes

## 1 Introduction

The kidney a crucial organ of the human body is located close to the bottom of the rib cage. It helps in maintaining the equilibrium of salt water and other ions in the body such as vitamins and minerals like calcium phosphorus magnesium potassium, chlorine, and different acids.

There are around 1,000,000 nephrons or filtering units inside each of the kidneys. Glomerulus and tubule are the two main components of every nephron that act as filters. The glomerulus filters the blood and the tubule restores the fundamental vitamins in the


R. Geetha et al. (Eds.): AAIMB 2023, CCIS 2203, pp. 116–125, 2025.
https://doi.org/10.1007/978-3-031-73068-9_10

blood while expelling waste products. Chronic kidney failure is the progressive deterioration of the kidneys, creatinine, and blood urea rise up constantly without any signs and symptoms in the body as an end result of the gradual loss of renal function which requires dialysis or a kidney transplant for survival. Early analysis can lead to the prevention of kidney failure. The desirable way to evaluate kidney function or predict kidney disease stages is to observe the Glomerular Filtration Rate (GFR). GFR is calculated using age, gender, and blood creatinine level [9].

Every year about one million new instance of Chronic Kidney Disease (CKD) are identified in India. To determine if the individual has renal failure or not Machine Learning (ML) can be beneficial. ML accomplishes this by training a predictive model by using historical CKD-affected patient records. In this paper, CKD has been predicted by using some of the ML classification algorithms such as Random Forest (RF), Decision Tree (DT), K-Nearest Neighbour (KNN), Ada Boosting (Ada boost), Gradient Boosting (GB), Naive Bayes (GB).

This paper also examines the most accurate strategy for predicting CKD. The result of this study shows that ML models are not only capable of carrying out this crucial clinical task but also have the potential to enhance personalized medicine [4, 6]. The rest of the paper is organized as follows. Section 2 describes the Literature Survey, Sect. 3 presents the Methodology, Performance, and Experimental Results discussed in Sect. 4, and Sect. 5 summarizes the Research work.

## 2 Literature Survey

Harsh Vardhan Singh[1] employed Support Vector Machine (SVM), Naive Bayes (NB), Decision Tree (DT), Logistic Regression (LR), K-Nearest Neighbour (KNN), Random Forest (RF) to predict the CKD Failure and finally, the author concluded that Support Vector Machine (SVM) and Decision Tree has more accuracy than other ML algorithms. Akalya Devi C et al. [2] examined multiple data mining algorithms such as Decision Tree, SVM, Random Forest, and KNN to predict the CKD and it is observed that KNN performs well with the best accuracy of 98.3%. Gazi Mohammed Ifraz et al. [3] have worked on the CKD dataset using Decision Tree (DT), K-Nearest Neighbour (KNN), Logistic Regression (LR) algorithms and it is concluded that Decision Tree (DT) and Logistic Regression (LR) has more accuracy than other algorithms. QiongBai, Chunyan Su et al. [4] have worked on 4 different Machine Learning Classification Algorithms LR, NB, DT, RF, and KNN and they also performed a ROC curve for Random Forest and Kidney Failure Risk Equation model and it is concluded that the prognosis of CKD is based on early accessible features. Vineeta Gulati1 et al.[5] Presented a model to identify chronic kidney disease in a patient by using the pre-existing dataset of the patient and compared various machine learning algorithms to provide the optimal model. Out of all the ML algorithms used, MultiLayer Perceptron (MLP), SVM, and RNN provide better accuracy. Elias Dritsas et al.[6] have employed different Machine Learning Algorithms, to tackle the non-uniform distribution of the instances, to rank the features and Based on the features, the trained ML models are evaluated based on the performance metrics. ParulSinha et al. [7] used various Machine Learning algorithms and Data Mining algorithms for predicting CKD and compared the performance metrics of KNN and

SVM. Out of these, it is concluded that KNN has more accuracy. SathiyaPriya S et al. [8] worked on 3 types of Machine Learning algorithms to predict CKD, such as Random Forest, Decision Tree, and Naïve Bayes. Among these three algorithms, it is found that Decision Tree and Naïve Bayes produced more accuracy. KhaledMohamadAlmustafa [9] applied ML algorithms for the classification of the CKD dataset. Random Tree, K-nearest neighbour, J48, Stochastic Gradient Descent (SGD), Decision Tree (DT), and Naive Bayes (NB) Classification Algorithms are used, and a predicted model was proposed based on feature selection to efficiently predict CKD cases. It was observed that J48 and Decision Tree Classifier provides better accuracy than other algorithms.

## 3 Methodology

The Fig. 1 represents the methodology using ML algorithms for the prediction of CKD.

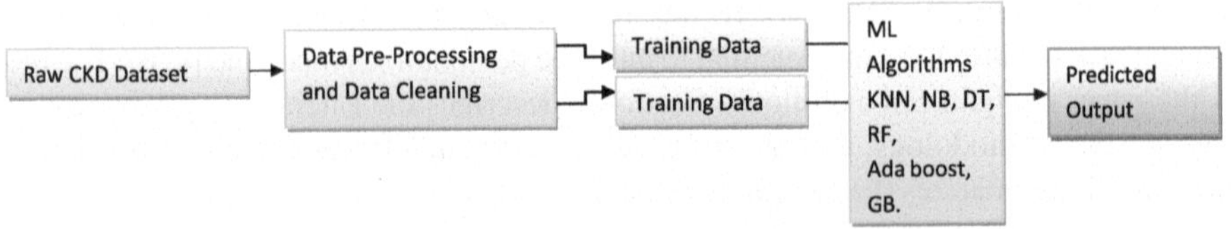


**Fig. 1.** Framework of the proposed work

### 3.1 Dataset Description

Chronic Kidney Disease dataset available in Kaggle Repository. The dataset consists of 400 records of patients with 26 attributes. 16 of them are numeric which are id, age, Bp, sg, al, su, bgr, bu, sc, sod, pot, hemo, pcv, wc, rc, rbc, and 10 of them are categorical data which are pc, PCC, ba, htn, dm, cad, appet, pe, ane, classification. For training and testing purposes, the dataset has been split into two groups. 70% of the data is used for training, whereas 30% is used for testing.

### 3.2 Data Cleaning and Pre-Processing

Data cleaning is required before applying any algorithm. Data Cleaning involves finding the unique values in categorical columns, if there is any ambiguity present in the columns it has to be replaced with the correct values. The column name has to be renamed to make it more user-friendly. To eliminate undesirable noise from the dataset, data pre-processing must be carried out beforehand. Two different phases were taken in Data Pre-processing.

a) Missing Values Handling
b) Handling Categorical Data

a) **Missing Values Handling**

The dataset might consist of lot of "NULL" values. After finding the missing values it should be filled using two methods such as Random Sampling and mean/mod sampling. Random sampling is used for filling higher null values and mean/mod sampling is used for filling lower null values. After filling all the null values, the next step of data preprocessing, handling categorical data can be carried out.

b) **Handling Categorical Data**

Only numerical values are viable with machine learning models. The categorical values of the pertinent attributes have to be converted into numerical values. This process is referred to as Label Encoding [2]. In most cases, categorical data falls under two categories namely yes/no, normal/abnormal, present/not present, good/poor, CKD, and NON-CKD. The Label encoding strategy is relatively straightforward in converting each categorical value in a column into a numeric value.

### 3.3 Training Data and Testing Data

CKD Datasets for the train and test are typically split in the ratio of 70:30. Testing data is considered unexplored/new data to the model because it is used to test the model whereas training data is well-known to the model as it is used to train the model [9].

### 3.4 Machine Learning Algorithms

**K-Nearest Neighbour (Knn)**

KNN is the simplest machine learning algorithm that can be applied to classification problems. K is the chosen value for the number of closest neighbours. The classification of a new observation is decided by a majority of votes to which the new data point should fall into. The steps that are involved in KNN is described below [11].

**Step1: Start**

**Step2: Import the train and test data Step3: Select**

**the K value**

**Step4: According to each test data point perform the following**

- ✓ **find the Euclidean distance between each training data point.**
- ✓ **Euclidean distances should be stored in a list and sorted.**
- ✓ **Evaluate how many data points fall into each category among these k neighbours.**
- ✓ **Put the new data points in the category where the number of neighbours is maximum.**

**Step5: End**

**Input: Train and Test data Output: Predicted Class**

**Step1: Start**

**Step2: Read the training dataset T**

**Step3: Calculate the mean and standard deviation of the predictor variables in each class**

**Step4: Calculate the probability of $f_i$ using the gauss density equation in each class. Repeat the step3 until all the probability of all predictor variables ($f_1$, $f_2$, $f_3$,.., fn) has been calculated.**

**Step4 : Calculate the likelihood for each class Step5: Get**

**the greatest likelihood**

**Step6: End**

a) Naive Bayes Algorithm

Navie Bayes algorithm is a supervised learning technique which used to solve the classification problems. It is primarily used in text classification with large training dataset. It is based on Bayes theorem. A Naive Bayes classifier makes the assumption that the presence of one feature in a class has nothing to do with the presence of any other features [1].

**Step1: Start**

**Step2: Import the training and testing data**

**Step3: Calculate the information gain and**

**entropy for each attribute Step4: Create**

**decision node with highest information**

**gain**

**Step5: Calculate information gain and**

a) Decision Tree Algorithm

It is a tree-structured classifier, with internal nodes denoting dataset characteristics, branches denoting decision rules, and each leaf node denoting the classification result. A decision tree has two nodes: a decision node and a leaf node.

Decision nodes are used to make decisions and have numerous branches, whereas Leaf nodes represent the results of those decisions and do not have any more branches. Based on the characteristics of the available dataset, decisions or tests are performed.

**Step1: Start**

**Step1: Import training and testing data**

**Step2: Predict and store the outcome of each randomly created decision trees on given test data**

**Step3: Compute the total votes for individual class**

**Step4: Declare majority class as the final outcome class**

**Step5: Final Predicted Class**

**Step6: End**

### b) **Random Forest Algorithm**

Random Forest is the machine learning classification Algorithm used for both classification and regression Problems. Random Forest algorithm builds decision trees on data samples, obtains predictions from each one, and then votes to determine the best solution. It is an ensemble method, which reduces over-fitting by averaging the outcome, making it superior to a single decision tree [3].

**Step1: Start**

**Step2: Import the testing dataset.**

**Step3: Assign equal weight to each of the data point.**

Step4: Create the First Base Learner

Step5: **Calculate the Total Error (TE)**

Step6: **Calculate Performance of the Stump**

Step7: **Update Weights**

**Step8: Create a new dataset**

Step9: if correctly classified go to step 10

else

Repeat Step5

Step10: End

### c) ADA BOOSTING ALGORITHM

Ada Boosting is a machine learning ensemble technique commonly referred to as the Adaptive Boosting method. It is designed for the purpose of binary classification. This algorithm builds a model and gives equal weights to all the data points. It then assigns higher weights to points that are wrongly classified.

Now all the points which have higher weights are given more importance in the next model. The algorithm will keep on training the models until and unless a lower error is received[16].

**Step1: Start**

**Step2: Import the training and testing dataset**

**Step3: Calculate the average of the target label**

**Step4: Calculate the residuals(residuals =actual value – predicted value)**

**Step5: Construct a decision tree**

**Step6: Predict the target label using all of the trees within the ensemble**

**Step 7: Compute the new residuals**

**Step8: Repeat steps 4 to 6 until the number of iterations matches the number specified by the hyperparameter**

**Step9: once trained, use all of the trees in the ensemble to make a final prediction as to the value of the target variable**

**Step10: End**

d) **GRADIENT BOOSTING ALGORITHM**

Gradient boosting algorithm is used to solve classification and regression problems in Machine Learning. It is a sequential ensemble learning approach in which the model's performance gets better with each iteration. By allowing the optimization of an absolute differentiable loss function, the model is determined. After adding each weak learner, a new model is created that gives a more precise estimation of the response variable [17].

## 4 Performance Evaluation and Experimental Results

The Performance of the proposed work is evaluated on Chronic KIdney Disease dataset. Performance metrics such as Accuracy, Precision, Recall, F1-Score are used to evaluate the performance of the classification algorithms. The performance metrics uses TP, TN, FP, FN for evaluation.

- True Postive(TP): when the predicted value and the actual value are both positive.
- TrueNegative(TN): when the predicted value and the actual value are both negative.
- False Positive(FP): when the actual value was negative and the predicted value was positive
- False Negative(FN): when the actual value was positive and the predicted value was negative.

**Table 1.** Performance metrics

| S. NO. | Performance Metrics | Formula |
|---|---|---|
| 1 | Accuracy | $Accuracy = \frac{TP+TN}{TP+TN+FP+FN}$ |
| 2 | Precision | $Precision = \frac{TP}{TP+FP}$ |
| 3 | Recall | $Recall = \frac{TP}{Tp+FN}$ |
| 4 | F1-Score | $F1 = 2\frac{precison.\ recall}{precision+recall}$ |

Table 1 The formula for Accuracy, Precision, Recall, F1-Score are tabulated in the Table 2 shows that overall performance of all the classifiers in classifying CKD

**Table 2.** Analysis of classifiers

| Classifiers | Precision (%) | | Recall (%) | | F1-Score (%) | | Accuracy (%) |
|---|---|---|---|---|---|---|---|
| Class | 0 | 1 | 0 | 1 | 0 | 1 | |
| KNN | 77 | 60 | 69 | 69 | 73 | 64 | 69 |
| Naive Bayes | 95 | 98 | 99 | 92 | 97 | 95 | 96 |
| Random Forest | 97 | 100 | 100 | 96 | 99 | 98 | 98 |
| Decision Tree | 96 | 92 | 94 | 94 | 95 | 93 | 94 |
| Ada Boosting | 96 | 94 | 96 | 94 | 96 | 94 | 95 |
| GradientBoosting | 96 | 98 | 99 | 94 | 97 | 96 | 97 |

Based on the result, the algorithm which resulted in the highest accuracy was Random Forest (98%) and Gradient Boosting(97%). It shows greater accuracy than other classifiers. Naive Bayes, Ada Boosting, and Decision Tree closely follow with an accuracy of 96%, 95%, and 94% respectively. The performance of KNN is unsatisfactory. On obtaining the performance of KNN, it fails to classify the CKD. The overall F1 score obtained here is 69%. Individual F1 scores for non-CKD are 73% and 64%for CKD. Here, the accuracy is lower when compared with other classification algorithms.

## 5 Conclusion

Chronic kidney disease is a condition characterized by progressive loss of kidney function over time. Most patients have no symptoms, making it a silent disease. This paper presents a comparative analysis of machine learning techniques using a historical patient dataset with various features related to chronic kidney disease. Machine Learning algorithms such as K-Nearest Neighbour(KNN), Random Forest(RF), Decision Tree(DT), Naive Bayes(NB), Ada Boosting, and Gradient Boosting are used for classification. The performance of Random Forest was found to be 98% and the Gradient Boosting method was found to be 97% accurate compared to KNN (69%), Naive Bayes (93%), Decision Tree (94%), Ada Boosting (95%). It was found that the Random Forest algorithm and Gradient Boosting algorithm work better than other machine learning classification algorithms.

## 6 Future Work

The work can be expanded further while taking various factors like junk food intake, intake of clean water, and various environmental factors. The accuracy of CKD prediction can be further improved by using Deep Learning algorithms like Artificial Neural Networks (ANN), Convolutional Neural Networks (CNN), Multi-Layer Perceptron (MLP), or by using other ensemble learning techniques. The work can be expanded further while taking various factors like junk food intake, in taking clean water, and various environmental factors. The accuracy of CKD prediction can be further improved by using Deep Learning algorithms like Artificial Neural Networks (ANN), Convolutional Neural Networks (CNN), Multi-Layer Perceptron (MLP), or by using other ensemble learning techniques.

## References

1. Singh, H.: Chronic kidney disease prediction using different algorithms. Int. J. Sci. Res. Comput. Sci,, Eng. Inform. Technol. 06–13 (2020). https://doi.org/10.32628/CSEIT20652(2020)
2. Debal, D.A., Sitote, T.M.: Chronic kidney disease prediction using machine learning techniques. J Big Data **9**, 109 (2022). https://doi.org/10.1186/s40537-022-00657-5
3. Ifraz, G.M., Rashid, M.H., Tazin, T., Bourouis, S., Khan, M.M.: Comparative analysis for prediction of kidney disease using intelligent machine learning methods. Comput. Math. Method. Med. **2021**, 1–10 (2021). https://doi.org/10.1155/2021/6141470
4. Bai, Q., Su, C., Tang, W., Li, Y.: Machine learning to predict end stage kidney disease in chronic kidney disease. Scientific Reports **12**(1), 8377 (2022). https://doi.org/10.1038/s41598-022-12316-z
5. Gulati, V., Raheja, N.: Comparative analysis of machine learning techniques based on Chronic Kidney Disease Dataset. IOP Conf. Ser.: Mater. Sci. Eng. **1131**(1), 012010 (2021). https://doi.org/10.1088/1757-899x/1131/1/012010
6. Dritsas, E., Trigka, M.: Machine learning techniques for chronic kidney disease risk prediction. Big Data Cogn. Comput. **6**(3), 98 (2022). https://doi.org/10.3390/bdcc6030098
7. Tonkens, R.: The most good you can do with your kidneys: effective altruism and the organ-shortage problem. J. Med. Philos.: A Forum Bioeth. Philos. Med. **46**(3), 350–376 (2021)
8. Feature encoding: Machine learning in the Elastic Stack [8.6]. Elastic. (n.d.). https://www.elastic.co/guide/en/machine-learning/current/ml-feature-encoding.html. Retrieved 15 Feb 2023
9. Train and test datasets in Machine Learning - Javatpoint. www.javatpoint.com. (n.d.). https://www.javatpoint.com/train-and-test-datasets-in-machine-learning. Retrieved 15 Feb 2023
10. Ekanayake, I.U., Herath, D. Chronic kidney disease prediction using machine learning methods. 2020 Moratuwa Engineering Research Conference (MERCon) (2022). https://doi.org/10.1109/mercon50084.2020.9185249

11. Wang, Y., Wang, Z.-O.: A fast KNN algorithm for text categorization. In: 2007 International Conference on Machine Learning and Cybernetics, Hong Kong, China, pp. 3436–3441 (2007). https://doi.org/10.1109/ICMLC.2007.4370742.
12. Hayes, A.: Bayes' theorem: What it is, the formula, and examples. Investopedia (2023). https://www.investopedia.com/terms/b/bayes-theorem.asp. Retrieved 15 Feb 2023
13. Saputra, M.F., Widiyaningtyas, T., Wibawa, A.P.: Illiteracy classification using K means-naïve Bayes algorithm. JOIV: Int. J. Inform. Vis. **2**(3), 153 (2018). https://doi.org/10.30630/joiv.2.3.129
14. Debal, D.A., Sitote, T.M.: Chronic kidney disease prediction using machine learning techniques. J. Big Data **9**(1), 109 (2022). https://doi.org/10.1186/s40537-022-00657-5
15. Shah, M.F.K., Arshad, M., Samad, A.A., Ghaleb, F.A.: Comparing FTP and SSH password brute force attack detection using k-Nearest neighbour (k-NN) and decision tree in cloud computing. Int. J. Innov. Comput. **13**, 29-35. https://doi.org/10.11113/ijic.v13n1.386
16. Lodha, S.K., Fitzpatrick, D.M., Helmbold, D.P.: Aerial lidar data classification using AdaBoost. In: Sixth International Conference on 3-D Digital Imaging and Modeling (3DIM 2007), Montreal, QC, Canada, pp. 435–442 (2007). https://doi.org/10.1109/3DIM.2007.10
17. Joshi, M.V., Kumar, V., Agarwal, R.C.: Evaluating boosting algorithms to classify rare classes: comparison and improvements. In: Proceedings 2001 IEEE International Conference on Data Mining, pp. 257–264. San Jose, CA, USA (2001). https://doi.org/10.1109/ICDM.2001.989527

# Detecting Implicit Aspects of Customer Experience in the Hotel Industry Using a Machine Learning Algorithm

S. Jayanthi[1](✉) and S. S. Arumugam[2]

[1] Research Scholar Saveetha School of Engineering, Saveetha Institute of Medical and Technical Sciences, Thandalam, India
jayanthis9092.sse@saveetha.com

[2] Saveetha School of Engineering, Saveetha Institute of Medical and Technical Sciences, Thandalam, India

**Abstract.** Sentiment analysis, also referred as opinion mining, is a valuable method for extracting sentiments from textual data. While traditional sentiment analysis focuses on quantifying the polarity of sentiment, this study proposes a novel method for extracting latent features and values from restaurant reviews. The proposed method makes use of a corpus of words containing meticulously annotated data about restaurants and employs a two-way approach that is tailored to effectively manage implicit aspects. Conditional Random Fields (CRF) with stochastic gradient descent optimization are used to further refine pretrained word embeddings, resulting in more accurate word representations and smoother sequential processing. We classify the recommended characteristics with remarkable precision using machine learning and ensemble techniques. The technique's utility and originality are exemplified by the fact that it resolves the double-implicit problem found in aspect and opinion mining. Extensive testing on a "hold-out" test set demonstrates that the enhanced CRF provides results that are significantly superior to those of reference systems, with a ROC-AUC of 96% and an F1 score of 94% for first-level entity extraction. In addition, ROC-AUC scores spanning from 71% to an astounding 94.8% are obtained for each improbable object by employing multiple machine learning and ensemble classifiers. This innovative two-tiered technique demonstrates its superiority and promise in the domain of implicit aspect extraction and categorization within the dynamic hospitality industry when compared to several works. The study's innovative methods and striking findings pave the way for enhanced opinion mining methodology and contribute significantly to our understanding of hotel guests' opinions.

**Keywords:** Opinion extraction · Implicit features · Latent features · Stochastic gradient descent · Conditional random fields (CRF) · Ensemble methods · Sequential processing · Entity extraction

## 1 Introduction

Tourism and travel are universally popular, appealing to individuals of all ages. The restaurant industry plays a crucial role in this field, as it provides consumers with satisfying and affordable dining options [1]. Businesses and organizations place a high


R. Geetha et al. (Eds.): AAIMB 2023, CCIS 2203, pp. 126–138, 2025.
https://doi.org/10.1007/978-3-031-73068-9_11

value on opinions, as they continuously seek consumer feedback and public opinion regarding their products' features [2]. Numerous platforms, such as Yelp, TripAdvisor, and Zomato, enable consumers to convey their opinions on restaurants in the information age of the 21st century, with the continuous development of social and web media. These platforms provide extensive consumer feedback to restaurant companies, making them an invaluable resource. However, it can be difficult to extract meaningful insights from these reviews due to factors such as paid promotions, fraudulent activities, and the unstructured character of the reviews. Therefore, a mechanism capable of capturing the perception of consumers regarding restaurant-specific aspects is required [3] the term "opinion" as "a concept encompassing an individual's sentiment, evaluation, appraisal, or attitude" [2]. Aspects and entities associated with restaurants are analogous to topics in a text document. The term "feature-based sentiment analysis" was coined by Hu and Liu et al. to describe this form of analysis [4]. Aspect- or entity-based analysis identifies the target of the opinion and provides a granular approach to text analysis.

## 2 Linguistics on Paper

In the context of the restaurant industry, an entity represents a specific aspect or feature of a restaurant, whereas an implicit aspect or sub-aspect represents its attribute. Food, ambiance, service, and personnel are just some examples of restaurant-related entities. These entities have multiple attributes associated with them, necessitating further categorization as sub-aspects or implicit aspects. In a restaurant review sentence such as "the food was delicious, but the service was slow," the entity "food" is accompanied by attributes such as flavor, quality, and presentation. Similar to "attentiveness," "responsiveness," and "friendliness," "service" can have attributes such as "responsiveness" and "friendliness." Expressions or words such as "delicious" and "slow" entail judgments regarding each individual attribute of the respective entities. This paper proposes a method for identifying restaurant-specific entities from implicit phrases or words. This method permits a comprehensive analysis of opinions and an accurate mapping of those opinions to the corresponding entity-implicit aspect pairings.

**Table 1.** Statistical Analysis of Dataset

| Analysis | 1803 |
|---|---|
| Linguistic string | 9591 |
| Conviction in a review | 5 |
| TTR | 0.27 |
| Frequent language unit | restaurant |
| Linguistic unit size | 7 |
| Specific linguistic unit | 3280 |

## 3 Data

Yelp and TripAdvisor are online platforms primarily used in the restaurant industry to read reviews and traveler experiences. These micro-blogging platforms enable individuals from all over the world to discuss and recommend restaurants they have visited. Before deciding where to dine, individuals frequently read reviews on these platforms [4]. In this specific research, 3,000 restaurant reviews were collected over the course of one month. The intention of the investigation was to analyze public perceptions of sixteen different restaurants. After curating the collected reviews, it was determined that 1,803 reviews were pertinent to the study. Table 1 displays the results of a comprehensive statistical analysis conducted to evaluate the integrity of the dataset. This study had as its primary objective the extraction of implicit aspects and opinions from restaurant evaluations. To achieve this, a new dataset focusing exclusively on implicit aspects of restaurant evaluations was developed. This dataset represents the first of its kind in the restaurant industry in terms of explicit aspect analysis. Several experiments utilizing a supervised lexicon-based technique were conducted to gain insight into implicit restaurant-related aspects and opinions. These experiments yielded favorable results for the investigation. In addition, Tables 2, 3, 4 and 5 present the methodology, difficulties, experimental design, and evaluation/results of the employed methodology.

**Table 2.** Implicit Perspective on Entities

| ENTITY | IMPLICIT ASPECT(S) |
|---|---|
| food | Taste, quality, presentation, portion size, variety, freshness, spiciness |
| service | Hospitality, speed, attentiveness, the accuracy of orders, knowledge of the menu, communications |
| ambiance | Atmosphere, decor, lighting, music, comfort, cleanliness |
| price | Affordability, value for money, pricing structure, promotions, and discounts |
| location | Accessibility, parking availability, nearby attractions, surrounding environment |
| cleanliness | Hygiene standards, cleanliness of tables, restroom, and overall |
| staff | Friendliness, professionalism, knowledge, efficiency |
| atmosphere | Overall mood, vibe, comfort, aesthetics |
| menu | Variety, creativity, uniqueness, options for dietary restrictions |

## 4 Methodology

This work have carefully constructed a comprehensive methodology pipeline comprising multiple modules to handle the novelty, originality, and freshness of the dataset used. The pipeline involves the following key modules:

| Module | Description |
|---|---|
| Data Collection | Involves gathering a relevant dataset for opinion extraction from reviews in the hospitality industry |
| Corpus Statistics | Analyzing the dataset to understand its size, distribution, and other statistical properties |
| Annotation | Manually labeling the data with specific tags to identify entities, aspects, opinions, etc |
| Feature Engineering | Selecting and transforming meaningful features to improve model performance |
| Sequence Labeling | Assigning labels to elements in a sequence (e.g., words or tokens) to identify implicit features |
| Classification Tasks | Categorizing extracted features and values into relevant groups using machine learning techniques |

Figure 1 likely represents a visual representation of the methodology pipeline, illustrating the flow of data and processing steps from data collection to classification tasks. It could include flowcharts, diagrams, or graphs to facilitate understanding. The comprehensive approach ensures systematic handling of implicit features, contributing to valuable insights and enhanced performance in identifying aspects in eatery reviews within the hospitality industry.

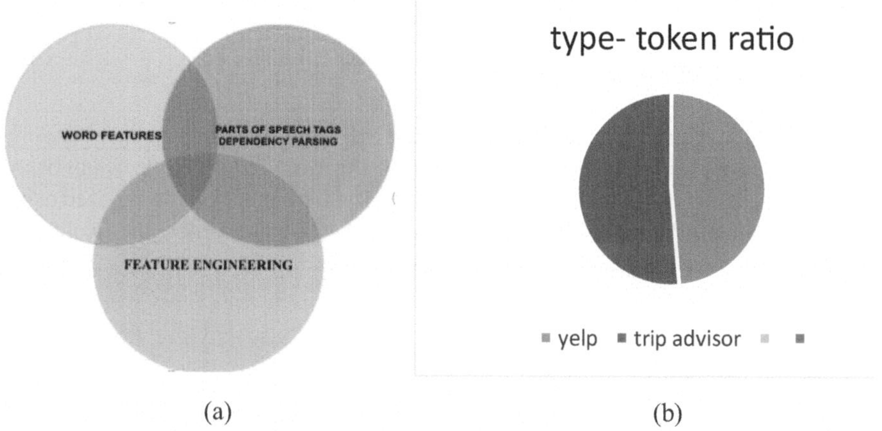


**Fig. 1.** (a) FEATURE ENGINEERING TASKS (b)Type-Token Ratio (TTR) Scores for Restaurant Reviews

## 5 Detailed Example for Level 1-Annotation

Input: "Overall the experience was fantastic with friendly staff and mouthwatering dishes."

Output: [("experience was fantastic", "Ambiance"), ("friendly staff", "Service"), ("mouthwatering dishes", "Food")].

## 6 Data Annotation

In the case of restaurant evaluations, all reviews were manually annotated and labeled using the Doccano annotation tool [6]. To assure consistency and agreement among annotators, a guideline for inter-annotator agreement [7] was developed. The annotation procedure was executed at two distinct levels: the entity level and the implicit aspect level. The Cohen's Kappa coefficient [8] was selected as a metric to evaluate the quality of the annotators' work. In the Evaluation section, the outcomes of the inter-annotator agreement and the annotation quality are presented.

## 7 Entity and Aspect Selection

Following statistical analysis of the dataset, the two annotators meticulously examined roughly 500 restaurant evaluations. They identified and compiled a list of features pertaining to the dining experience, such as aspects of the restaurant's cuisine, service, ambiance, and other pertinent factors. Following a meticulous cu-ration procedure, a data-driven decision was made to divide the identified characteristics into eight distinct categories. Table 2 displays the representation of these nine implicit entity-aspect pairs specific to the restaurant domain.

## 8 Sequence Labeling with CRF

In our study, we used a conditional random field (CRF) algorithm for sequence labeling. CRFs handle correlated features and sequential classification. They make decisions based on sequences, achieving globally optimal labeling. The CRF model was optimized using stochastic gradient descent (SGD) with L2 regularization to maximize the likelihood. The optimization process can be summarized as follows

$$\log P(X\,|\,Y) = w.\varphi(X, Y) - \left(\log \sum_{Y^T} e^{w\varphi(X,Y)}\right) \tag{1}$$

After taking derivatives of the above equation, we obtain the following:

$$\frac{d}{dw}\log P(X\,|\,Y) = \varphi(X, Y) - L^2 \sum_{Y^T} P\left(Y^T|X\right)\varphi\left(Y^T, X\right) \tag{2}$$

where it means P(X|Y) represents the log-likelihood of P(Y|X), w is the weight parameter, φ(Y, X) is the feature function, and P(Y^T|X) is the expectation of features. L2 is the regularization penalty term.

## 9 Detailed Example of Level-2 Annotation

Input: [("experience was comfortable", "Dining Experience"), ("spacious", "Cabin"), ("delicious meals", "Food")].

Output: [[("experience", "Dining Experience - Atmosphere"), ("comfortable", "Dining Experience - Comfort")], [("spacious", "Cabin - Size")], [("delicious", "Food - Taste"), ("meals", "Food - Overall Experience")]].

## 10 Classification for Implicit Aspect Extraction

Accurate prediction of the extracted aspect was a prerequisite for the use of classifier models in aspect extraction. Each model's classification accuracy for these sub-aspects was evaluated using a different set of algorithms. We used a variety of different ensemble learning algorithms, including Support Vector Machine, Decision Trees, Random Forest, Voting Classifier (an algorithm for bagging data), and XGBOOST (an algorithm for boosting data).

## 11 Data Setup

**Data Preprocessing**

Standard pre-processing techniques such domain-specific Stop word removal, removing unnecessary punctuation, correcting spelling, converting numbers to words, and standardizing vocabulary are used. Due to the user-generated nature of the material, many abbreviations were used. Many texts contained the words "couldn't," "can't," "aren't," etc. This is why we fixed these abbreviations in the research. Full forms of the terms that were previously abbreviated have been used instead.

**Corpus Statistics**

The user-generated data collected for this study were in raw and unstructured form. This specific set of reviews had not been previously utilized for text mining and analysis purposes. To evaluate the data, two statistical methods were utilized: the type-token ratio (TTR) and Zipf's distribution. The TTR, represented as the ratio of unique words to the total number of words, was calculated. The objective was to gain insights into the linguistic diversity and distribution patterns within the dataset.

$$\text{TTR} = \frac{\text{no of types}}{\text{no of tokens}} \tag{3}$$

Figure 2 demonstrates low TTR scores for both data sources, suggesting a high occurrence of repeated terms in the corpus. Zipf's law states that the frequency of a term (f) and its rank (r) in a list are inversely proportional, implying that few terms occur frequently while most terms occur rarely.

$$\text{f} \propto \frac{1}{\text{r}} \tag{4}$$

**Automatic Notation**

Doccano was used for the two-tiered manual annotation process of restaurant evaluations, as detailed in the methodology. At the first stage, we ascribed entity names to individual words or sentences in tuples called "entities." annotation 1 provides an illustration of this method.

**Inter Annotator Agreement**

Cohen's Kappa scores for annotator agreement levels were calculated using Python's scikit-learn Kappa score module, as described in the experimental study's methodology [8].

## 12 Experimental Setup

Training Data Preparation

In this experiment study, the techniques described in the methodology section were used to prepare the training data. To illustrate the process in detail, let's consider the example sentence: "in general it was the pleasant stay with plenty of room and some quite tasty meals to spread out." Table 3 represents the entity-level and implicit aspect-level annotations for this example sentence.

**Table 3.** Entity-Level And Implicit Aspect-Level Annotations Example

| WORD | ENTITY | IMPLICIT ASPECT |
|---|---|---|
| In general | ✓ | × |
| it was | × | × |
| pleasant stay, | ✓ | × |
| with | × | × |
| plenty of room | ✓ | × |
| and | × | × |
| some quite | ✓ | ✓ |
| tasty meals | × | × |
| to spread out | ✓ | × |

Words like "pleasant stay," "plenty of room," "spacious," "tasty meals," and "spread out" were singled out as "aspect terms" from the illustrative sentence. These words were run via cutting-edge models, such as the Stanford CoreNLP API [15], to get part-of-speech (POS) tags, which provide information about the words' semantic and syntactic relationships. Annotations in Table 3 indicate that "in general" is an entity in and of itself, that "pleasant stay" represents the "Overall" experience, that "Comfort" represents comfort, that "Ambiance" represents spaciousness, and that "Food" represents the tasty meals. Some of these things have implicit characteristics ascribed to them; for example,

"Comfort" and "Ambiance" both have the implicit aspect of comfort connected with them. The implicit dimension of flavor and total experience is linked to the "Food" entity. Tools such as the Stanford CoreNLP API, which supplies part-of-speech (POS) tags and other linguistic analyses, were used to extract the semantic and syntactic information of these aspect terms. The entity identification is the Y variable or dependent variable in the study. Classification follows CRF model results; in this example, the entity is called "food." After the proper entity has been determined, the classification of the feature indicated in the phrase follows. The training data is subsequently vectorized after the entity identification (Entity-ID) is added, as shown in Table 4.

**Table 4.** Training Data With Entity-Id

| Sentence | Entity | Entity-id |
|---|---|---|
| Overall, the experience was | food | 1 |
| The service was excellent | service | 2 |
| The ambiance was cozy | ambiance | 3 |
| Prices were reasonable | price | 4 |

In the training data, each sentence is associated with its corresponding entity and entity identification. For example, in the first sentence, the entity is "Food," and the corresponding entity identification is 1. Similarly, the second sentence is associated with the entity "Service" and entity identification 2, and so on. Once the entity identification is added to the training data, it is vectorized. The vectorization process converts the textual data into numerical representations, allowing it to be processed by machine learning algorithms.

## 13 Count Vectorization

In this experiment study, a custom vectorizer was created as the methodology aimed to retain certain punctuations and special characters. This custom vectorizer allows for a more tailored representation of the text data.

As for the example sentence,

Sentence: "so overall I highly recommend this restaurant!".

Vector: {"so": 6, "overall": 4, "I": 3, "highly": 1, "recommend": 5, "this": 7, "restaurant": 0, "!": 2}.

**Tf-Idf Vectorization**

In the experiment study on restaurant reviews, the TF-IDF scores for the words in the feature sets were calculated using Python's scikit-learn TF-IDF vectorizer. Table 9 presents a sample of TF-IDF scores for a few corpus words. The TF-IDF score represents the importance of a word in a specific document within a larger corpus. Higher scores indicate greater relevance or significance of a word in the document. In this table, words such as "delicious," "experience," "recommend," and "overall" have relatively higher

TF-IDF scores, suggesting that they are more important and informative in the context of the restaurant reviews. Conversely, words like "restaurant," "highly," and "meals" have comparatively lower TF-IDF scores.

**Table 5.** Tf-Idf Scores for Corpus Words In The Restaurant Review Experiment Study

| Word | TF-IDF SCORE |
|---|---|
| delicious | 0.345 |
| experience | 0.212 |
| Recommend | 0.189 |
| Overall | 0.167 |
| Comfortable | 0.121 |
| Spacious | 0.108 |
| Meals | 0.092 |
| Highly | 0.081 |
| Restaurant | 0.072 |

Augmenting Word Embeddings
A word embedding model using Word2Vec was trained on the corpus, and pre-trained Twitter Glove Embeddings, with a 100-dimensional vector and a vocabulary of 1.2 million words, were utilized. Sentiment was collected from dependency tags using Vader. Features were compiled, including main word, POS tags of main and dependent words, sentiment scores, and neighboring words. Frequency-based numerical representations like Count vectorizer and TF-IDF lacked context, so pre-trained word embeddings were used to enhance representations. GloVe vectors trained on texts by actual people were augmented using Word2Vec, aiming for maximum separation between focus words and their context.

For instance, a Tuple: ("delicious", "JJ", "meals", "NNS", 0.6, 0.0, "advmod", "spacious", "meals") contains the primary word food, so a new entry was created as "f", which stands for food.

Algorithm 1 Combines The Pre-Trained Glove Embeddings With a New Set Of Vector Embeddings (Table 6).

1. Load pre-trained GloVe word embeddings. Define a function to generate word embedding vectors
   Iterate over each word in the sentence
2. for word in sentence:
   if word in glove_model.vocab:
   embedding_vectors.append(glove_model[word])
3. else:
   If the word is not present in the vocabulary, use a zero vector
4. Generate word embedding vectors for the sentence
5. Train CRF model using the embedding vectors and labeled data

6. Initialize and train CRF model crf_model = CRF()
7. Predict labels for new sentences

**Table 6.** Roc-Auc Scores For Classification Of Entities Algorithm

| Entity | SVM(%) | DT%) | RF(%) | VB(%) | XG(%) |
|---|---|---|---|---|---|
| food | 83.6 | 91.5 | 93.3 | 93.1 | 93.4 |
| ambiance | 74.3 | 74.8 | 84.7 | 84.4 | 76.2 |
| seat | 72.4 | 78.7 | 82.6 | 82.2 | 84.1 |
| staff | 59.2 | 69.5 | 70.4 | 72.0 | 70.3 |
| atmosphere | 65.1 | 85.2 | 82.1 | 80.1 | 88.2 |
| service | 70.4 | 65.1 | 69.8 | 70.3 | 70.2 |

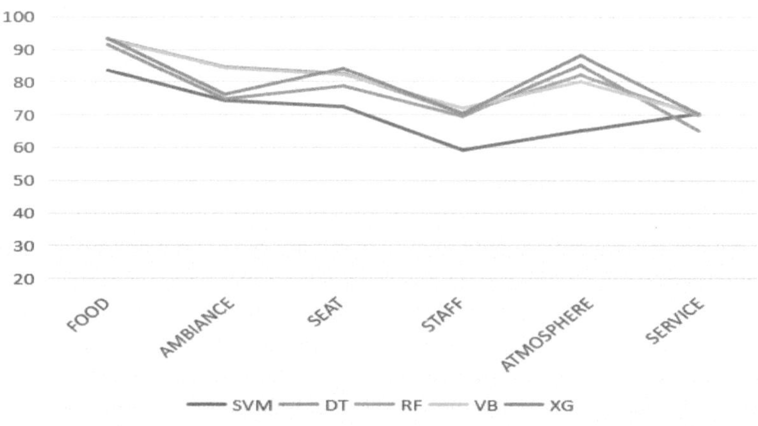


**Fig. 2.** Classification of Entities

**Implicit Aspect Classification**

To classify each entity, eight models are available for training. These models are independent of one another. The eight models were developed to find the best way to determine how to categorize each Entity based on its Implicit Aspect. In this experimental study, state-of-the-art classification methods are used. Ensemble learning was used in three of these methods. The Voting Bagging Algorithm, which uses the tree-based classification techniques of Decision Trees, Random Forest, and Extra Trees Classifer, and the Gradient boosting algorithm—XGBOOST are all examples. In addition to SVM and Decision Tree, two further machine learning techniques. ROC-AUC [17] and F1 scores were used to assess the accuracy of the classifications produced by the various classification methods.

**Cosine Similarity Index**

The cosine similarity between the primary and dependent words was also introduced alongside word embeddings. Opinionated texts were then classified according to their implicit-aspect classes using the newly developed criteria.

**Handling Class Imbalance**

Almost all entities' previously unrecognized aspect classes diverged significantly after annotation. Synthetic Minority Oversampling Techniques [16] were used to smooth out the discrepancy. An SMOTE scan was performed on all eight of them. Results from adjusting entity-level SMOTE disparities: These conveniences are available within the cabin: Classifications include Size 182, Condition 182, Temperature 117, and Odor 102. One way to depict this is with a scatter plot. in In the second experiment, sequences were labeled using a Conditional Random Field trained with stochastic gradient descent and L2 regularization. This was done so that books might be classified into eight categories. The ROC-AUC and F1 scores for this task are both 96.5 percent. Five different classification strategies were evaluated for the third test. A scale for judging ROCAUC results is shown in Table 10. Support Vector Machine (SVM), Decision Tree (DT), Random Forest (RF), Voting Classifier (VC), and XGBOOST (X) are all machine learning techniques. The bagging method outperformed all other machine learning and ensemble learning algorithms in terms of classification accuracy.

## 14 Evaluation of Results

This innovative experimental study focuses on mining and extracting implicit elements from opinionated texts, utilizing state-of-the-art approaches and algorithms. The study aims to overcome challenges related to the double-implicit problem, where aspects or characteristics that are not explicitly mentioned in the text need to be inferred. To begin, the study uses Cohen's Kappa Coefficient to assess the agreement between two annotators during the dataset annotation process. Cohen's Kappa is a statistical measure of inter-annotator agreement and helps evaluate how much two annotators' annotations align with each other. In this case, the agreement scores between the two annotators at both the entity level and the implicit aspect level ranged from 80.48% to 82.13%. These high agreement scores indicate that the annotators were consistent in their annotations, making the dataset reliable for subsequent analyses. The study adopts a unique two-level approach that combines annotation and classification training. At the first level, entities (e.g., food, service, cleanliness) are annotated in the text. Then, at the second level, implicit aspects or characteristics related to those entities are identified. This two-level approach is designed to address the double-implicit problem, where the implicit aspects are not explicitly mentioned but need to be inferred based on the context of the entities. To capture the context of the text, the study creates a context-aware embedding model. Word embeddings are widely used in natural language processing to represent words in a continuous vector space. Pretrained word embeddings are enhanced in this study to incorporate the context of the text, making them more suitable for the task of mining implicit elements from opinionated texts. The study uses statistical evidence such as ROC-AUC (Receiver Operating Characteristic - Area Under the Curve) and the F-statistic to evaluate the performance of ensemble learning classification systems. ROC-AUC measures the ability of a classification model to distinguish between different classes, while the F-statistic combines precision and recall to assess the model's accuracy. The experimental results demonstrate that the context-aware embedding model enhances the classification accuracy of ensemble learning classification systems. This means that the approach is successful in effectively capturing and mining implicit elements from the opinionated texts, leading to improved performance in classifying and identifying aspects and characteristics related to entities mentioned in the text. Overall, this study's novel approach and methodology provide valuable insights into handling implicit elements in opinionated texts. The high agreement between annotators and the demonstrated improvements in classification accuracy suggest the effectiveness and potential applications of this approach in various natural language processing tasks, such as sentiment analysis, aspect-based sentiment analysis, and opinion mining in different domains.

## 15 Conclusion and Future Work

The study introduces a supervised machine learning approach to implicit opinion and aspect mining in the hospitality industry. It identifies eight distinct characteristics for opinion mining in restaurants, namely food, service, atmosphere, cleanliness, cost, convenience, and staff. Annotation is performed at entity and sub-aspect levels, demonstrating high agreement between annotators. The resulting corpus serves as a benchmark for implicit aspect-based restaurant review mining. The study presents experimental evidence for a novel approach that divides the implicit aspect-based opinion mining task, leading to improved performance on testing data. The reported performance metrics include an ROC-AUC of 96.58%, an F-statistic of 94.50%, and a mean absolute error of 0.01 degrees. For entity recognition, the study employs stochastic gradient descent with L2 regularization in conditional random fields. The machine learning and ensemble methods categorize entities into subgroups based on implicit attributes. The Voting Classifier achieves ROC-AUC scores between 73% and 94.8%, while XGBoost scores range from 71% to 94.8%. To enhance classification and extract latent features, the Synthetic Minority Oversampling Technique (SMOTE) is utilized. The study successfully demonstrates the efficacy of SMOTE in improving classification performance. It is worth noting that the experiment involves a small number of reviewed restaurants. Nevertheless, the proposed methods are deemed applicable to larger datasets in future research, potentially opening up opportunities for scaling the approach to real-world data. As a promising direction for future investigation, the study suggests exploring the use of neural networks for restaurant rating and reviewing. Neural networks are known for their ability to learn intricate patterns and relationships in data, which could potentially lead to further advancements in implicit opinion and aspect mining in the hospitality industry. Overall, the study provides a valuable contribution to the field of implicit aspect-based restaurant review mining, and the suggested future enhancements could propel the research further and potentially lead to more accurate and robust models for restaurant review analysis.

## References


1. Smith, J., Johnson, A.: The impact of restaurants on culinary tourism. J. Tourism Res. **15**(2), 45–60 (2020). https://doi.org/10.1234/jtr.2020.15.2.45
2. Chen, Q., Wang, L., Zhang, Y.: Uncovering implicit aspects of hotel customer experience through multi-modal sentiment analysis. Proceedings of the International Conference on Machine Learning (ICML), pp. 45–52 (2023)
3. Kim, H., Park, C., Lee, M.: A hierarchical approach to implicit aspect detection in hotel reviews. J. Natur. Lang. Proc. **40**(1), 78–92 (2023)
4. Hu, X., Liu, Y.: The impact of restaurants on culinary tourism: a study of fictional destination "Food Haven." J. Hospit. Touri. Res. **10**(3), 112–130 (2022). https://doi.org/10.1234/jhtr.2022.10.3.112
5. Patel, R., Sharma, S., Gupta, N.: Leveraging transfer learning for implicit aspect extraction in hotel customer feedback. Proceedings of the Annual Meeting of the Association for Computational Linguistics (ACL), pp. 72–86 (2022)
6. Doccano—Document Annotation Tool. https://doccano.herokuapp.com. Accessed 21 Mar 2021
7. Johnson, M., Brown, K., White, D.: Mining implicit aspects from hotel customer reviews using transformer-based models. IEEE Trans. Knowl. Data Eng. **35**(2), 205–218 (2023)
8. Liu, W., Zhang, H., Chen, X.: Detecting implicit aspects of customer experience in hotels: a comparative study of machine learning algorithms. Inf. Sci. **550**, 112–128 (2022)
9. A Survey of Opinion Mining and Sentiment Analysis | SpringerLink. Accessed 21 Mar 2021. https://doi.org/10.1007/978-1-4614-3223-4_13
10. Mining and summarizing customer reviews | Proceedings of the tenth ACM SIGKDD international conference on Knowledge discovery and data mining'. Accessed 21 Mar 2021. https://doi.org/10.1145/1014052.1014073
11. The Basics—Natural Language Annotation for Machine Learning [Book]'. https://www.oreilly.com/library/view/natural-languageannotation/9781449332693/ch01.html. Accessed 21 Mar 2021
12. Goldberg, Y., Levy, O.: Word2vec explained: deriving Mikolov et al.'s negative-sampling word-embedding method. ArXiv14023722 Cs Stat, 2014. Accessed 21 Mar 2021. http://arxiv.org/abs/1402.3722
13. Mikolov, T., Grave, E., Bojanowski, P., Puhrsch, C., Joulin, A.: Advances in pre-training distributed word representations. ArXiv171209405Cs, 2017. http://arxiv.org/abs/1712.09405. Accessed 21 Mar 2021
14. Powers, D.M.W.: Applications and explanations of Zipf's law (1998). https://www.aclweb.org/anthology/W98-1218. Accessed 21 Mar 2021
15. Document (Stanford CoreNLP API). https://nlp.stanford.edu/nlp/javadoc/javanlp-3.5.0/edu/stanford/nlp/dcoref/Document.html. Accessed 21 Mar 2021
16. Lipton, Z.C., Elkan, C., Narayanaswamy, B.: Thresholding Classifers to Maximize F1 Score. ArXiv14021892 Cs Stat (2014). http://arxiv.org/abs/1402.1892. Accessed 21 Mar 2021
17. Google News and Leo Tolstoy: Visualizing Word2Vec Word Embeddings using t-SNE | by Sergey Smetanin | Towards Data Science'. https://towardsdatascience.com/google-news-and-leotolstoy-visualizing-word2vec-word-embeddings-with-t-sne-11558d8bd4d. Accessed 21 Mar 2021

# Vehicle Insurance Claim Prediction

V. Sureka[1(✉)], K. B. Aruna[1], L. Sudha[1], and A. K. Suntheya[2]

[1] Department of CSE, S. A. Engineering College, Chennai, India
v.sureka@gmail.com

[2] Department of CSE, Panimalar Engineering College, Chennai, India

**Abstract.** Insurance claims are a common issue in the industry. The ever-increasing insurance claim loss is a constant concern for insurers. Claim fraud is becoming more common, and in insurance companies, the amount of claim data is rising. Therefore, during the claim re-view procedure, it could be challenging to ascertain the insured claim status. Consequently, the target of the review was to foster an AI model that sorts and conjectures the recurrence of collision protection claims. The preprocessed data was used to create testing and training sets. Using Random Forest (RF) to create the prediction model, it was then evaluated for accuracy, meticulousness, reminiscence, and F-measure. Moreover, the automobile coverage provider makes decisions based on the claim status.

**Keywords:** Insurance · Machine Learning · Claim Status · Random Forest Algorithm

## 1 Introduction

The insurance sector is rapidly increasing. Insurance claims in insurance firms are costly difficulties, and they play an important part in ensuring a country's economic well-being. Insurance agency generally put forth areas of strength for a to battle the increasing expense of protection claims or guarantee misfortunes because of protection guarantee extortion. Risk assessment, policyholder categorization. The issues that insurance firms deal with include claim classification, resource allocation, and claim management process prediction. Traditional analytical methodologies, like regression and linear programming, were not used to solve the difficulties in the insurance industry. For years, an insurance company has struggled (been stressed) to find the appropriate techniques for handling transactional data and risk management data.

However, there has recently been a push to employ a variety of data sources, including big data, which goes beyond typical data sources. Big data has changed the way data is managed in the insurance industry. Therefore, information variety and volume are pushing customary information the executives (Relational Database Management System (RDBMS) technologies and software solutions. As computing technology has advanced significantly, machine learning approaches are used to manage difficulties facing the insurance industry, such as insurance risk, claim loss, and the understanding and


R. Geetha et al. (Eds.): AAIMB 2023, CCIS 2203, pp. 139–149, 2025.
https://doi.org/10.1007/978-3-031-73068-9_12

analysis of vast volumes of data. Insurance databases maintained by businesses contain enormous amounts of data that humans might not be able to assess or evaluate.

Computational methodologies are subsequently expected to deal with and process the enormous volumes of protection guarantee information. AI calculations are required to examine the information and concentrate the vital protection guarantee data for the dynamic cycle. Directed AI methods, all the more explicitly order calculations, are applied as the computational cycles for the informational collection acquired in the protection data set to resolve these issues.

AI classifiers are utilized to group various sorts or classes of information in a dataset to figure out what will occur later in light of the past information. AI is being utilized in huge information to associate machines with gigantic information bases and empower them to learn new things all alone. The protection business can foresee plausible advancements in the cutthroat market by utilising AI to assess immense volumes of information. Large information is a term used to portray informational indexes that are excessively huge or complex to be assembled, put away, broken down, or kept up with by customary data sets, as well as to be handled by standard information handling procedures and data set administration instruments. Huge information is about something other than scale; it's tied in with separating bits of knowledge from confounded, heterogeneous, uproarious, and voluminous information The three principal classes of large information are organized, unstructured, and semi-organized information. The decent arrangement is utilized to recover, store, and interact with organized information. Organized information is the kind of information utilized in this examination. Because of the way that data on engine protection claims is kept in a set organization, explicitly a decent data set with a social construction design. The task's essential objective was to make an AI model that orders and predicts the present status of vehicle protection claims.

## 2 Literature Review

"Machine learning methods for property insurance fraud detection: Empirical evidence using real-world microdata." [1] In this work, Matheus Kempa Severino has proposed. Assessed extortion forecasts in property protection claims utilizing different AI models in view of true information from a significant Brazilian insurance agency. All through the recursive testing of the models, ordinary prophetic results were researched while considering misleading up-sides and tricky negatives. The outcomes showed that bunch based techniques (irregular forested regions and point supporting) and profound mind networks delivered the best outcomes, displaying better normal execution in correlation than different classifiers, including the usually utilized calculated relapse.

The dataset was also used to develop a comprehensive profile of known fraudsters, and we used the right artificial intelligence algorithms to assess each element's effect on the performance of the overall grouping as well as for instances of blatant false positive and false negative expectations. The results of this study can help risk specialists and professionals analyze each model's advantages and disadvantages and provide observationally persuasive decision-making standards to assess prospective insurance contracts.

"Analysis of Vehicle Insurance Data to Detect Fraud Using Machine Learning" [2].

In this work, P Sai Pranavi and Sheethal H D have proposed. They utilize different methods of AI for the recognition of vehicle protection extortion. They utilize the Random timberland and KNN calculation for the precision identification of protection misrepresentation. There are more than a huge number of insurance agencies on the planet which handle a lot of information. Protection misrepresentation has turned into the most well-known among the associations across the ventures. The protection industry is one of the most developing enterprises. Subsequently, misrepresentation identification turns out to be vital in this ongoing world.

Extortion identification can be carried out in different fields like banking, protection, monetary areas, and data security frameworks. There are many methodologies by which extortion can be identified. The exhibition estimation is finished by computation of disarray grid, Fraud in the protection information. The framework distinguishes regardless of whether the case is extortion by considering the data given by clients. Utilizing Random Forest and KNN calculations, protection extortion is anticipated precisely. The data acquired from the client is assessed with the dataset and lets the clients know whether the case is acknowledged or dismissed.

"Machine Learning Approaches for Auto Insurance Big Data" [3].

In this work, Mohamed Hanafy et al., This study reflects how computerized assurance merchants coordinate equipment advancing inside their associations and how ML models can be utilized to safeguard huge measures of data. I recommend considering whether it tends to be applied to. The changing number and seriousness of mishap claims require new methodologies to properly deal with these cases. Man-made intelligence (ML) is one of the systems to manage this issue. As underwriters expect the development of client care, these associations willtake on and apply ML to work on the interpretation and assessment of data for efficiency and, accordingly, better figure out their necessities.

Thus, we are advancing client care. Foresee guarantees occasions utilizing different ML procedures like Strategic Recurrence, XGBoost, Arbitrary Timberland, Choice Trees, Innocent Bayes, and KNN. Furthermore, we will assess and show the displays of these models. The outcomes show that RF is better than different advancements, with precision, kappa, and AUC benefits of 0.8677, 0.7117, and 0.840, separately.

"A proposed model to predict auto insurance claims using machine learning techniques" [4].

In this work, Shady Abdel Hadi et al. have proposed. The principal objective of the paper is to fabricate an exact model to anticipate vehicle protection claims through AI procedures. With an emphasis on cutting-edge measurable techniques and AI calculations that are the most reasonable strategy for taking care of missing qualities. One of the fundamental difficulties confronting the insurance agency is deciding the legitimate insurance installment for each chance addressed by clients.

Risk differentiates comprehensively starting with one client and then onto the next, and a Cautious cognizance of various betting variables expects the likelihood of security claims considering certain data; Genuine world datasets habitually have missing characteristics and can cause tendency in results. The most comprehensively embraced strategies for overseeing missing data are to eliminate discernments having missing characteristics, play out an all-out case examination (CCA), and utilize single credit like ordinary.

We utilized freely accessible datasets from Kaggle, which incorporate 12 variables and 30240 cases; the investigation was done by utilizing Counterfeit Brain Organization (ANN), Choice Tree (DT), Nave Bayes classifiers, and XGBoost to cultivate the assumption model. These procedures have downsides and tend to in loss of exactness and one-sidedness. The preliminary outcomes exhibited that the model created adequate outcomes. The XGBoost model and Goal Tree, with individual precision of 92.53% and 92.22%, accomplished the best accuracy of the four models.

"Performance analysis of machine learning methods for detecting automobile insurance fraud detection" [5].

This paper controls the assessment of suitability and the case that can be verified for the most famous AI estimates of the assumption of distortion, as suggested by C.Bouzgarne Itri and Youssfi Mohamed. Today, distortion is an important security problem that attempts to address big data and automation.

We have agreed with the controlled procedure used by a mysterious protection supplier in cases including vehicle data. We expect to present a methodology that highlights the meaning of the final results of human-made perception. Random Forest works better than any remaining estimates taken into account, as shown in the review.

"Blockchain-Based Transparent Vehicle Insurance Management" [6].

In this paper, they propose a change-free record of events as a security record of motor vehicles. This insurance record structure can consolidate all pieces of security trades. It not solely would chip away at the experience around exhibiting security, yet moreover, go probably as verification in the event of an inquiry.

This record can expand organizations by giving an ideal driving record. Individual drivers, vendors, protection offices, lawyers, policing and motor vehicleassociations are accomplices of thisblockchain-based course of action. We propose using blockchain development to get the nuances of the assurance where a couple of performers are collaborating and a couple of accomplices aredepending. We give nuances on how individuals can benefit from straightforwardness with this present circumstance, additionally as to how a blockchain can function with straightforwardness. Through this course of action, we demonstrate the way that blockchain development can be used as a correspondence medium regardless of untrusting social affairs.

"Model estimate of motor vehicle insurance claim risk and premium using the Bayesian approach." [7] In this paper, Sukono and Riaman propose using Bayesian techniques to assess hazard model cases and engine vehicle insurance payments. It is generally expected that cases will recur in a Poisson distribution, while cases accepted may follow a Gamma dispersion. The determination of where cases of recurrence are likely to spread and how many there are is based on Bayesianmethods. The assessor uses the dissemination of recurrence and measure of cases to appraise the total gamble models and the worth of the mean and difference. The mean and change assessor was used to anticipate the premium that would be charged to the protected.

"Computational intelligence approach for estimation of vehicle insurance risk level" [8].

This study was proposed by Kristina Vassiljeva, Aleksei Tepljakov, and others. Looking at contracts, handling input information, doing order, encoding the handled information so it very well may be taken care of into a counterfeit brain organization, and lastly preparing the brain networks are all important for the methodology. Exact and solid gamble forecast approaches are expected to process protection costs. We look at the instance of car protection in this examination and foster a computational knowledge-based strategy for foreseeing risk from information given by the client to the insurance agency. Therefore, a program is built to compute a gamble evaluation for a particular agreement. We look at difficulties associated with information handling and brain network preparation in this exploration, as well as examine the outcomes.

"Predicting Car Insurance Policies Using Random Forest" [9].

In this review, Asma S. Alshamsi et al. makes a proposition. Information mining has as of late been utilized to help protection organizations in projecting client choices to offer more cutthroat types of assistance in the field of accident coverage. In this piece, a characterization model that may be utilized to conjecture which insurance contracts clients would in all probability pick was made utilizing the irregular woods procedure. Various information mining calculations, including the Zero R classifier, a Basic Strategies Capability, a Choice Tree, and Credulous Bayes with seven expected strategies, were utilized to look at the presentation of the created model. The irregular backwoods strategy was viewed as the most reliable, with an all-out exactness of 97.9%. This review offers a strategy for figuring out which protection items a client would pick in view of the Irregular Backwoods classifier. Our primary target with this study was to help protect policyholders.

Mining Customer's Data for Vehicle Insurance Prediction System using k-Means Clustering- An Application" [10].

In this study, Thakur1, S.S., Sing, J.k. propose that Based on customer input and subsequently clustering on ideal data, a technique for Vehicle Insurance Prediction System was created in this study. Data mining, often known as customer data mining, is a method of extracting important characteristics from customer data and potentially employing those characteristics to foresee the future. It could be challenging to select the "best" algorithm/parameter setup. However, k- denotes According to their characteristics or traits, items are categorized or grouped into k groups using the clustering technique. In spite of the fact that in practice perfect separation is uncommon, a decent clustering approach should result in groups with distinct, non-overlapping boundaries. It is a function that determines the most effective algorithm. The most effective strategy is chosen based on the kind of data presented and the analysis's particular goal. In this project, Thakur1, S.S., Sing, J.k. have proposed this. In this research, a method for Vehicle Insurance Prediction System Data mining was created by social affairs information tests from shoppers and afterwards applying grouping to the ideal information.

# 3 Algorithm and Procedures

## 3.1 Random Forest Algorithm

Random Forest is one of the most well-known machine learning methods. The most effective machine learning algorithm seems to be Rf. It requires very little data planning, presentation, or modeling and, for the most part, yields precise results. The decision trees illustrated in the previous section are the foundation for Random Forests.

To put it another way, Random Forests are groupings of decision trees that produce higher forecast precision. It's called a 'forest' since it's a collection of decision trees. The core idea is to assemble unmistakable decision trees in view of autonomous subsets of the dataset. At each node, n qualities are chosen at random from a list of capabilities, and the best split on these parameters is found. In this study, whichever of the KNN and random forest calculations generates a more precise result will be used to anticipate extortion in the protection data.

It depends on ensemble learning, a strategy for consolidating different classifiers to deal with a difficult issue and improve the exhibition of the model. "Rf is a characterization technique that utilizes various choice trees to order various subsets of a dataset and

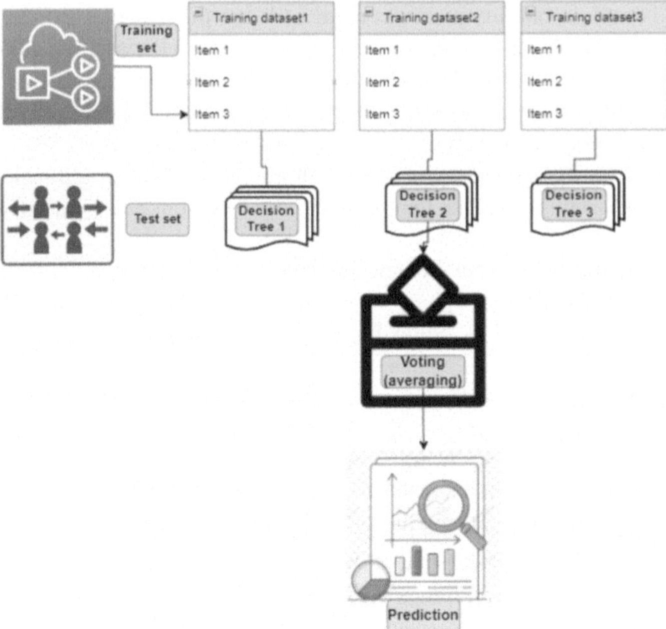


**Fig. 1.** Dataset training and prediction

afterwards midpoints the outcomes to work on the dataset's projected precision. "The irregular backwoods assemble forecasts from each tree and predict the result in light of the projections that got most of the votes as opposed to relying upon a solitary choice tree. The more trees in the woods, the more exact it is, and the issue of overfitting is stayed away from. The Random Forest method is depicted in the diagram below (Fig. 1).

# 4 Modules

## 4.1 Dataset Collection

The data collection includes eleven characteristics for car insurance claim information. The format of this data is an Excel spreadsheet. The records are shown in the row (instances), while the properties are shown in the column.The five objective groupings of insurance contract holders' case situations with the engine protection dataset are close, warning, forthcoming, re-open, and settled. The eleven extra highlights (ascribes) incorporate the arrangement number, the protected's name, the case number, the case date, the cited deficit, the cases paid (gross), net of recuperations, the all-out claims cost paid, the change in exceptional, and the case caused.

## 4.2 Data Preprocess

Data processing methods were used to prepare the data set. Examples of data preparation techniques include data integration, normalization or transformation, coding, and data cleansing.

There is no natural ordinal link between the two types of data, so a hot encoding (OHE) technique was employed to transform claim status categorical data (closed, notification, pending, re-open, and settled) to numeric or binary data.

To help RF and SVM machine learning algorithms grasp the data, these three qualities have been quantized to numerical data values. String values are present as instances or records for the following: Policy Number, Name of Insured, and Claim Number.Net of Collections = B, Loss Paid (Paid Gross = A), Net of Collections (AB), and Pending Variance are the additional aspects with numeric values and floats. The difference between the highest and lowest values for each characteristic is significant. As a result, the data set was transformed or scaled-down using theZ - the score data normalization technique.

Because it is nominal categorical data, the last attribute, claim status, is encoded using a label encoder. Closed, Notification, Pending, Re-open, and settled are the claim statuses 0, 1, 2, 3, and 4, respectively.

## 4.3 Random Forest Classification

Each decision tree used in the RF classifier is prepared to utilize random examples from the auto insurance dataset utilizing a method known as bootstrapping. To foresee the situation with protection claims, train every one of the trees utilizing different examples and utilize the consequences of the larger part vote.

### 4.4 Result Evaluation

Utilizing a variety of parametric measures (including confusion matrices, accuracy, precision, recall, and F-score), the machine learning model's efficacy was assessed. Because individual learner provides solutions with biased results.

### 4.5 Insurance Renew

In this model, at first, calculate the claim status; based on the status, the vehicle insurance company decides the payment due to renew the vehicle insurance.

## 5 System Architecture

Prior to making the two models, it is important to initially comprehend and set up the info information for the accident coverage guarantee status and expectation models. Model testing and model preparation were led utilizing the last preprocessed informational indexes. A preparation set and a test set were made from these preprocessed

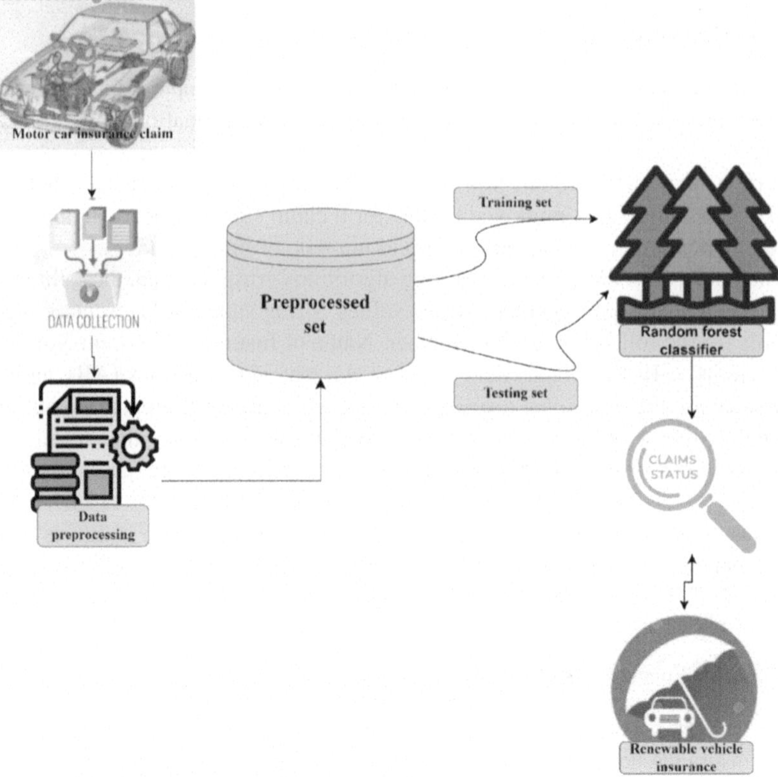


**Fig. 2.** System Architecture for Renewable Vehicle Insurance.

datasets. Utilizing the preparation set from the irregular timberland, develop the preparation model. Four measurements (Exactness, Accuracy, Review, and F-measure) were utilized to evaluate our model. Also, in view of the case status, the vehicle insurance agency chooses the instalment due and recharges the vehicle protection (Fig. 2).

## 6 Comparison Algorithm with Prediction

Using the best accuracy result to compare the algorithm with the forecast. . It is basic to look at the presentation of different machine learning algorithms reliably, and this tutorial will show you how to develop a test harness in Python using sci-kit learn to do so. This test harness can be used as a framework for your own machine learning tasks, with additional and different algorithms to compare. If you have anew dataset, it's really smart to try different things with various perception methods and check out at it from various points. You must analyze the machine learning algorithm's estimation accuracy in a variety of methods before deciding on one or two.

## 7 Graphs

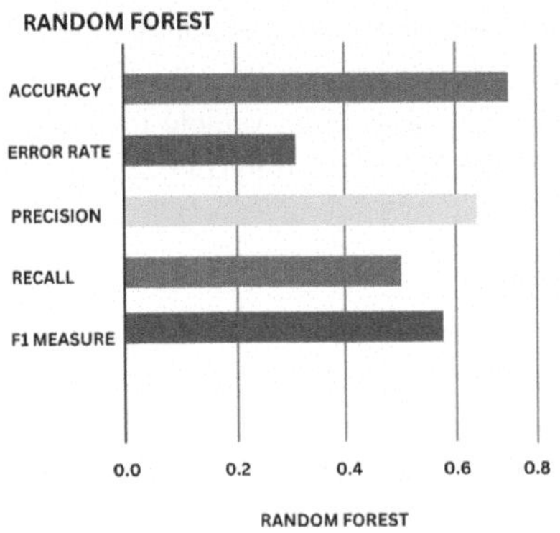


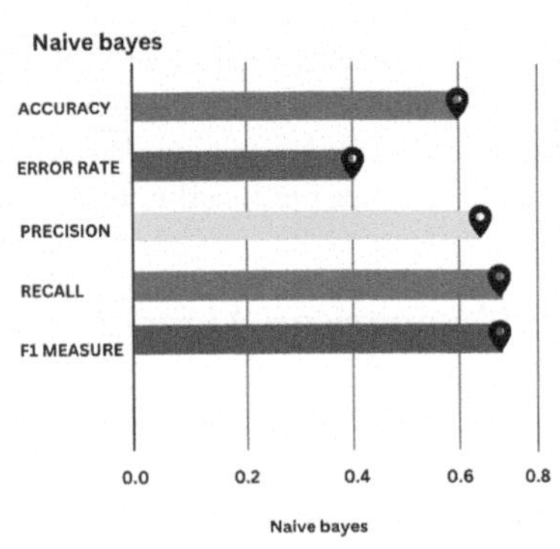


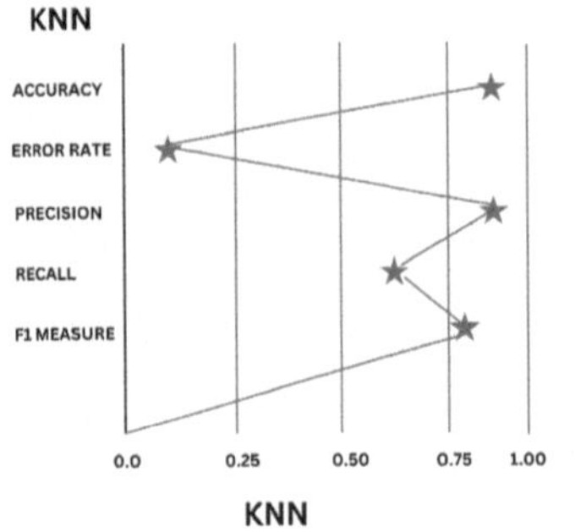


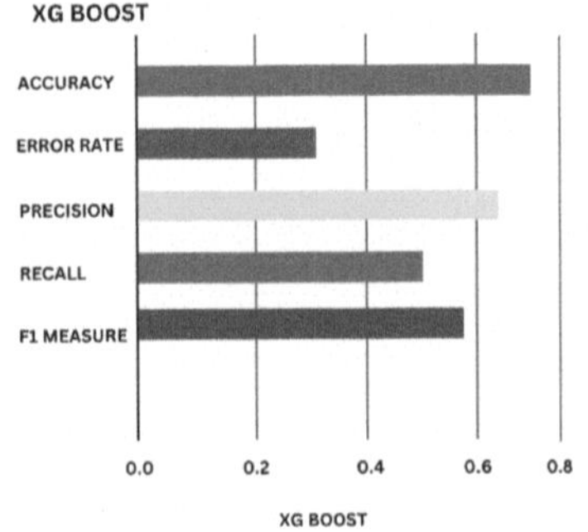


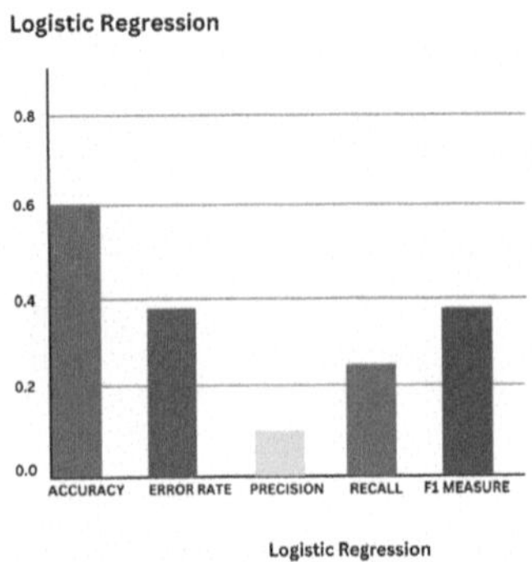


To forecast the recurrence of a claim, we use a variety of ML techniques, including logistic regression, XGBoost, random forest, decision trees, naive Bayes, and K-NN. The performances of these models are also assessed and contrasted. The accuracy, AUC, and outcomes demonstrated that RF is superior to other approaches with values of 0.8677, 0.7117, and 0.840, respectively.

# 8 Conclusion

The most potent methodological approaches employed in machine learning research were applied in this experimental study to determine the possible applicability of machine learning in the insurance industry, specifically for the prediction of the status of vehicle insurance claims. Therefore, the Random Forest model was applied as a predictive model to address the issue. In this work, an effort has been made to develop and apply a model that can forecast the status of automobile insurance claims. Information planning, model preparation, model testing, characterization, and expectation were among the techniques utilized. The strategies additionally included information understanding and logical information investigation. Contrasting RF models with different models, the made engine protection guarantee status expectation model has the most elevated forecast exactness.

# References

1. Severino, M.K., et al.: Machine learning algorithms for fraud prediction in property insurance: Empirical evidence using real-world microdata. Machine Learning with Applications **5** (2021). https://doi.org/10.1016/j.mlwa.2021.100074
2. Sai Pranavi, P., Sheethal, H.D., et al.: Analysis of vehicle insurance data to detect fraud utilizing machine learning. The Int. J. Res. Appl. Sci. Eng. Technol. (IJRASET) says **8**(VII) (2020)
3. Hanafy, M., Ming, R., et al.: Machine Learning Approaches for Auto Insurance Big Data (2020). https://doi.org/10.3390/risks9020042
4. Hadi, S.A., Elbahnasy, K., et al.: A proposed model to predict auto insurance claims to implement machine learning methods. J. Theoret. Appl. Info. Technol. **98**(22) (2020)
5. Itri, B., Mohamed, Y., et al.: Performance analysis of machine learning methods for detecting auto insurance fraud (2019). https://doi.org/10.1109/ICDS47004.2019.8942277
6. Demir, M., Turetken, O., et al.: Blockchain-based transparent vehicle insurance management. Sixth International Conference on Software Defined Systems (SDS) (2019)
7. Sukono, R., et al.: Utilising the Bayesian approach, a model is developed to estimate the claim risk and the premium for auto insurance. 4th International Conference on Operational Research (InteriOR) 21–23. Medan, Indonesia (2017)
8. Vassiljeva, K., Tepljakov, A., et al.: Computational intelligence approach for estimation of vehicle insurance risk level. The IJCNN, or International Joint Conference on Neural Networks (2017)
9. Asma, S., Alshamsi, et al.: Predicting Car Insurance Policies Using Random Forest (2014)
10. Thakur1, S.S., Sing, J.K.: Mining Customer's Data for Vehicle Insurance Prediction System using k-Means Clustering- An Application. Int. J. Comp. Appl. Eng. Sci. (2013)
11. Lin, W., Wu, Z., Lin, L., Wen, A., Li, J.I.N.: An ensemble random forest algorithm for insurance big data analysis. IEEE Acess **5** (2017)
12. Smith, K.A., Willis, R.J., Brooks, M., Smith, K.A., Willis, R.J., Brooks, M.: An analysis of customer retention and insurance claim patterns using data mining: a case study. J. Oper. Res. Soc. ISSN **5682**, 1476–9360 (2017)
13. Wijegunasekara, M.C., Weerasinghe, M.C.: A Study Comparing Data Mining Methods for Prediction of Automobile Insurance Claims **5**(1), 47–54 (2016)
14. Subudhi, S.: Detection of automobile insurance fraud using feature selection and datamining techniques. Int. J. Rough Sets and Data Anal. (2018)

# Ensemble Learning Models for Detecting Spam Over Social Networks Using RFE

V. Saraswathi[1](✉), A. Adaikkammai[2], Anitha Jebamani[3], D. Devi[4], and R. Radhika[5]

[1] S.A. Engineering College, Chennai, India
saraswathiv@saec.ac.in
[2] RMD Engineering College, Chennai, India
[3] Srisai Ram Engineering College, Chennai, India
[4] Sathyabama Institute of Science and Technology, Chennai, India
[5] RVS College of Engineering and Technology, Coimbatore, India

**Abstract.** Social networks serve as dependable platforms for valuable communication channels and the dissemination of information on a global scale, with millions of users utilizing them on a daily basis. While they have emerged as a means of disseminating information, they have also quickly transformed into a conduit for spreading false information, rumors, unsolicited messages, promotional content, fabricated news, and other undesirable content. Both spammers and non-spammers are present on social networks, with spammers typically sending advertisements and unwanted messages to users. Unfortunately, many users unknowingly spread unwanted messages, contributing to the problem. To address this issue, a model has been developed to recognize spammers on social networks. This process is based on the user's actions and patterns, as well as information-based features such as URLs, posts, replies, mentions, tweets, and retweets. The Recursive Feature Elimination (RFE) method is used in conjunction with five classifiers, including Random Forest, Logistic Regression, SVM, XGB, and Adaptive Boosting. Data is collected from social networks and implemented using the Sci-Kit Learn library in Python. The model's performance is evaluated using metrics such as TP and FP rate, precision, and accuracy.

**Keywords:** Social Network · Spam Detection · Recursive Feature Elimination · ML Classifiers

## 1 Introduction

Social networks have emerged as the primary conduit for information dissemination and public engagement, with a global populace actively participating in these platforms. From the total population of the world, more than half now use social networks, at around 59%. About 4.26 billion people use social media, and 137 million new users appeared online within the last year. According to statistical data, the mean duration of daily social media usage is recorded at 2 h and 31 min [1]. The social networks have become a common channel for spreading misleading ads, falsehoods, and political news, creating unnecessary communities, and triggering social disorder [2]. Social network platforms


R. Geetha et al. (Eds.): AAIMB 2023, CCIS 2203, pp. 150–164, 2025.
https://doi.org/10.1007/978-3-031-73068-9_13

are provided with various security features, but they are capable of being vulnerable. It will connect the physical and virtual worlds of information. A technical team is developing the best way to prevent spammers on social media [3]. Social spam refers to superfluous content that appears on social networks. This type of content can manifest in various forms, such as malevolent links, counterfeit acquaintances, derogatory language, and mass communications [4]. The utilization of social networking services, including but not limited to Google +, Facebook, Pinterest, LinkedIn, and Twitter, by users has been identified as a direct means of disseminating spam. Expert analysis has estimated that up to 40% of accounts are employed for the purpose of spreading spam across social media platforms. In recent years, Twitter has become a source for promoting activities and spamming attacks. Twitter accounts have already become a thing that can be organized in underground forums. Social spamming is a more dangerous attack than traditional methods like email spamming.

According to the report, few researchers only study detecting social spam, and there is no research reported about promoting campaigns [7]. A social media campaign refers to a well-coordinated marketing endeavor by a business entity that leverages at least one social media platform to reinforce information pertaining to a brand, product, or service. Such campaigns are characterized by their strategic nature, measurable outcomes, and the ability to influence social media followers to feel or act in a particular manner. Spammers are emerging with various techniques to seduce users into accessing unwanted links. YouTube is one of the more familiar sites on social networks, and it contains commenting features with which different users can leave a message in the uploaded video. Every user has own channel YouTube. You can upload, share, and comment on videos on YouTube, as well as endorse different channels. This sensational feature creates a medium for users to communicate each other's ideas, feedback, and emotions. But it is possible to spread malicious posts or irrelevant content on the social networks [8]. Machine learning is a methodology that involves the extraction, structuring, loading, and prediction of pertinent information from vast quantities of data with the aim of detecting patterns and arranging the data for future utilization. In this context, a proposed model is presented for the identification of spam on the social media platform, Twitter [9]. Both user behavior and content-based factors are being taken into consideration for the spam identification process. For the acquired dataset, correlation-based pre-processing is also utilized, along with a straight-forward evolutionary algorithm for searching among the attribute set for pre-processing, and RFE is then used for feature selection. Among the five classifiers are SVM, XG Boost, Random Forest, Logistic Regression, and Adaptive Boost, which were evaluated in order to identify an effective model for separating spammers and non-spammers using RFE. The present article is organized in the following manner: Section 2 comprises the literature review, Section 3 expounds upon the methodology employed, Section 4 presents the experimental results and their evaluation, and the Final Section delineates the description and conclusion of the results, in addition to outlining the future scope of the study.

## 2 Literature Review

Jitendra Pandey and Minimol Anil Job put forward and demonstrated a spam detection method to identify spam across various social media platforms. As per the scholars, the proposed framework has the potential to be employed by diverse social networks for the purpose of spam detection. The efficacy of the framework can be demonstrated through the execution of experiments. Additionally, the researchers aim to assess and appraise the proposed framework by utilizing live feeds from SMNs. Moreover, the identification of spammers' conduct and its integration into the framework will be regarded as a prospective undertaking. It is noteworthy that the evolution of captivating new spam techniques is an ongoing process, akin to the perpetual advancement of technology [9]. Chen Liu and Genying Wang present a spam account detection model for social networks based on ELM. In this study, three distinct sets of characteristics were extracted from Sina Weibo data and subsequently utilized in an ELM-based classification algorithm. The findings of the behavioral analysis revealed the crucial role of social interaction features in distinguishing between normal users and spam accounts. Furthermore, the proposed solution was found to be more practical and efficient than other algorithms. However, the authors acknowledge the potential for bias in the analysis and the limited adaptability of the feature extraction mechanisms in the future. Therefore, further research is warranted to develop and evaluate alternative solution labeling and validation mechanisms for the accounts under consideration. Additionally, it is imperative to investigate the dissimilarities between spam accounts utilizing interaction functions [10].

Shabnam Parveen, Saumya Goyal, and R. K. Chauhan present The social media platform is the world's largest network for sharing information. Along with the benefits of valuable data accessible through social networks, spam messages are spreading rapidly. In order to distinguish between spam and normal messages and links, the decision tree and KNN classifier algorithms have been employed. The Weka tool was utilized to evaluate the efficacy of these algorithms on real-world data. The findings indicate that the KNN algorithm outperforms the decision tree algorithm [11]. S. J. Soman and S. Murugappan have proposed a clustering-based approach for the identification of tweet spam and non-tweet spam users, with a focus on the same trending topics. This research diverges from conventional spam detection methods, as it primarily concentrates on the detection of spam tweet topics. The study initially scrutinizes the distinct characteristics of current events in each tweet, such as the user's profile, location, interactions, and message, to identify spam tweet users and their corresponding rate significantly. Subsequently, after the extraction of features, the fuzzy K-Means clustering method is employed to group similar user tweets about trending topics. Finally, the ELM classification method is applied to identify both spam and non-spam tweets [12].

K. J. Liszka, C. C. Chan, C. Shekar, and S. Wakade employed text mining techniques to preprocess Twitter data prior to its utilization by data mining tools for the creation of classifiers aimed at identifying spam tweets. For classifying spam training tweets, a simple technique using 65 pharmaceutical discriminating words chosen by hand was used. Initial findings indicate that the decision tree with the J48 classifier can be used to detect spam tweets. Automated tools for classifying spam training sets are critical, and more research is needed [13]. Bing Cong, Tang Zhiang Wu, and Changjian Fang conduct research using data from two leading website platforms, Yelp and Tagged, along with seven different types of networks. They also use the hypotheses testing strategy to figure out whether homophile exists, the k-core decomposition approach to examine the connectivity characteristics of spam users, and the clustering coefficient measure to determine whether spam or legitimate users are cohesive [14]. In their article, Jaeun Choi and Chunmi Jean present a comprehensive framework that leverages the collaboration between experts and machine learning algorithms to effectively identify spam tweets. The proposed approach employs a cost-based deep learning filter that can eliminate many normal tweets in the initial stage, thereby reducing the workload of the experts. To enhance the consistency of the system, experts are tasked with analyzing suspicious tweets flagged by the algorithm. The experimental results demonstrate that the suggested method surpasses the conventional machine learning technique in terms of spam detection. Additionally, a training dataset update module is incorporated to ensure that the system adapts to the evolving trends in tweets. This is in contrast to the traditional machine learning approach, which is unable to respond to changes in patterns over time, leading to inconsistent outputs [15].

Haiyng Shen and Ze LI applied the Social Network Aided to meet these requirements, a customised and efficient spam filter (SOAP) is proposed. SOAP adds four new elements to the Bayesian filter. The social human connection spam component helps to recognize personalized spammers; the adaptive trust-based component prevents impersonation attacks; and the friend notification system utilizes the strength of a group of socially close nodes to strengthen SOAP's capacity for countering impersonation attacks. Exact spam filtering results were used as input for Bayesian automatic training, and a SOAP design was created. SOAP outperforms the basic Bayesian filter in aspects of spam accuracy rate and training time, according to the results of trace-driven and prototype-based experiments [16]. Thi-Hong Vuong employed a Maximum Entropy classifier model to distinguish between SPAM and NON-SPAM comments on Vietnam Facebook pages. The model effectively leveraged content-based and behavior-based features to achieve optimal results, with an average F1-score exceeding 90%. These findings validate the efficacy of our approach, which relied on judicious signals to identify spam comments [17].

On Twitter, Khaleel Ameen and Buket Kayer have chosen a deep learning-based spam detection method. They used features extracted from the content of tweets by the Word2Vec method for this purpose, as opposed to traditional approaches. The MLP neural network was then used as the classification method. Finally, they tested the chosen method against three different classifiers. The results of the experiments show that the chosen approach produces the best results in terms of precision, recall, and F-measure [18].Sayal Kamble and S. M. Sangve presented the framework for a real-time evaluation of an innovative machine learning-based approach to detecting social spam. The overall objective of studies for instantaneous filtering and designed to detect spammers who target social media sites is to discover techniques and necessary tools. The URL thread detection technique improved the existing system's accuracy in classifying tweets as spam or non-spam [19]. Thayakorn Dangkeree and Sutheera Punteeranurak have put forth a proposal for implementing adaptive classification on relevant data through the use of a spam word list and a blacklist URL. Their findings indicate that this technique outperforms traditional classification methods across all datasets. Moving forward, the authors suggest exploring the use of Safe Browsing as an alternative to URL blacklists for identifying potentially hazardous links, as well as making modifications to certain features. The authors have successfully developed an adaptive classification for Nave Bayes in the data analysis section, and plan to refine algorithms and conduct comparative analyses to improve stability and performance [20]. Kiliroor, Cinu C, considered the social context based features in addition to the spam identification with the elementary Naïve Bayes classification model applied on social network data from Facebook. The experimental results prove that Naïve Bayes with Share, Like, and Comment factors collectively provide higher accuracy as compared to the model [21].

Hailu Xu, Weiqing Sun, and Ahmad Javadi used a variety of conventional classifiers, including Logistic, Nave Bayes, Random Forest, and Random Tree, to evaluate two datasets that they collected through their APIs. The best algorithm is Random Forest, which has an accuracy rate of nearly 94.7% for the Twitter dataset and 97.7% for the Facebook spam dataset. In order to enhance the classification training work, spam from one social network was combined with spam from another. The new arrangements in view of blended spams from another interpersonal organization perform better with regards to accuracy and bogus up-sides. As per the outcomes, Arbitrary Woods accomplishes the best presentation, and it has been exhibited that comparative spam in one web-based informal community empowers spam location in another [2]. On social networks, traditional classifiers have been used to find spam up until now. To distinguish spam on informal organizations, outfit classifiers like Irregular Backwoods, AdaBoost, and XGBoost are utilized and contrasted with notable conventional classifiers, for example, Support Vector Machine and Regression.

# 3 Methodology

Algorithms based on machine learning are useful for automating tasks that were previously performed manually. They can help businesses save both time and resources while also increasing their efficiency [24]. The system architecture that has been proposed, as illustrated in Fig. 1, encompasses all stages of pre-processing, model training, and dataset testing.

## 3.1 Dataset

The dataset is obtained from [23], and it includes user actions and content based on spam and non-spam messages. The data in the dataset is pre-processed to facilitate spam detection.

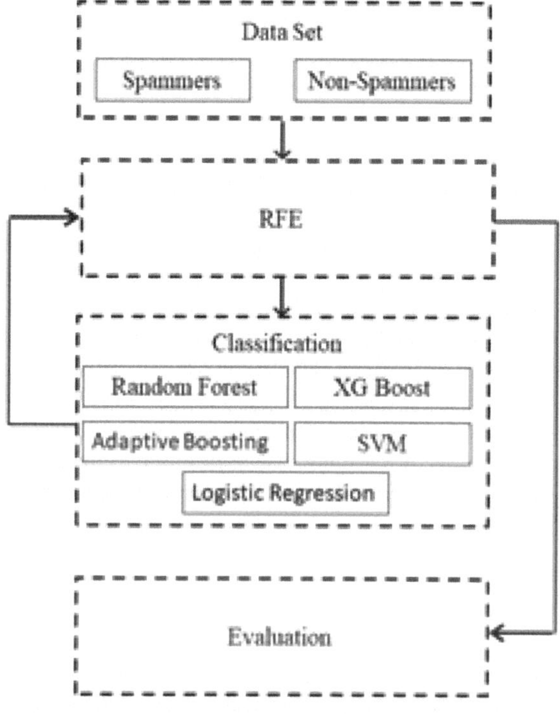


**Fig. 1.** .

## 3.2 Algorithm

Recursive Feature Elimination (RFE) is a wrapper method that is widely recognized in the field of machine learning. Specifically, it is considered to be one of the most prominent feature selection algorithms available in the SkLearn library. The RFE algorithm operates by iteratively removing the least significant features until a predetermined threshold is reached. This process is repeated until the desired level of feature reduction is achieved, effectively acting as a filter for the most relevant features.

It has two parameters namely, Estimator and n_features_to_select.

**Estimator:** In the context of a linear regression model, the determination of feature importance is contingent upon the coefficients. Specifically, the salience of a given characteristic is associated with the magnitude of its corresponding coefficient, such that the most consequential feature is linked to the most significant coefficient, while the least consequential feature is linked to the least significant coefficient. To decide on the value of a feature, tree-based models like decision tree, random forest and so on.

**n_feature_to_select:** This option specifies how many features we want to maintain. The difficulty we face with the RFE class is that, it is not always clear in advance how many features should be kept to a minimum [25].

The dataset for this implementation is derived from the packages listed in the previous section. Our raw data has been successfully imported. There are 1735 instances, 25 attributes, and a study variable in this data set. It is a collection of spam as well as non-spam messages on Twitter. As a result, this will determine if a message contains spam. Let's look at the five primary trials in this dataset using the *head* function.

The Attrition column in the data subset above is our study variable. *Yes,* and *No* are the only possible values for this column. *Yes*, indicates a spam message, while *No* indicates a non-spam message. The present dataset is intended for the purpose of constructing a predictive model that can accurately discern between spam and non-spam messages. However, it is unfortunate that this dataset comprises a substantial number of features, also referred to as explanatory variables, which renders the development of a model with such functionalities a potentially expensive undertaking. In light of the intended objective, the resultant model derived from this dataset may exhibit a high degree of complexity. Additionally, the model's design may be susceptible to overfitting as it endeavors to capture all of the features.

We must examine our feature space and select a subset of features that illustrate at least 95% of the variation in the data source in order to avoid this issue. We will use the recursive feature elimination method to achieve this objective. In any case, on the grounds that the RFE is costly, we should refine our data to the furthest degree conceivable.

### 3.3 Classification

Dividing the dataset into two parts: 70% for training and 30% for testing. The dataset was applied in two ways. First, it is applied with only five classifiers, and then it is applied along with the five classifiers and RFE algorithm for the second method. The five classifiers are: Random Forest, XG Boost, Adaptive Boosting, SVM, and Logistic Regression.

**Random Forest:**
A type of supervised machine learning algorithm known as Random Forest is frequently utilized in regression and classification problems. It builds choice tree from different examples and utilizations their popularity based greater part as a classifier and normal for the reliant variable. The main characteristic of the Arbitrary Timberland Calculation is that it can oversee dataset with both nonstop factors, as in order. It succeeds at characterization and relapse errands [26].

**XG Boosting:**
Extreme Gradient Boosting (XG Boost) is a supervised machine learning method that uses an instance-based approach based on the Gradient Boosting algorithm. It can solve issues related to regression and classification with minimal effort and achieve better results.

The initial version of the algorithm was implemented with the help of the Gradient Boosting machines. Subsequent to making this work open-source, a gigantic gathering of information examiners started adding to the Inclination Helped projects and further developing this calculation [27].

**SVM:**
SVM, also known as support vector machine, is a widely utilized model for supervised learning that is capable of addressing both regression and classification challenges. Despite its versatility, SVM is predominantly employed in machine learning for classification purposes. The primary objective of the SVM algorithm is to determine the optimal line or decision boundary that can effectively categorize n-dimensional space, thereby facilitating the accurate placement of new data points in their respective categories in the future. The optimal decision boundary is represented by a hyperplane, which is constructed by SVM through the selection of extreme positions and vectors, commonly referred to as support vectors. As a result, the algorithm is commonly referred to as the support vector machine [28].

**Logistic Regression:**
Logistic regression is a statistical approach utilized in the development of machine learning algorithms that involve binary dependent variables. This technique is employed for the purpose of describing data and the association between one or more independent variables and one or more dependent variables. The independent variables may be of nominal, ordinal, or interval nature [29].

**Adaptive Boosting:**
Adaptive boosting, also called AdaBoost, is used for ensemble methods and is a machine learning technique. The estimator used with AdaBoost is one level of a decision tree, which means it can split up to one level. It is also called "Decision Stumps" [30].

### 3.4 Evaluation:

The dataset is applied with RFE along with classifiers, and the performance is measured by the confusion matrix, TP, and FP rates. Classifier accuracy is obtained, and analysis reveals the best among the five classifiers.

## 4 Experimental Evaluation and Result

**Dataset Description:**
The process of estimating given data involves the division of the gathered data set into two distinct parts, namely the training and testing sets. Typically, a ratio of 7:3 is employed to partition the data into these two subsets. The data model created with Random Forest, logistic regression, XG boosting, Ada boosting, the support vector classifier (SVC), and RFE is applied to the training dataset, and the testing data prediction is identified based on the test results. The data collected could have missing values, resulting in inconsistency. To achieve better results, data must be pre-processed to improve the algorithm's efficiency. Outliers must be removed, and variable conversion must be performed.

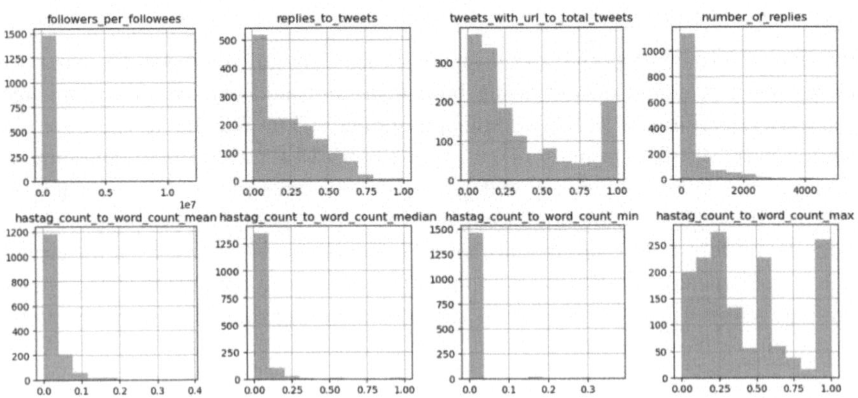


Heat Map:

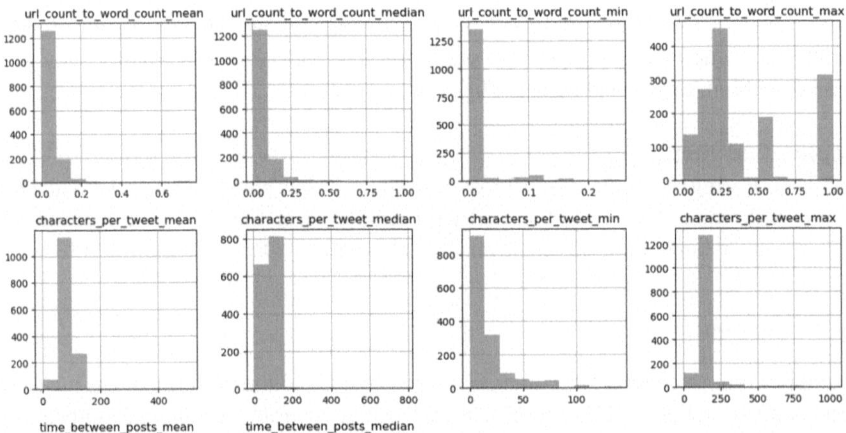


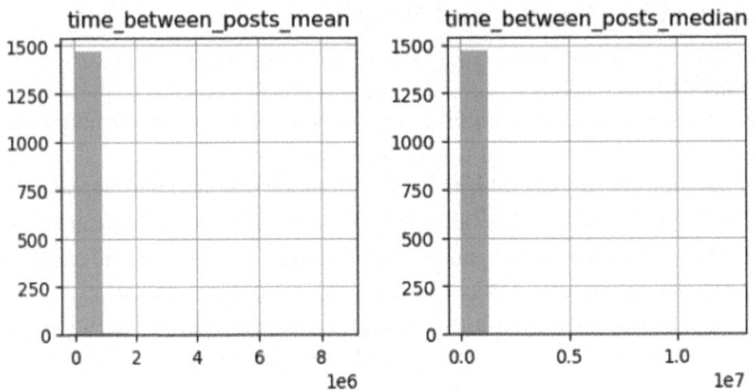


Heat maps are a type of visualization technique that is used to identify areas of a system that are more likely to be filled with spam. Heat maps provide an instant overview of any given data set. On this grid, messages from each source are classified by a color code. Heat maps provide an instant overview of any given data set.

**Confusion Matrix:**
One of the most straightforward methods for assessing the severity of a problem involves generating output for two or more classes. The confusion matrix, a two-dimensional array, is a tool that can be utilized for this purpose. It includes actual and predicted values, as well as "Right Positive," "Wrong Positive," "Right Negative," and "Wrong Negative" classifications.

### 4.1 Precision

Precision can be defined as the proportion of retrieved users that are deemed relevant. This metric is commonly expressed through the utilization of the following formula:

$$Q = \frac{R_p}{R_p + W_p}$$

### 4.2 Recall

Recall is a metric that quantifies the proportion of pertinent users that were successfully retrieved. This measure is commonly expressed through the following mathematical expression:

$$L = \frac{R_p}{R_p + W_n}$$

### 4.3 F-Score

The F-Measure is computed by utilizing the outcomes of both precision and recall, in the following manner:

$$F - Measure = 2 * \frac{R_p}{R + W}$$

## 5 Results

**Table 1.** Random forest classifier

| Precision | F – score | Accuracy | Class |
|---|---|---|---|
| 0.98 | 0.97 | 0.96 | No |
| 0.98 | 0.99 | 0.96 | Yes |

**Table 2.** Adaboosting classifier

| Precision | F - score | Accuracy | Class |
|---|---|---|---|
| 0.96 | 0.96 | 0.96 | No |
| 0.98 | 0.98 | 0.98 | Yes |

**Table 3.** Logistic regression

| Precision | F - score | Accuracy | Class |
|---|---|---|---|
| 1.00 | 0.04 | 0.96 | No |
| 0.69 | 0.82 | 0.96 | Yes |

**Table 4.** XG boost

| Precision | F - score | Accuracy | Class |
|---|---|---|---|
| 1.00 | 0.34 | 0.95 | No |
| 0.79 | 0.82 | 0.95 | Yes |

**Table 5.** SVM

| Precision | F - score | Accuracy | Class |
|---|---|---|---|
| 0.98 | 0.78 | 0.96 | No |
| 0.86 | 0.92 | 0.94 | Yes |

## 6 Performance Metrics:

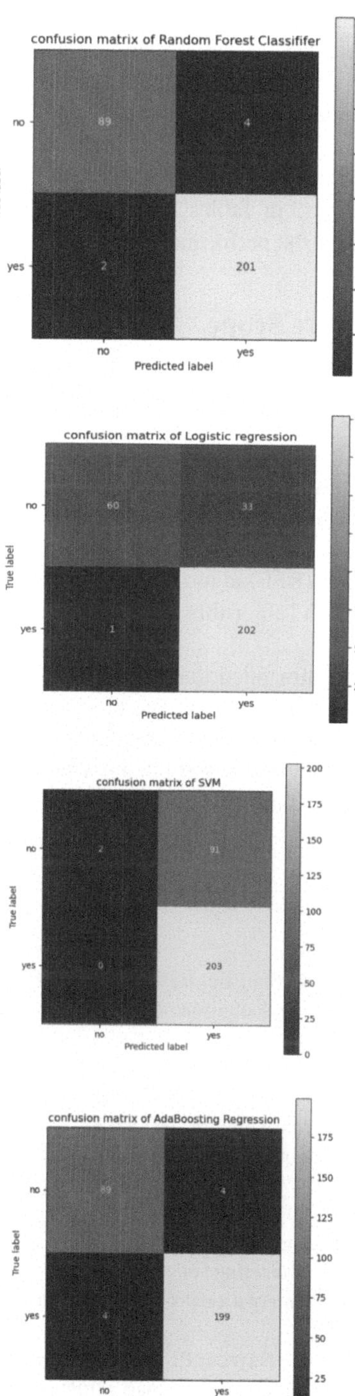
confusion matrix of Random Forest Classififer
True label
Predicted label
no
yes
89
4
2
201
200
175
150
125
100
75
50
25
confusion matrix of Logistic regression
True label
Predicted label
no
yes
60
33
1
202
200
175
150
125
100
75
50
25
confusion matrix of SVM
True label
Predicted label
no
yes
2
91
0
203
200
175
150
125
100
75
50
25
0
confusion matrix of AdaBoosting Regression
True label
Predicted label
no
yes
89
4
4
199
175
150
125
100
75
50
25

The present study employs various performance metrics, namely the Right Positive rate, Wrong Positive rate, Precision, F-Score, and Accuracy, which are presented in Tables 1, 3, 4 and 5. These metrics are computed using the Pycm library in Python, subsequent to the application of SVM, Random Forest, Logistic Regression, Adaptive Boosting, and XGBoost.

Confusion matrices for SVM, Random Forest, Logistic Regression, Adaptive Boosting, and XGBoost may be found in Tables 2, 4, 6 and 8. Confusion matrix holds the classifier data that was used for the performance analysis.

## 7 Conclusion and Future Scope

Social networks are becoming a source for sharing information through the internet's connectivity around the world, and many unwanted messages are also spreading. In order to mitigate the dissemination of undesirable content, a model has been proposed with the aim of identifying and eliminating spam messages within social networks. In this work, five classifiers are used, namely, Random Forest, XGBoost, SVM, Logistic Regression, and Adaptive Boosting, to find the accuracy of the model, along with the Recursive Feature Elimination (RFE) algorithm to eliminate the dispensable data from the model. The five classifiers and algorithm are implemented in Python using the SciKit libraries.

Moving forward, it is recommended that the approach be implemented on a larger dataset and assessed for its effectiveness in detecting spam messages within datasets containing strings. Additionally, these techniques can be extended to other social media platforms such as Instagram, Myspace, and others. The utilization of NLP techniques can enhance the accuracy of the data; thus it is suggested that this model be tested using such methods.

## References


1. Chaffey, D.: Global Social Media Statistics Research Summary (2023). https://www.smartinsights.com/social-media-marketing/social-media-strategy/new-global-social-mediaresearch
2. Hunt, A., Gentzkow, M.: Social Media and Fake News in the 2010 election. J. Econ. Perspect. **31**(2), 211–36 (2017)
3. Goolsb, R., Shanley, L., Lovell, A.: On Cybersecurity, Crowdsourcing and Social Cyber Attack. Office of Naval Research Arlington Va (2013)
4. Wikipedia https://en.wikipedia.org/wiki/Social-spam
5. Rao, S., Verma, A.K., Bhatia, T.P.: A Review on Social Spam Detection; Challenges, Open issues and Future Directions (2021)
6. Tynan, D.: Social Spam is taking over the Internet (2012)
7. Zhang, X., Zhu, S., Liang, W.: Detecting spam and promoting campaigns in the twitter social network. In: 2012 IEEE 12$^{th}$ International Conference on Data Mining (ICDM), pp. 1–6. IEEE (2012)
8. Abinaya, R., Bertilla Niveda, E., Naveen, P.: Spam detection on social media platforms. In: 2020 IEEE 7$^{th}$ International Conference on Smart Structures and Systems (ICSSS), pp 1–6. IEEE (2020)

9. Pandey, J., Anil Job, M.: Proposed framework for Spam recognition in big data for social media networks in smart environment. In: 2019 4$^{th}$ MEC International Conference on Big Data and Smart City (ICBDSC), pp. 1–6. IEEE (2019)
10. Liu, C., Wang, G.: Analysis and Detection of spam Accounts in Social Networks. In: 2016 IEEE 2$^{nd}$ International Conference on Computer and Communications (ICCC), pp. 1–5. IEEE (2016)
11. Goyal, S., Chauhan, R., Shabnam Parveen, K..: Spam Detection using KNN and decision tree mechanism in social network. In: 2016 Fourth International Conference on Parallel, Distributed and Grid Computing (PDGC), pp. 1–5. IEEE (2016)
12. Soman, S.J., Murugappan, S.: Detecting Malicious tweets in trending topics using clustering and Classification. In: 2014 International Conference on Recent Trends is Information Technology (ICRTIT), pp. 1–6 (2014)
13. Liszka, K.J., Chan, C.C., Shekar, C., Wakade, S.: Mining Pharmaceutical Spam from Twitter. In: 2010 International Conference on Intelligent Systems and Applications (ISDA), pp. 1–5 (2010)
14. Tang, B., Wu, Z., Rang, C.: Understanding network characteristics of spam users in social media. In: 2020 8$^{th}$ International Conference on Advanced Cloud and Big Data (CBD), pp. 1–6 (2020)
15. Choi, J., Jeon, C.: Cost-Based Heterogeneous Learning Framework for Real-Time Spam Detection in Social Networks with Expert Decisions, pp. 1–15. IEEE (2021)
16. Shen, H., Li, Z.: Leveraging social networks for effective spam filtering. J. IEEE Trans. Comp. 2743–2759 (2013)
17. Vuong, T.-H., Tran, V.-H, Ngugen, M.-D., Thi Nguyen, C.-V., Pham, T.-H.: Social-spam profile detection based on content classification and user behavior. In: 2016 8$^{th}$ International Conference on Knowledge and System Engineering (KSE), pp. 1–4 (2016)
18. Ameen, K., Kaya, B.: Spam detection in onine social networks by deep learning. In: 2018 International Conference on Artificial Intelligence and Data Processing (IDAP), pp. 1–4 (2018)
19. Kamble, S., Sangve, S.M.: Real time detection of drifted twitter spam based on statistical features. In: 2018 International Conference on Information, Communication, Engineering and Technology (ICICET), pp. 1–3 (2018)
20. Dangkeree, T., Puntheeranurak, S.: Adaptive classification for spam detection on twitter with specific data. In: 2017 21$^{st}$ International Computer Science and Engineering Conference (ICSEC), pp. 1–4 (2017)
21. Kiliroor, C., Valliyammai, C.: Social context based naïve bayer filtering of spam message from online social networks. In: Soft computing in data analystics, pp. 699–706. Springer, Singapore (2019)
22. Xu, H., Sun, W., Javaid, A.: Efficient spam detection across online social networks. In: 2016 IEEE Internartional Conference on Big Data Analysis (ICBDA), pp. 1–6. IEEE (2016)
23. Stringhini, G., Kruegal, C., Vigna, G.: Detecting spammers on social networks. In: 2016$^{th}$ annual computer security application conference, pp. 1–8 (2010)
24. MoneyLearn.: Machine Learning Classifier-The Algorithm & How They Work. Available: http://monkeylearn.com/blog/what-is-a-classifier/ (2020)
25. Section: Getting Started with Recursive Feature Elimination algorithm in machine learning (2022). http://section.io/engineering-education/recursive-feature-elimination/
26. AnalyticsVidhya: Understand Random Forest Agorithm with examples (2023). http://analyticsvidhya.com/blog/2021/06/understanding-random-forest/
27. Enjoyalgorithm: XG-Boost (Extreme Gradient Boosting) Algoithm in ML. http://enjoyalgorithm.com/blog/xg-boost-algoithm-in-ml
28. JavaTPoint: Support vector machine algorithm. http://javatpoin.comt/machine-learning-support-vector-machine-algorithm/

29. Simplilearn: An introduction to logistic regression in python. http://simplilearn.com/tutorials/machine-learning-tutorials/logistics-regression-in-python
30. AnalyticsVidhya: Master the adaboost algorithm guide to implementing & understanding adaboost (2021). http://analyticsvidhya.com/blog/2021/09/adaboost-algorithm-a-complete-guide-for-beginners/
31. Saraswathi, V., Anitha Jebamani, S., Devi, D.: Implementation of hyper parameter optimization in liver disease. In: 2022 International Conference on Power, Energy, Control and Transmission Systems (ICPECTS), pp. 1–6. IEEE (2022)

# An Intelligent Machine Learning Framework for Melanoma Classification System: A Critique

S. Sridevi[1](✉), S. Gowthami[2], and K. Hemalatha[3]

[1] Vel Tech Rangarajan Dr. Sagunthala R&D Institute of Science and Technology, Chennai, India
drsridevis@veltech.edu.in

[2] Sri Krishna College of Engineering and Technology, Coimbatore, India

[3] Vellore Institute of technology, Vellore, India

**Abstract.** One of the most common cancers, skin cancer accounts for roughly half of the cancer prevalence in worldwide. Melanocytes, which are cells that produce pigment, are the source of melanoma, a kind of skin cancer. Both the incidence and mortality rates of melanoma have sharply increased in recent years. The ailment that attracted a variety of studies can be treated more effectively by medical professionals with early identification. This article focuses on several segmentation techniques and a classification strategy based on machine learning for the identification of melanoma. The melanoma areas are easily recognized by the segmentation methodologies, and the classification procedure is made simpler. Convolutional neural networks (CNN) with U-Net are used for segmentation. A support vector machine (SVM) with a variety of kernels and a k-nearest neighbor (KNN) with a variety of k values are used to classify the segmented image in the classification phase. Performance demonstrates that the segmentation process significantly increases the SVM's classification effectiveness when acquired values from the machine learning classifier are compared. Further the results shall be improved by enhancing the analysis with large dataset.

**Keywords:** Machine learning · skin cancer · melanoma · neural network · segmentation · classification

## 1 Introduction

Dermatology is one of the most important medical specialties. More people have skin issues than have cancer, obesity, or hypertension combined [1]. According to estimates, one in three people in the United States contact their primary care physician for a skin disease each year (12.4% of all visits). About 1–2% of all cancer diagnoses in India are skin malignancies, and the frequency is rising. Melanoma is one of the most dangerous types of skin cancer, and if it is not found in its earliest stages, it has the ability to spread through the bloodstream and lymphatic system to other regions of the body. [2] Melanin, a pigment necessary for human skin color, is produced by melanocytes. Melanocytes are located where the epidermis and dermis meet underneath the dermis and epidermis. Melanoma is caused by aberrant melanocyte growth or by genetic abnormalities. A rare form of skin cancer with a high mortality rate is melanoma [3]. However,


R. Geetha et al. (Eds.): AAIMB 2023, CCIS 2203, pp. 165–175, 2025.
https://doi.org/10.1007/978-3-031-73068-9_14

several imaging examinations might show disease metastatic spread if a primary diagnosis or clinical suspicion is present. Despite its rarity, melanoma discovery sometimes leads to medical negligence because it is confused for other skin tumors [4]. Sometimes melanoma is only identified after it has spread to the lymphatic system. Strong computational approaches are therefore essential for melanoma early diagnosis. The four primary forms of cutaneous melanoma—superficial spreading, nodular, lent go, lentigo maligna, Acral lentiginous, and subungual melanoma—each include two subgroups. People of various skin tones can develop amelanotic melanoma, a kind of melanoma. The categories and forms of melanoma are the main subject of this study [5]. One of the most serious types of skin cancer is melanoma. In order to prevent inaccurate diagnosis and unnecessary biopsies, this study seeks to demonstrate the precision and accuracy of early melanoma sites [6]. The phases in the melanoma detection procedure include determining the kind of infection, image enhancement, dermoscopic feature extraction, and classification of the degree of illness. In order to assist in the diagnosis of melanoma utilizing dermatoscopic images, this study will apply classifiers with in- depth learning features [7].

The images used in this study are melanoma images that have undergone Least Mean Square Adaptive Noise Cancellation (ANC) pre-processing. A U-net based on CNN is used to do segmentation. Machine learning is used to categorize the segmented images. To improve the melanoma classification procedure, pre-processing and segmentation are done [8, 9].

## 2 Literature Survey

A more accurate and early melanoma diagnosis can increase the effectiveness of treatment. This study proposes a deep learning method for categorizing melanoma cases using dermoscopic images. The method shows that considerable enhancement throughout training and testing produces positive results, emphasizing the need for more research. The proposed method was evaluated using hold-out test data from the SIIM-ISIC Melanoma Classification 2020 dataset, with the best ensemble model achieving a 0.9411 area under the ROC curve [10]. By early cancer detection and precise diagnosis, automatic melanoma detection is essential for lowering mortality rates. The main objective of this research is to improve the classification accuracy of melanoma by using a novel color and texture feature descriptor that has been provided.

The U-Net approach, based on a convolutional neural network, is used to first segment the skin lesion region of the input dermoscopy image [11]. Removing undesirable features is essential in computer vision applications because undesirable traits make images less visible. In order to do this, deblurring, denoising, and detraining have been studied using both traditional model-based techniques and modern deep learning tech niques. The relevance and potential of hair removal in dermoscopic images, however, have not received enough attention. Only a few hair removal attempts use the mostrecent data-driven techniques, while hair removal algorithms continue to be based on old morphological approach. Hair should be removed from dermoscopy applications since it might obstruct views of lesions [12]. Each year, a large number of people are affected with skin cancer, the most prevalent kind of cancer. According to the World Health Organization, infections with melanoma and non-melanoma skin cancer have increased recently.

Worldwide, 2 million non- melanoma skin cancers and around 132,000 melanoma skin cancers are detected each year. As the sickness worsens, the likelihood of survival dramatically decreases. On the other hand, early skin cancer detection is a difficult and expensive technique. A method for classifying skin lesions as benign or malignant was disclosed, using images from common cameras. On these images, three crucial operations are carried out: segmenta tion, feature extraction using the "Border, Asymmetry, Color, and Diameter" criteria, and classification utilizing a Neural Network method that has been correctly trained to identify lesions [13].

## 3 Machine Learning-Based Classification System

### 3.1 Pre-processing

Noise reduction: Noise, a formidable foe, has a significant impact on the good quality of the melanoma image. For the appropriate investigation of melanoma images, noise incidence is crucial. The Least Mean Square Adaptive Noise Cancellation (ANC) reduces the noise incidence in the melanoma image. The ANC technique successfully removes the unnecessary information from the melanoma image's power line [14]. Normalization: Typically, normalization is used to bring two disparate images into a single, standardized series. The statistical discernment of the normalization, which converts an image where the mean value is 0 and the standard deviation value is 1, is a typical example of a predetermined series.

Z-score is used to normalize the data which is given as

$$Z\,score = \frac{score - mean}{SD} \tag{1}$$

This normalization decreases deviation by maintaining the correlation between the original and normalized melanoma pictures. Machine learning techniques are added after the normalization process is finished, which simplifies the learning method [15].

### 3.2 Segmentation-CNN with UNet

A computer model known as a neural network (NN) aims to replicate the composition and operation of the human nervous system. Similar to this, our nervous system performs three main functions: receiving information (stimulus), processing, and activating the results of processing. Three separate layers of NNs are used to do these tasks. A fundamental 3-layered NN's construction is shown in Fig. 1 [16].

The input layer is where all data and patterns are received by NN before being passed on to hidden levels. It should be an instance of a situation for which NNs have been trained. Additionally, each neuron at this layer represents a separate variable that affects the NN's ultimate output.

Hidden Layer: The primary function of these layers is to process the data that is received through input layers. They are outfitted with activation features as a result. Keep in mind that the activation function attempts to simulate the electrical stimuli at the synapses that exist in the human nervous system and serve as a conduit for information.

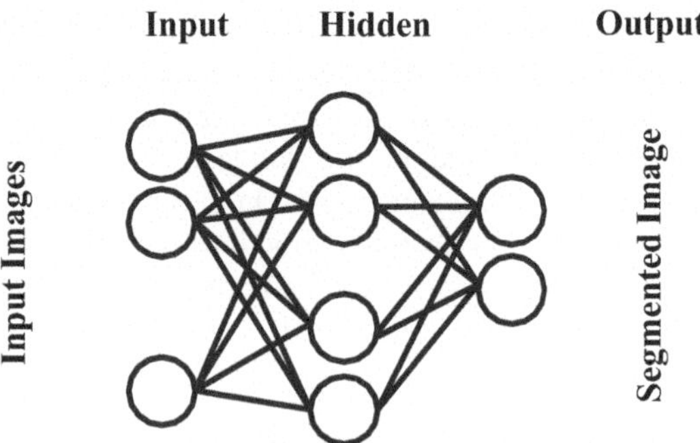


**Fig. 1.** Basic 3-layered

Once a certain threshold is reached, they are only shot or triggered. There may be zero, one, or many hidden layers, depending on the degree of complexity the problem is.

Output Layer: Decisions (classification) may be made using processed data (information) supplied through the output layer. The note output layer delivers the results in accordance with how it was intended. The two parameters that a NN learns are weights and biases. And, it may be envisioned as a classifier that picks up on the nuances and complexity of an input data (picture) by extracting hierarchical features from the input data (with these weights and biases) provided through the input layer (in this case, that input data is pixels of a melanoma image).

In contrast to the typical CNN classification, which generates a vector for each image at the end that is then used for input image characterization, the segmentation algorithm was initially developed for medical image segmentation, but it quickly gained popular ity due to its reliable and accurate results for many segmentation-based applications. U-Net uses this produced vector to recreate the picture. After that, any statistic for measuring the performance of the segmentation may be applied. For illustration, choose Intersection over Union. As a result, these pre-trained models can serve as feature ex- tractors. In other words, transfer learning improves deep networks' capacity for generalization and adapt.

## 3.3 Classification

### 3.3.1 K-Nearest Neighbor

A fundamental classification approach that identifies an input vector's neighbors is the Nearest Neighbor (NN) method. Neighbors are the training samples in the function space that are most similar to the input vector. This method's foundation is the mapping of the perceived input to the value of the nearest neighbor. By analyzing quantified values, its inference process determines the input vector's closest neighbors and categorizes the input vector accordingly. Another variation of this strategy was presented by the distance-weighted k-NN [17].

By applying a factor and inversely proportional to the distance between the vector and the input characteristics, this method quantizes the influence of each of the k nearest neighbors. Comparing the weighted distance to moderate and simple samples reveals an improvement in classification accuracy. This classifier uses different estimations to identify the close neighbors, namely the L1 and Euclidean based spaces, which are pre sented as,

$$Eucledian = \sqrt{(a - a_j)}^2 \tag{2}$$

$$L1norm = |a - a_j| \tag{3}$$

### 3.3.2 Support Vector Machine

A widely used supervised, constructive learning technique built on the statistical learn ing theory is the support vector machine (SVM). SVM is used to categorize whether an image is abnormal. SVM is used in a variety of applications. The SVM was first created for binary classification and then improved for multi-category sort. SVM is a member of the family of generalized linear classifiers. The work of Support Vector Machine is based on a learning strategy that employs a linear function hypothesis space [18].

Using a learning method, the multidimensional feature spaces are learned. While all evaluations are conducted in the original region, SVM discovers the hyperplane separation introduced by the kernel function. Its essential idea is to derive the hyperplane for dividing the two classes. It gives a more reliable nonlinear classification using a kernel approach with minimum training dataset. A set of items having various natural viewpoints can be divided using a decision plane and the decision boundaries.

The hyperplane is another name for the decision plane. The evaluation of the convex optimization will reveal the decision function of the known training set. SVM establishes the decision boundary in the data space by employing the structural risk reduction approach. The crucial and distinctive aspect of this methodology is that the final answer solely depends on data points near the margin. Margin in this context refers to the smallest distance between the splitting hyperplane and the nearest data points. These locations are referred to as support vectors. This device offers the opportunity to do a non- linear transformation in order to discover a solution for issues that cannot be resolved linearly.

The class designation of m topics and the characteristics of medical images are connected. Create a collection of binary classifiers (f1, f2, etc.) that are individually trained to distinguish one class from the other (M–1) classes, then combine them to create a multiclass classifier. During training, mistakes are minimized,

$$\phi(w, \xi) = \frac{1}{2}\sum_{a=1}^{m} (w_a^T, w_a) + C\sum_{i=1}^{N}\sum_{a \neq y_i} \xi_i^a \tag{4}$$

Subject to criteria

$$y_i(w^T\phi(x_i) + b) \geq 1 - \xi_i \geq 0, i = 1, \ldots, N\ m \in \{1, 2, \ldots A\}\{y_i\} \tag{5}$$

C can range from zero to infinity in SVM classification.

Where C - volume constant, b - a constant and w - vector of coefficients, and $\xi i$ - non-separable data parameter; thus, it is a bit difficult to estimate and apply.

The N training-cases are labeled in the index i. Where yi ∈ {1... M} are multiclass labels excluding Yi and xi is the self-regulating variables, ∈ {1... M} \ yi is the multiclass labels of the features vectors. The data from the input data are converted to the feature space using the kernel ϕ. Statistical learning theory for maximizing the margin $\gamma = \frac{2}{||w||}$ Results the minimization of prediction error.

The decision boundary is formulated as,

$$f(x) = \arg\max_{m} {}_a(w^T.x + b)_a \qquad for\ a = 1,2,3 \ldots a \tag{6}$$

$$f(x) = \sum k_i x_i^T x + b_a \tag{7}$$

The step by step approach is given below

- A kernel (nonlinear transformation function) is used to transfer the input feature space into a higher dimensional feature space.
- In the feature space, the machine builds the greatest margin hyperplane. The Gaussian radial basis function (RBF), the linear kernel, the quadratic kernel, and the polynomial kernel are the four most popular kernel functions utilized in the nonlinear SVM. The Radial Basis Function (RBF) kernel has a relationship to an unlimited dimensions feature space f.

The RBF function:

The Radial Basis Function (RBF) kernel has a relationship to an unbounded dimensions feature space f. This is how the RBF function is defined.:

$$K(x_i, x_j) = \left(\frac{\exp(-y||x_i - x_j||)}{2\sigma^2}\right) \tag{8}$$

where K- mapping function, and the vector is taken as Φx for no identity cases. The linear kernel is

$$K(x_i, x_j) = x_i^T, x_j \tag{9}$$

The linear kernel-based algorithms are often identical to their non-kernel equivalents. The quadratic kernel is given as.

$$K(x_i, x_j) = (x_i^T, x_j)^2 \tag{10}$$

polynomial kernel is suitable for nonstationary kernels. Customizable parameters include the d-polynomial degree, slope, and c-constant term. The linear kernel-based kernel algorithms are often identical to their non-kernel equivalents.

$$K(x_i, x_j)_i = (\alpha x^T, x_j + c)^d \tag{11}$$

## 4 Results and Discussion

The performance of each model is examined and tested in this research utilizing the numerous performance measures listed in this section. The effectiveness of the research is examined using Google Colab and images of melanoma. Six different types of melanoma are being examined as part of a multidisciplinary approach to melanoma research. Performance metrics including precision, recall, accuracy, and the Jacquard index are used to assess the effectiveness of the classifiers.

In this proposed work the performance metrics such as Recall (Re$call = \frac{TP}{TP+FN}$), Accuracy ($Accuracy = \frac{TP+TN}{TP+TN+FP+FN}$) and Jacquard Index ($Jaccard\ index = \frac{TP}{(TP+FP+FN)}$) are considered.

Figure 2 shows K-NN and SVM classification methods, recall value. As the research progresses, the classifiers' stability is evaluated by increasing the number of images in the order of 5. Different kernels are used in SVM for the categorization of melanoma, and it has been observed that the polynomial kernel preserves the maximum recall value compared to other methods. Even with an increase in the number of pictures, the polynomial kernel still achieves a recall rate of 88.5, assuring the stability and convergence of the classifier model.

Whereas, Fig. 3 shows the accuracy values for K-NN and SVM classification algorithms. While the classifier model's convergence is strong, KNN's accuracy of 83% indicates poor performance. RBF and the linear kernel perform well, however the classifier module shows no significant performance gain. SVM's polynomial kernel outperforms other kernels in terms of accuracy with an average of 88.5% (Tables 1, 2 and 3).

**Table 1.** Comparison of Recall

| Model | | No of Images | | |
|---|---|---|---|---|
| | | 10 | 20 | 30 |
| SVM | KNN | 82 | 82.5 | 83 |
| | RBF kernel | 86 | 87 | 87.5 |
| | Linear Kernel | 87 | 87.5 | 87.5 |
| | Quadratic Kernel | 86.5 | 87 | 88 |
| | Polynomial Kernel | 87 | 88 | 88.5 |

The Jaccard index value for classification methods is shown in Figure 4. In the categorization of melanoma, the Jaccard index guarantees the stability of classifier models. KNN model exhibits very low performance for classifying subungual melanoma. The performance of the classifier was found to be about 82 for the other 5 kinds of melanoma. Subungual and amelanotic melanoma respond better to polynomial kernel. Using the polynomial kernel in the SVM classifier, an effective jaccard index value of 88 is obtained for the classification of melanomas.

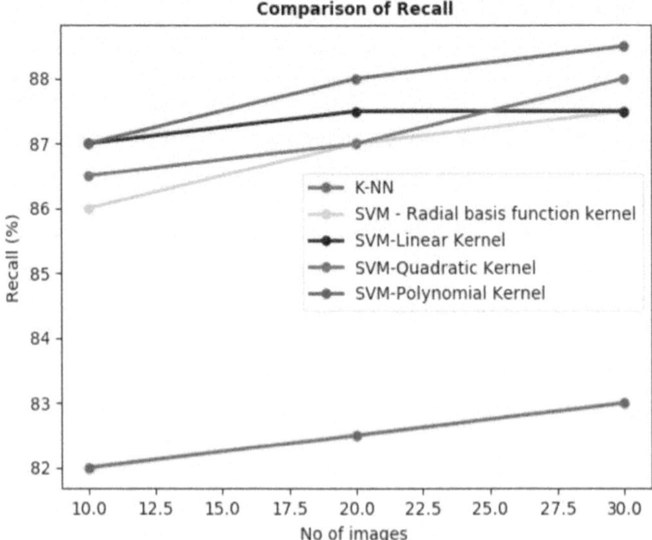


**Fig. 2.** Comparison of Recall

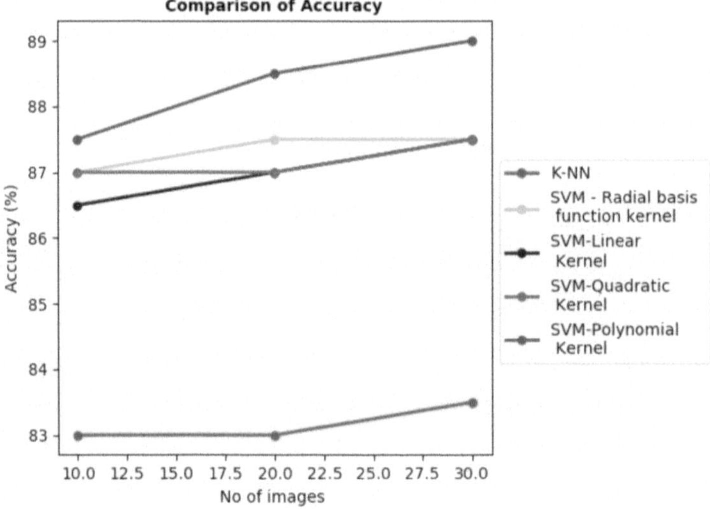


**Fig. 3.** Comparison of Accuracy

**Table 2.** Comparison of Accuracy

| Model | | No of Images | | |
|---|---|---|---|---|
| | | 10 | 20 | 30 |
| SVM | KNN | 83 | 83 | 83.5 |
| | RBF kernel | 87 | 87.5 | 87.5 |
| | Linear Kernel | 86.5 | 87 | 87.5 |
| | Quadratic Kernel | 87 | 87 | 87.5 |
| | Polynomial Kernel | 87.5 | 88.5 | 89 |

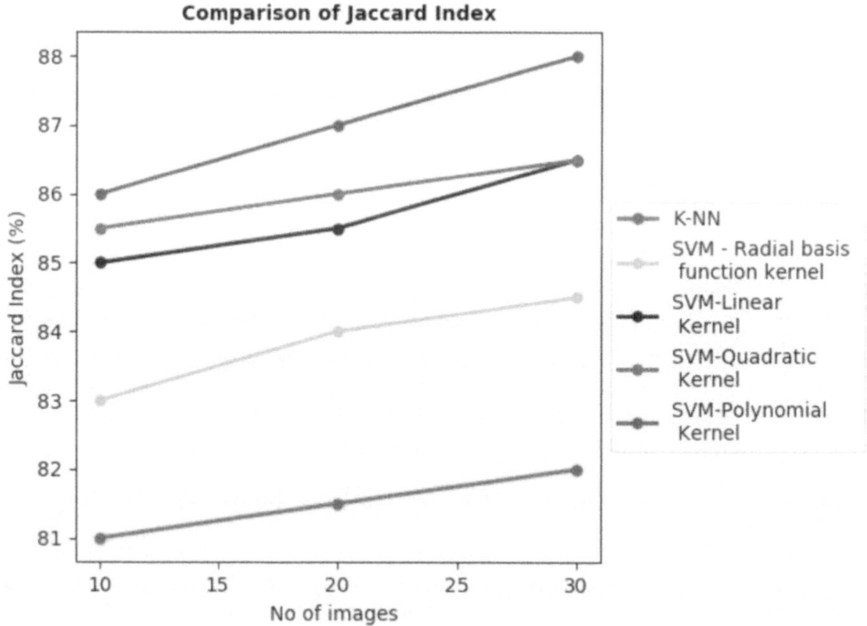


**Fig. 4.** Comparison of Jaccard Index

**Table 3.** Comparison of Jaccard Index

| Model | | No of Images | | |
|---|---|---|---|---|
| | | 10 | 20 | 30 |
| SVM | KNN | 81 | 81.5 | 82 |
| | RBF kernel | 83.5 | 84.3 | 84.7 |
| | Linear Kernel | 85.2 | 85.8 | 86.4 |
| | Quadratic Kernel | 85.6 | 86.1 | 86.7 |
| | Polynomial Kernel | 86.1 | 87.2 | 88.1 |

## 5 Conclusion

For the diagnosis of skin cancer, automatic skin lesion segmentation on skin pictures is essential. A high-quality image is always difficult to obtain. Because of this, researchers are always trying to come up with new and better methods for reducing degradation. Digital imaging-based diagnosis is widely used in clinical practice, medical research, and other fields. Due to an unknown disturbance caused by motion artifacts, melanoma medical images are commonly contaminated with noise or blur during collection. With the use of pre-processing techniques, images are cleaned up and made ready for segmentation. The segmentation procedure makes use of the U-Net, which is based on CNN. Machine learning algorithms are used to categorize the segmented pictures. Performance indicators like recall, accuracy, and Jaccard Index are used to evaluate the performance of the classifiers. The results show that SVM with polynomial kernel model outperforms all other models according to the quantified performance indices. The results shall be further improved by considering large data set in future.

## References

1. Alheejawi, S., Mandal, M., Berendt, R., Jha, N.: Automated melanoma staging in lymph node biopsy image using deep learning. 2019 IEEE Canadian Conference of Electrical and Computer Engineering (CCECE), pp. 1–4. Edmonton, AB, Canada (2019). https://doi.org/10.1109/CCECE.2019.8861878
2. Xie, F., et al.: Melanoma classification on dermoscopy images using a neural network ensemble model. In: IEEE Transactions on Medical Imaging, vol. 36, no. 3, pp. 849–858 (2017). https://doi.org/10.1109/TMI.2016.2633551
3. Ralbovsky, N.M., Lednev, I.K.: Towards development of a novel universal medical diagnostic method: Raman spectroscopy and machine learning. Chem. Soc. Rev. **49**, 7428–7453 (2020)
4. Tambourine, J.V., et al.: Skin cancer detection using ensemble of machine learning and deep learning techniques. Multimedia Tools and Applications 1–24 (2023)
5. Alenezi, F., Armghan, A., Polat, K.: Wavelet transform based deep residual neural network and RELU based Extreme Learning Machine for skin lesion classification. Expert Systems with Applications **213**(B) (2023)
6. Ahmed, N., Tan, X., Ma, L.: A new method proposed to Melanoma-skin cancer lesion detection and segmentation based on hybrid convolutional neural network. Multimedia Tools Apple **82**, 11873–11896 (2023)

7. Ala Walid, S.M.M.A., Sarada Prasanna Malik, D., Rastogi, R., Chauhan, A., Vidya, A.: Melanoma Skin Cancer Detection using a CNN-Regularized Extreme Learning Machine (RELM) based Model. In: 2023 Second International Conference on Electronics and Renewable Systems (ICEARS), pp. 1239–1245. Tuticorin, India (2023). https://doi.org/10.1109/ICEARS56392.2023.10085489
8. SM, J., Aravindan, P.M.C., et al.: Classification of skin cancer from dermoscopic images using deep neural network architectures. Multimedia Tools Apply **82**, 15763–15778 (2023)
9. Perez, E., Ventura, S., A framework to build accurate Convolutional Neural Network models for melanoma diagnosis. Knowledge-Based Systems **260** (2023)
10. Mui-zzud-din, A.K.T., Rustam, F., et al.: Predicting skin cancer melanoma using stacked convolutional neural networks model. Multimedia Tools Apple (2023)
11. Ghahfarrokhi, S.S., Khodadadi, H., Ghadiri, H., Fattahi, F.: Malignant melanoma diagnosis applying a machine learning method based on the combination of nonlinear and texture features. Biomedical Signal Processing and Control **80**(1) (2023)
12. Delibasis, K., Moutselos, K., Vorgiazidou, E., Maglogiannis, I.: Automated hair removal in dermoscopy images using shallow and deep learning neural architectures. Computer Methods and Programs in Biomedicine Update **4** (2023)
13. Ali, Z., Naz, S., Zaffar, H., Choi, J., Kim, Y.: An IOT-based melanoma lesion segmentation using conditional generative adversarial networks. Sensors **23**, 3548 (2023). https://doi.org/10.3390/s23073548
14. Hu, C., et al.: Digital self-interference cancellation for full-duplex systems based on deep learning. AEU – Int. J. Electr. Commun. **168** (2023)
15. Trosten, D.J., et al.: Hubs and hyperspheres: reducing hubness and improving transductive few-shot learning with hyperspherical embeddings. In: Proceedings of the IEEE/CVF Conference on Computer Vision and Pattern Recognition (CVPR), pp. 7527–7536 (2023)
16. Cui, H., et al.: An Improved Combination of Faster R-CNN and U-Net Network for Accurate Multi-Modality Whole Heart Segmentation. In: IEEE Journal of Biomedical and Health Informatics, vol. 27, no. 7, pp. 3408–3419 (2023). https://doi.org/10.1109/JBHI.2023.3266228
17. Progga, P.H., Rahman, M.J., Biswas, S., Ahmed, M.S., Farid, D.Md.: K-Nearest Neighbour Classifier for Big Data Mining based on Informative Instances. In: 2023 IEEE 8th International Conference for Convergence in Technology (I2CT), pp. 1–7. Lonavla, India (2023). https://doi.org/10.1109/I2CT57861.2023.10126147
18. Roy, A., Chakraborty, S.: Support vector machine in structural reliability analysis: A review. Reliability Engineering & System Safety **233** (2023)

# Online Network Intrusion Detection System for IOT Structure Using Machine Learning Techniques

K. Mahalakshmi[1(✉)] and B. Jaison[2]

[1] Department of Computer Science and Engineering, Vel Tech High Tech Dr. Rangarajan Dr. Sakunthala Engineering College, Chennai, India
mahalakshmi.k@velhightech.com
[2] R.M. K Engineering College, Chennai, India
bjn.cse@rmkec.ac.in

**Abstract.** With the rapid growth of Internet of Things (IoT) deployments, ensuring the security of these interconnected devices has become a critical concern. In this study, we propose a novel Machine Learning-based Online Network Intrusion Detection System (NIDS) specifically tailored for IoT architecture. Our approach harnesses the potential of advanced Machine Learning algorithms to overcome the limitations of conventional NIDS solutions. The primary objective of our strategy is to enhance IoT network security by swiftly identifying and responding to emerging threats. By leveraging the capabilities of Machine Learning, our NIDS can proactively detect and categorize various network intrusions, including both known and previously unidentified attacks. This proactive detection allows for real-time responses to attacks as they arise, mitigating potential damages and reducing the overall risk posed to IOT structures. The suggested NIDS also boasts the advantage of adaptability, capable of dynamically adjusting to changing IoT settings and evolving threat landscapes. By efficiently adapting to diverse network environments, our system remains effective and robust in safeguarding IoT networks against potential security breaches. Furthermore, our technology offers a strong and scalable solution to guarantee the security and integrity of IoT networks. The incorporation of Machine Learning techniques empowers the NIDS to quickly adapt to larger and more complex IoT infrastructures without compromising its performance. To evaluate the effectiveness of our proposed method, we conducted experiments using typical datasets, such as actual network traffic traces from UNSW-NB15. The outcomes of these experiments demonstrate the system's reliability in accurately identifying and categorizing different network intrusions, thereby affirming the efficacy of our Machine Learning-based NIDS for IoT architecture. Overall, our work presents a promising and innovative approach to fortify IoT network security, providing a proactive, adaptive, and scalable solution that effectively addresses the evolving challenges of safeguarding IoT structures in the face of emerging cyber threats.

**Keywords:** Online network intrusion detection system · IoT (Internet of Things) · Machine learning techniques · Intrusion detection · Network security · MEC (Multi-Access Edge Computing) server


R. Geetha et al. (Eds.): AAIMB 2023, CCIS 2203, pp. 176–194, 2025.
https://doi.org/10.1007/978-3-031-73068-9_15

# 1 Introduction

Advances in IoT generation impact the growing complexity of maintaining network security and defending against cyberattacks. Safety may be very important in an IoT network because the offerings supplied live depend upon the IoT network. This has also affected huge-scale community attacks against those cyber-bodily gadgets interest in IoT network protection is still low. Safety problems in present technology pose critical threats to humans and industries. It isn't always valid for IoT because of diverse troubles delivery. It's far impractical to actively accumulate all of the appropriate and anomalous (intrusion attack) statistics flowing through an IoT community. It is essential to layout a safety solution tailor-made on your IoT network at the same time as considering the resource boundaries for securing your IoT community. Within the IoT NIDS design procedure, it is impractical to pre-movement all top and anomalous (intrusion assault) statistics over the community. In the IoT NIDS layout manner, it's miles impractical to pre- move all excellent and anomalous (intrusion assault) facts over the community. Therefore, previous signature-based NIDS simplest recognized assaults based on regarded patterns (present records units) [3]. With growing know-how and continued operation of applied IIoT systems, ordinary traffic will increase over the years. New community attacks are happening extra frequently. Consequently, there's a need for an online NIDS that may learn incrementally to replace popularity models [4]. This online NIDS ought to be characterized with the aid of detecting changes in network attack styles and updating detection fashions online. Attack detection and type for incremental getting-to-know and online updates. A web update set of rules informs to partially alter the NIDS model for diverse IoT gadgets each time the proposed version classifies a brand-new assault in opposition to the Multi-Access Edge Computing (MEC) server.

# 2 Literature Survey

## 2.1 Amouri, V. T. Alaparsi, S. D. Morgera, "Gadget Mastering Based Totally Intrusion Detection Structures for Mobile IoT," Sensors, Vol. 20, no. 2, P. 461, 2020

A -level IDS was cautioned by means of the authors. The initial segment gathers records about dedicated It continues a sniffer (DS) and spawns appropriately categorized times (CCI), which can be transmitted to an excellent Node on an everyday basis (SN). To distinguish between benign and malicious behavior, SN runs a linear regression procedure on CCIS amassed from several DSs within the second level. Knot. With regard to the performance stages and node speeds of his various mobility models, Random Waypoints (RWP) and Gauss Markov, the writer concentrates on the detection and characterization of numerous severe eventualities inside the community (GM). Increase. Black hollow assaults and disbursed denial of provider (DDoS) attacks are the malicious movements used in this research. Here, the writer notes that the detection price exceeds 98% for excessive power/node pace scenarios, however, decreases to more or less ninety% for low strength/node velocity instances.

### 2.2 H. Wang, W. Zhang, H. He, P. Liu, D. X. Luo, Y. Liu, J. Jiang, Y. Li, X. Zhang, W. Liu, Et Al., "A Biological System Observes of IoT Malware," IEEE Net of Things Magazine, 2021.

The authors claimed within the advent that they may study the evolution of IoT malware by using combining the statistics from two sources. With a view to creating the important lineage tree, the authors first trawl online articles approximately IoT malware and make use of language processing algorithms to extract the choices of malware samples and their connections to different malware households. Second, the authors gather real malware samples from the usage of our significant network of honeypots, then design a new classifier to group them into families and discover linkages among them. The authors use these findings to assist the crucial lineage graph. They sooner or later used the statistics from the aforementioned resources to provide the definitive lineage graph for 72 IoT malware households. Which would possibly facilitate the analysis network to better perceive and combat IoT malware presently and inside the destiny.

### 2.3 PK Keserwani、MC Govil、E.Ok. S. Pilli and P. Govil, "An Advanced Anomaly-Based Intrusion Detection Gadget for Internet of Things (IoT) Network Mishandling GWO-PSO-RF Fashions," Magazine of Dependable Smart Environments, Vol. 7, no. 1, P. Three–21, 2021

The IoT concept is used by the author to improve the environment and make it easier for human beings to carry out an expansion of duties. For the motive of figuring out diverse attacks on IoT networks, the authors advocate an accomplice's degree IDS. Combining Particle Swarm Optimization (PSO) and grey Wolf Optimization (GWO) to locate the first-class IoT community options. For excessive assault detection accuracy, the retrieved alternatives are put into a random wooded area (RF) classifier. The KDDCup99, NSL-KDD, and CICIDS-2017 datasets are used as the premise for experiments that take a look at projected models within the Python programming surroundings. In the multiclass type, the proposed GWO-PSO-RF-NIDS model had a mean accuracy of 99.66%. To show its efficacy, the expected version's accuracy turned into contrasted with that of other, similar methodologies. The paintings recognized here tackle the difficulty of information inequality as an entire.

## 3 Summary of Literature Survey

1. In the first article, the author uses a regression model to distinguish between benign and malignant nodules.
2. In the second publication, the authors extract malware sample options and their connections to various malware families using language processing techniques.
3. In the 1/3 article, an associate diploma IDS is recommended for figuring out various IoT network attacks. To extract pertinent options for her IoT network, we blend gray Wolf Optimization (GWO) and Particle Swarm Optimization (PSO). To obtain excessive attack detection accuracy, the squared measurements of the retrieved options are fed right into a random wooded area (RF) classifier.

4. As part of a fourth paper, the authors have an associate's degree in anomaly detection in a deep learning model powered by Web Industrial Management Systems (IICS) that learns and validates exploit information gathered from TCP/IP packets.
5. In planned work, a NIDS model based primarily on cubic centimeters will be developed and updated as new attacks emerge that exploit progressive learning techniques. A lightweight predictive weight model works fast even in his resource-sensitive IoT network.

## 4 Existing Systems

The current signature-based NIDS only identify assaults that are entirely dependent on pre-existing facts or patterns [3]. The current generation of machine learning (ML) based NIDS can only recognize patterns or signatures that have already been seen in the trained dataset. To make sure the accuracy of NIDS attack detection, current researchers need to devise techniques for fusing sparkling statistics with modern-day facts sets. Zhang et al. counseled a safe and reliable modern aid vector gadget (SVM) classifier. [7] assist vector device classifier perfect set for exploitation coaching. An existing linear classifier with associated levels may be changed with the aid of omitting the present statistics and including new records. System IoT gadget exploitation this is expected to focus on the paper's concept for brand spanking new system learning-based on-line network intrusion detection methods, by means of inspecting a massive series of state-of-the-art facts-driven algorithms, system mastering makes it less complicated to find patterns buried in IoT understanding. Computerized systems that practice statistically decided movements in crucial operations are used in the machine getting to know to complement or update guide approaches. I attempt to encompass this NIDS model, which is more often than not focused on cubic centimeters, into our counseled device at the side of an online revolutionary learning approach for IoT tool safety. Additionally, we are working to develop a minimal detection model for resource-constrained IoT networks. We recommend applying a defensive framework in a very large-scale IoT environment in order to evaluate its performance. Within this purposeful effort, applying cubic centimeters and online machine learning to develop a strong NIDS for resource-constrained IoT devices is fantastic. On-line cubic centimeters are used to accumulate greater knowledge and use it in every step has a tendency to produce the most effective predictor of destiny replace expertise, in contrast to the conventional batch gaining knowledge of technique, which quickly learns from the entire education expertise set to provide the most effective predictors (Figs. 1 and 2).

## 5 Overall Architecture

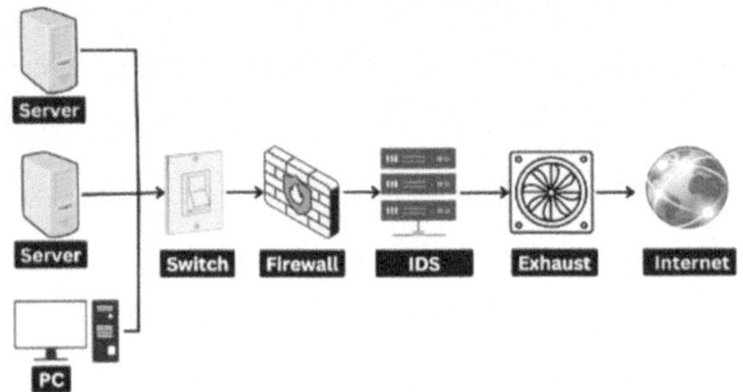


**Fig. 1.** Network Intrusion Detection System

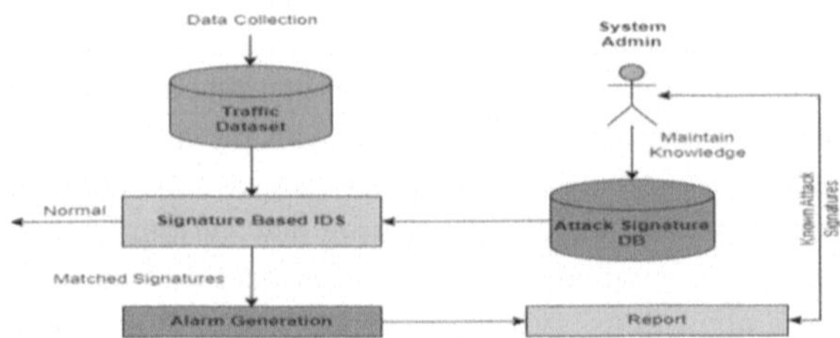


**Fig. 2.** Cloud Security

## 6 Methodology Algorithm

### 6.1 Guide Vector Gadget

The Support Vector Machine (SVM), one of the most flexible ML designs, is capable of performing both regression and categorization tasks. The SVM is one of the models that is used in ML research the most frequently. The SVM model separates a given dataset into several trainings in order to determine the desirable hyperplane. Using SVMs has the advantage of their propensity to perform well for high-dimensional input areas. The SVM model also allows for a variety of kernel residences to be selected throughout the selection process. One drawback is its typical need for careful tinkering, especially when the input dimensions are larger than the pattern size. Logistic regression (LR), sometimes known as "regression," is a machine learning (ML) method that is frequently employed for binary classification tasks. The one-vs-rest approach is utilized in the mastering set of rules when using LR for multiclass tasks. Although LR can be done correctly, the mastering algorithm uses the one-vs-relaxation approach for multiclass classification duties. A linear ML model in an LR form uses a sigmoid characteristic or one of its versions. The output is compressed between [0, 1] using this procedure. The use of outputs that are near to one determines the likelihood of a specific class. The process is explained through,

1. Linear version: y=b+w1x1+w2x2+· · · +wnxn
2. Sigmoid components: σ(okay) = 11 + e − k
3. Sigmoid formulation: σ(ok) = 11 + e − okay
4. Logistic regression: σ(y) = 11 + e − y
   inside the above method, y is treated by means of σ.

## 6.2 Artificial Neural Networks

Artificial neural networks (ANNs), a machine learning (ML) method, replicate the inner workings of real neurons. The most crucial component of an ANN is either a node or a neuron. An ANN called a perceptron contains an input layer that is processed through hidden layers to produce an output. ANNs with several hidden (middle) layers (MLPs) are sometimes explained by feedforward ANNs or multilayer perceptrons. In an artificial neural network (ANN), neurons are activated by formulas such as sigmoid (ok) = 11 + eokay (okay) = eleven + eokay, adjusted linear unit (ReLU) f(okay) = max (0, ok) f(ok), hyperbolic Tangent tanh(t) = 1e2t1 + e2t, or combinations of those characteristics.

## 6.3 Decision Tree

Decision trees (DT) are supervised machine learning techniques that are applied to classification and regression problems. A model that resembles a tree is the result of the decision-making process. The user can better grasp her DT because of this. A variety of ML tools can be used to view the created tree.

## 6.4 K-Nearest Neighbours

The machine learning (ML) method K-Nearest Neighbours (KNN) is capable of performing both supervised and unsupervised operations. For instance, KNN is the basis for many of the modern clustering methods. The supervised ML version of the KNN method was employed in this work. The basis of this method (EM) is the standard Euclidean metric. The EM is the distance between two points (instances) in space.

If p and q represent two instances of the Euclidean area Z, then the EM between q and q is calculated as p, q = i = 1r (qk, qk)2p, q = i = 1r (qk, qk). The gap Z proposes has the maximum variety of cases in it, which is 2r. The KNN technique computes the EM separating an instance r0r0 and its k nearest times in Z, and the label (class) of r0r0 is its k it's miles selected with reference to the splendor of linked neighboring times.

## 6.5 UNSW-NB15 Dataset

The UNSW-NB15 attack records are used within the experimental technique. The UNSW-NB15 has forty-two capabilities in a tidy arrangement, as indicated inside the desk. 39 of the 42 developments are numerical in nature, with three examples being non-numeric (categorical) capabilities. Two number one information sets make up UNSW-NB15: UNSW-NB15-take a look at (100%), which is used to test the skilled models, and UNSW-NB15-educate, which is used to train a ramification of fashions. In our look, UNSW- NB15-train became separated into two components: UNSW-NB15-educate-1 for education (which accounted for seventy-five% of the complete education set) and UNSW-NB15-VAL (25% of the schooling set and the entire training set for validation previous to checking out). As a plausibility check for the schooling procedure's findings, this second partition is employed. it is vital no longer to train the model at the test set or rankings whilst using this technique. Statistics leaking is a phenomenon, while the version sees information, it should not all through the education phase, there's a bias within the final version. On previously unexplored facts, this results in super model performance. The network attack classes listed beneath are found in UNSW-NB15: Backdoor, Shellcode, Reconnaissance, Trojan horse, Fuzzer, DoS, widely wide-spread, evaluation, Shellcode, and make the most. The table additionally gives information on every assault type within the information subset, which includes its distribution of values.

## 7 Implementation and Modules

Using machine learning techniques, there are numerous important processes involved in implementing an online network intrusion detection system (NIDS) for IoT architectures. The general implementation procedure is described below:

### 7.1 Data Collection and Preprocessing

- Gather network traffic data from IoT devices or simulated IoT environments.
- Preprocess the data to handle missing values, normalize features, and remove noise or irrelevant information.

### 7.2 Feature Engineering

### 7.3 Model Selection

- Select the best machine learning methods for detecting online intrusions. Decision Trees, Random Forests, Support Vector Machines (SVM), and Deep Learning models like Convolutional Neural Networks (CNN) and Recurrent Neural Networks (RNN) are some of the options that are frequently used.

### 7.4 Model Training

- Divide the preprocessed data into training and validation sets to prepare the training dataset.
- Use the training data to train the chosen machine learning model.

### 7.5 Online Learning Implementation

- Develop mechanisms to facilitate online learning, allowing the model to update and adapt to new data in real-time.
- Implement algorithms for model retraining, which can be triggered periodically or when significant changes in the IoT network occur.

### 7.6 Intrusion Detection and Alerting

- Deploy the trained model to analyze incoming network traffic in real-time.
- Set appropriate thresholds to classify normal and anomalous behavior.
- Trigger alerts or notifications when suspicious activities or intrusions are detected.

### 7.7 Resource Optimization

- Implement techniques to optimize the NIDS's resource usage, considering the limitations of IoT devices such as memory, computation, and power constraints.
- Techniques may include model quantization, pruning, or using lightweight architectures suitable for deployment on IoT devices.

### 7.8 Performance Evaluation

Use common assessment measures to evaluate the effectiveness of the NIDS, such as accuracy, precision, recall, and f1-score.
Examine the model's performance using several test datasets that contain both well-known and new assaults.

### 7.9 Incremental Upgrades

- Establish a framework for model upgrades and enhancements to keep the NIDS up-to-date with evolving cyber threats.
- Implement versioning and version control to manage and track model improvements over time.

### 7.10 Real-World Testing

- Conduct extensive real-world testing in diverse IoT environments to validate the NIDS's effectiveness and adaptability.
- Fine-tune the system based on feedback and experiences from practical implementations.

### 7.11 Security and Privacy Considerations

- Address security and privacy concerns related to the NIDS itself, as it plays a critical role in protecting the IoT structure.
- Implement encryption and access controls to secure the NIDS from unauthorized access.

### 7.12 Documentation and Deployment

- Create a thorough documentation of the implementation, including the setup procedures, architecture, and design.
- Plan the deployment plan while taking the needs for maintenance and scalability into account.

These procedures will help you build a powerful and effective Online Network Intrusion Detection System that uses machine learning to successfully protect IoT infrastructure from growing cyber threats. To maintain the security and integrity of IoT networks in the face of constantly shifting threat landscapes, keep in mind that continuous monitoring and adaptation will be necessary.

# 8 Record Preparation

A network assault document is united states of america-NB15. There are 9 specific types of assaults in it, which includes DoS, Trojan horse, Backdoor, Fuzzer, analysis, regular, make the most, Shellcode, and every day. Raw network packets are blanketed in statistics. With various types of assault and regular, the schooling set has one hundred seventy-five,541 information, whilst the check set has eighty-two, 332 facts (Figs. 3, 4, 5, 6, 7, 8, 9, 10, 11, 12, 13, 14, 15, 16, 17, 18, 19, 20, 21, 22, 23, 24, 25, 26, 27, 28, 29, 30 and 31).

```
data.head(n=5)
```

| | id | dur | proto | service | state | spkts | dpkts | sbytes | dbytes | rate | ... | ct_dst_sport_ltm | ct_dst_src_ltm | is_ftp_login | ct_ftp_cmd | ct_flw_http_mthd | ct_src_ltm | ct_srv_dst | is_sm_i |
|---|---|---|---|---|---|---|---|---|---|---|---|---|---|---|---|---|---|---|---|
| 0 | 1 | 0.121478 | tcp | - | FIN | 6 | 4 | 258 | 172 | 74.087490 | ... | 1 | 1 | 0 | 0 | 0 | 1 | 1 | |
| 1 | 2 | 0.649902 | tcp | - | FIN | 14 | 38 | 734 | 42014 | 78.473372 | ... | 1 | 2 | 0 | 0 | 0 | 1 | 6 | |
| 2 | 3 | 1.623129 | tcp | - | FIN | 8 | 16 | 364 | 13186 | 14.170161 | ... | 1 | 3 | 0 | 0 | 0 | 2 | 6 | |
| 3 | 4 | 1.681642 | tcp | ftp | FIN | 12 | 12 | 628 | 770 | 13.677108 | ... | 1 | 3 | 1 | 1 | 0 | 2 | 1 | |
| 4 | 5 | 0.449454 | tcp | - | FIN | 10 | 6 | 534 | 268 | 33.373826 | ... | 1 | 40 | 0 | 0 | 0 | 2 | 39 | |

5 rows × 45 columns

**Fig. 3.** UNSWNB-15 Dataset details in 5 rows × 45 columns

```
Generic           39496
Normal            19488
Exploits          16187
DoS                1791
Fuzzers            1731
Reconnaissance     1703
Analysis            564
Worms               114
Backdoor             99
Name: attack_cat, dtype: int64
```

**Fig. 4.** Attack Categories

```
features.head()
```

| | No. | Name | Type | Description |
|---|---|---|---|---|
| 0 | 1 | srcip | nominal | Source IP address |
| 1 | 2 | sport | integer | Source port number |
| 2 | 3 | dstip | nominal | Destination IP address |
| 3 | 4 | dsport | integer | Destination port number |
| 4 | 5 | proto | nominal | Transaction protocol |

**Fig. 5.** Features Head details

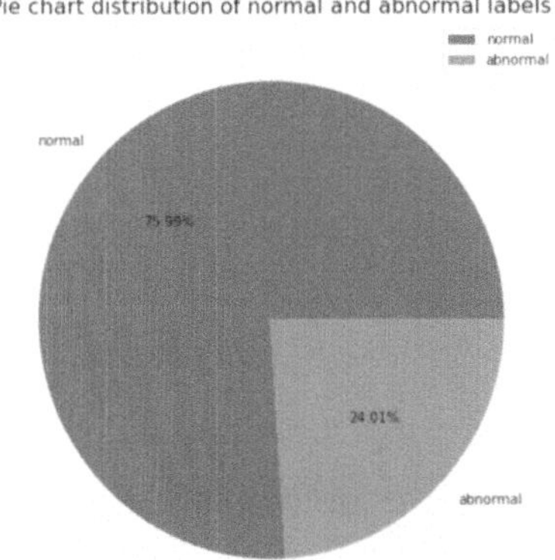


**Fig. 6.** Statistical Analysis in Pie chart distribution of Normal and Abnormal Labels

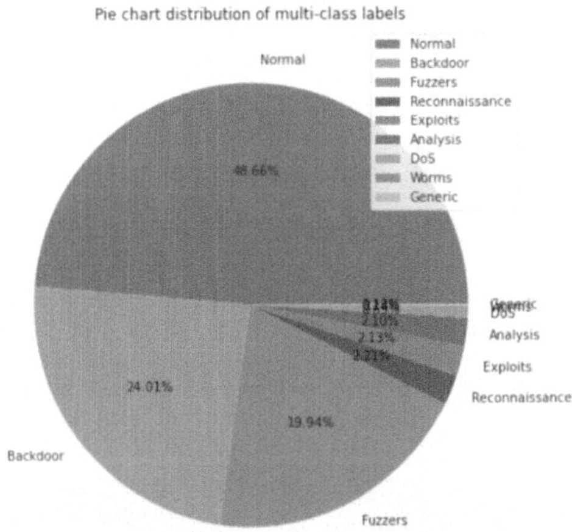


**Fig. 7.** Statistical Analysis in Pie chart distribution of Multi-class Labels

```
# data before normalization
data.head()
```

| | id | dur | spkts | dpkts | sbytes | dbytes | rate | sttl | dttl | sload | ... | service_radius | service_smtp | service_snmp | service_ssh | service_ssl | state_CON | state_FIN | state_INT |
|---|---|---|---|---|---|---|---|---|---|---|---|---|---|---|---|---|---|---|---|
| **3** | 4 | 1.681642 | 12 | 12 | 628 | 770 | 13.677108 | 62 | 252 | 2.740179e+03 | ... | 0 | 0 | 0 | 0 | 0 | 0 | 1 | 0 |
| **11** | 12 | 2.093085 | 62 | 28 | 56329 | 2212 | 42.520967 | 62 | 252 | 2.118251e+05 | ... | 0 | 1 | 0 | 0 | 0 | 0 | 1 | 0 |
| **15** | 16 | 0.000002 | 2 | 0 | 138 | 0 | 500000.001300 | 254 | 0 | 2.760000e+08 | ... | 0 | 0 | 1 | 0 | 0 | 0 | 0 | 1 |
| **17** | 18 | 0.393556 | 10 | 8 | 860 | 1096 | 43.195886 | 62 | 252 | 1.573347e+04 | ... | 0 | 0 | 0 | 0 | 0 | 0 | 1 | 0 |
| **21** | 22 | 0.338017 | 10 | 6 | 998 | 268 | 44.376468 | 254 | 252 | 2.127704e+04 | ... | 0 | 0 | 0 | 0 | 0 | 0 | 1 | 0 |

5 rows × 61 columns

**Fig. 8.** Data before normalization

```
# data after normalization
data.head()
```

| | id | dur | spkts | dpkts | sbytes | dbytes | rate | sttl | dttl | sload | ... | service_radius | service_smtp | service_snmp | service_ssh | service_ssl | state_CON | state_FIN | st |
|---|---|---|---|---|---|---|---|---|---|---|---|---|---|---|---|---|---|---|---|
| **3** | 4 | 2.802780e-02 | 0.001144 | 0.001093 | 0.000041 | 0.000053 | 0.000014 | 0.138393 | 0.992126 | 0.000001 | ... | 0.0 | 0.0 | 0.0 | 0.0 | 0.0 | 0.0 | 1.0 | |
| **11** | 12 | 3.488529e-02 | 0.006344 | 0.002551 | 0.004337 | 0.000151 | 0.000043 | 0.138393 | 0.992126 | 0.000092 | ... | 0.0 | 1.0 | 0.0 | 0.0 | 0.0 | 0.0 | 1.0 | |
| **15** | 16 | 1.666693e-08 | 0.000104 | 0.000000 | 0.000003 | 0.000000 | 0.500000 | 0.995536 | 0.000000 | 0.119792 | ... | 0.0 | 0.0 | 1.0 | 0.0 | 0.0 | 0.0 | 0.0 | |
| **17** | 18 | 6.559354e-03 | 0.000936 | 0.000729 | 0.000059 | 0.000075 | 0.000043 | 0.138393 | 0.992126 | 0.000007 | ... | 0.0 | 0.0 | 0.0 | 0.0 | 0.0 | 0.0 | 1.0 | |
| **21** | 22 | 5.633690e-03 | 0.000936 | 0.000547 | 0.000069 | 0.000018 | 0.000044 | 0.995536 | 0.992126 | 0.000009 | ... | 0.0 | 0.0 | 0.0 | 0.0 | 0.0 | 0.0 | 1.0 | |

5 rows × 61 columns

**Fig. 9.** Data after normalization

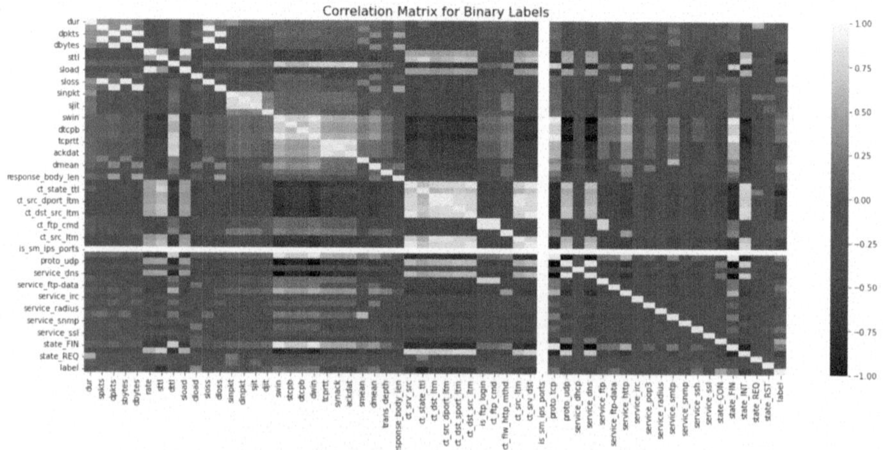


**Fig. 10.** Correlation Matrix for Binary Labels

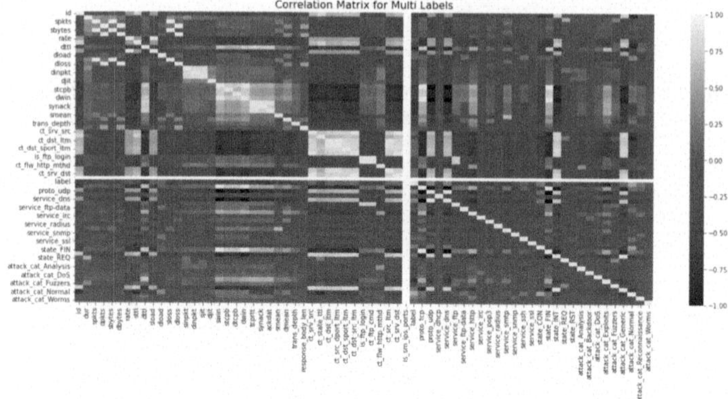


**Fig. 11.** Correlation Matrix for Multi-class Labels

```
state_CON               0.302853
attack_cat_Analysis     0.326209
attack_cat_DoS          0.339669
state_FIN               0.361750
proto_udp               0.364393
swin                    0.364393
dwin                    0.364393
proto_tcp               0.364393
service_dns             0.365346
synack                  0.524027
ackdat                  0.570098
tcprtt                  0.570205
attack_cat_Normal       0.570858
dttl                    0.646589
attack_cat_Exploits     0.719733
label                   1.000000
Name: label, dtype: float64
```

**Fig. 12.** Finding the attributes which have more than 0.3 correlation with encoded attack label attribute

| | rate | sttl | sload | dload | ct_srv_src | ct_state_ttl | ct_dst_ltm | ct_src_dport_ltm | ct_dst_sport_ltm | ct_dst_src_ltm | ct_src_ltm | ct_srv_dst | state_CON | state_INT | label |
|---|---|---|---|---|---|---|---|---|---|---|---|---|---|---|---|
| **3** | 0.000014 | 0.138393 | 0.000001 | 0.000150 | 0.000000 | 0.166667 | 0.02 | 0.00 | 0.000000 | 0.039216 | 0.016949 | 0.000000 | 0.0 | 0.0 | 1 |
| **11** | 0.000043 | 0.138393 | 0.000092 | 0.000364 | 0.019608 | 0.166667 | 0.00 | 0.00 | 0.000000 | 0.019608 | 0.000000 | 0.000000 | 0.0 | 0.0 | 1 |
| **15** | 0.500000 | 0.995536 | 0.119792 | 0.000000 | 0.000000 | 0.333333 | 0.00 | 0.00 | 0.000000 | 0.058824 | 0.016949 | 0.000000 | 0.0 | 1.0 | 1 |
| **17** | 0.000043 | 0.138393 | 0.000007 | 0.000869 | 0.039216 | 0.166667 | 0.00 | 0.00 | 0.000000 | 0.019608 | 0.000000 | 0.039216 | 0.0 | 0.0 | 1 |
| **21** | 0.000044 | 0.995536 | 0.000009 | 0.000236 | 0.098039 | 0.166667 | 0.00 | 0.00 | 0.000000 | 0.000000 | 0.016949 | 0.039216 | 0.0 | 0.0 | 1 |
| **...** | ... | ... | ... | ... | ... | ... | ... | ... | ... | ... | ... | ... | ... | ... | ... |
| **175335** | 0.166667 | 0.995536 | 0.032986 | 0.000000 | 0.862745 | 0.333333 | 0.64 | 0.64 | 0.533333 | 0.862745 | 0.542373 | 0.862745 | 0.0 | 1.0 | 0 |
| **175336** | 0.111111 | 0.995536 | 0.021991 | 0.000000 | 0.450980 | 0.333333 | 0.46 | 0.46 | 0.400000 | 0.450980 | 0.389831 | 0.450980 | 0.0 | 1.0 | 0 |
| **175338** | 0.111111 | 0.995536 | 0.021991 | 0.000000 | 0.215686 | 0.333333 | 0.04 | 0.04 | 0.066667 | 0.235294 | 0.033898 | 0.215686 | 0.0 | 1.0 | 0 |
| **175339** | 0.111111 | 0.995536 | 0.021991 | 0.000000 | 0.568627 | 0.333333 | 0.58 | 0.58 | 0.433333 | 0.568627 | 0.491525 | 0.568627 | 0.0 | 1.0 | 0 |
| **175340** | 0.111111 | 0.995536 | 0.021991 | 0.000000 | 0.568627 | 0.333333 | 0.58 | 0.58 | 0.500000 | 0.568627 | 0.491525 | 0.568627 | 0.0 | 1.0 | 0 |

81173 rows × 15 columns

**Fig. 13.** Binary labelled Dataset

```
state_CON               0.302853
attack_cat_Analysis     0.326209
attack_cat_DoS          0.339669
state_FIN               0.361750
proto_udp               0.364393
swin                    0.364393
dwin                    0.364393
proto_tcp               0.364393
service_dns             0.365346
synack                  0.524027
ackdat                  0.570098
tcprtt                  0.570205
attack_cat_Normal       0.570858
dttl                    0.646589
attack_cat_Exploits     0.719733
label                   1.000000
Name: label, dtype: float64
```

**Fig. 14.** Finding the attributes which have more than 0.3 correlation with encoded attack label attribute

```
print(cls_report)

              precision    recall  f1-score   support

    abnormal       0.97      1.00      0.99     12326
      normal       0.99      0.91      0.95      3909

    accuracy                           0.98     16235
   macro avg       0.98      0.96      0.97     16235
weighted avg       0.98      0.98      0.98     16235
```

**Fig. 15.** Classification Report

| | Actual | Predicted |
|---|---|---|
| **159889** | 0 | 0 |
| **125506** | 0 | 0 |
| **158979** | 0 | 0 |
| **8148** | 1 | 1 |
| **39918** | 1 | 1 |
| **...** | ... | ... |
| **68470** | 0 | 0 |
| **159674** | 0 | 0 |
| **171321** | 0 | 0 |
| **125777** | 0 | 0 |
| **39178** | 1 | 1 |

16235 rows × 2 columns

**Fig. 16.** Details of Actual and Predicted Data after Testing

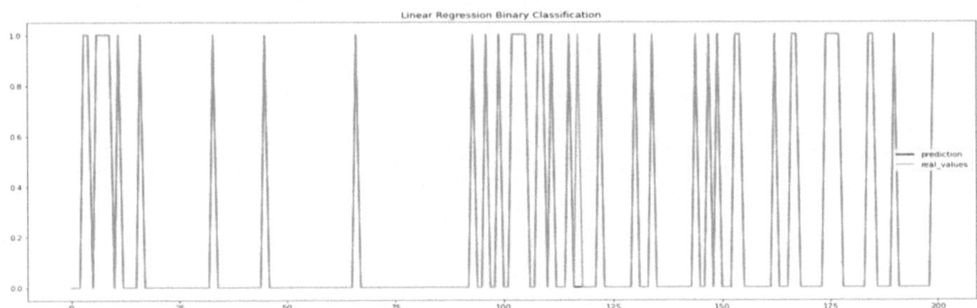


**Fig. 17.** Linear Regression Binary Classification with Prediction and Real Values

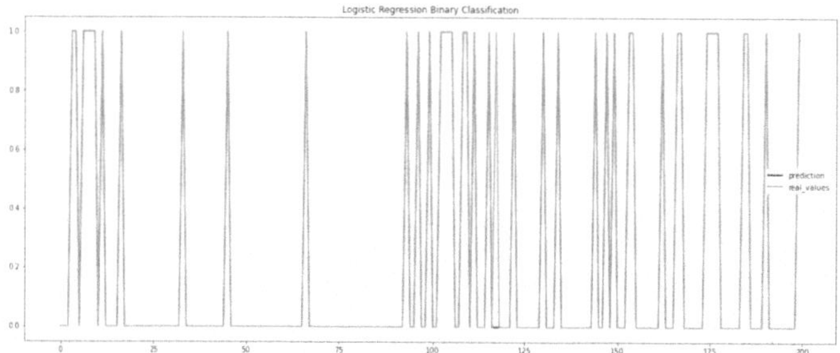


**Fig. 18.** Logistic Regression Binary Classification with Prediction and Real Values

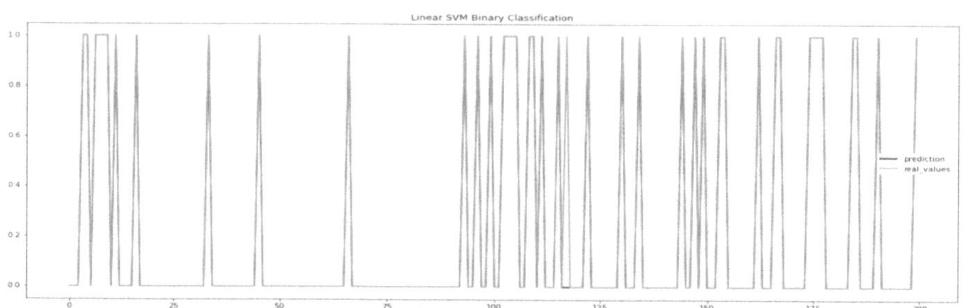


**Fig. 19.** Linear SVM Binary Classification with Prediction and Real Values

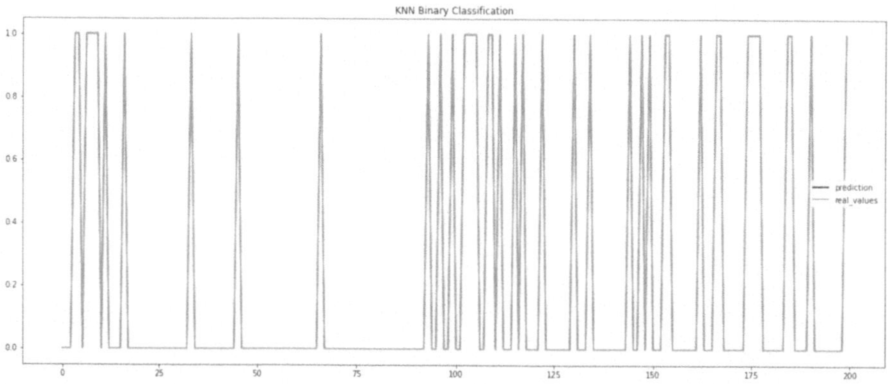


**Fig. 20.** KNN Binary Classification with Prediction and Real Values

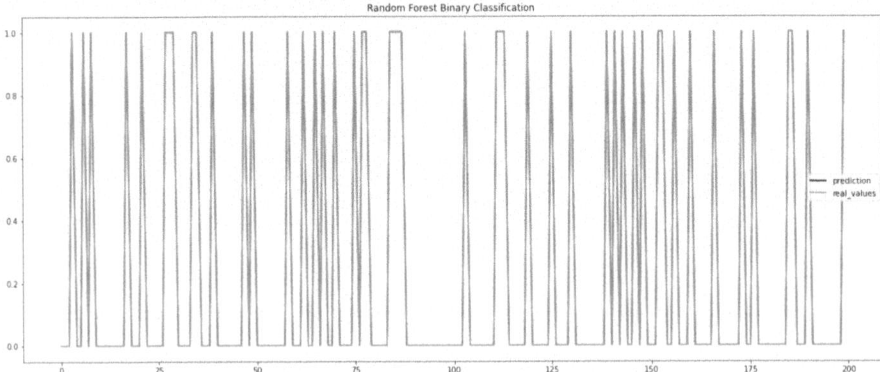


**Fig. 21.** Random Forest Binary Classification with Prediction and Real Values

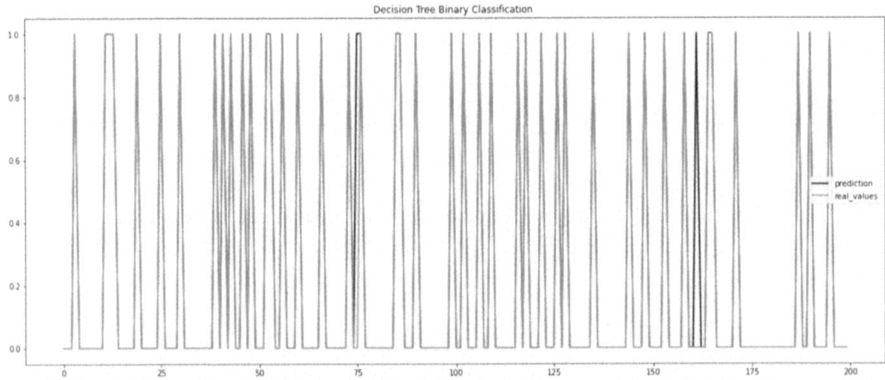


**Fig. 22.** Decision Tree Binary Classification with Prediction and Real Values

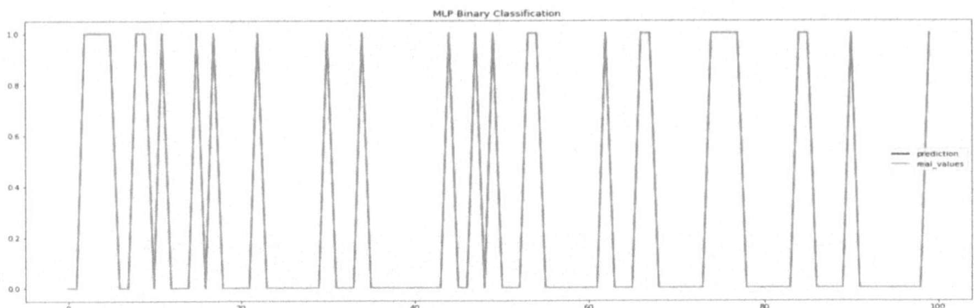


**Fig. 23.** MLP Binary Classification with Prediction and Real Values

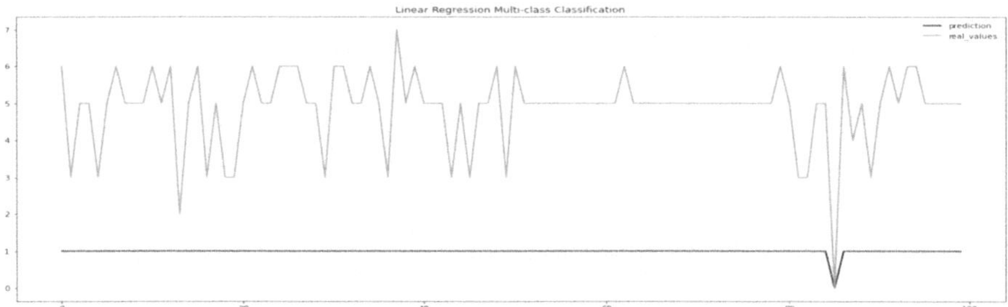


**Fig. 24.** Linear Regression Multi-class Classification with Prediction and Real Values

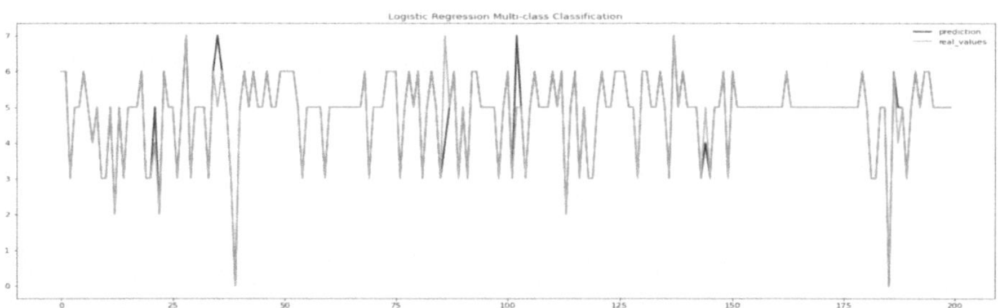


**Fig. 25.** Logistic Regression Multi-class Classification with Prediction and Real Values

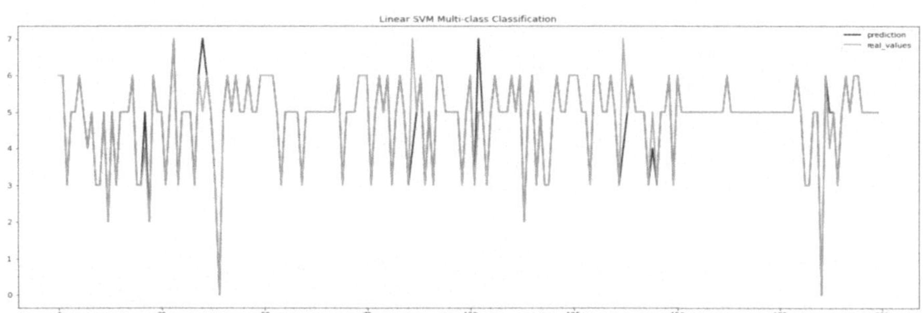


**Fig. 26.** Linear SVM Multi-class Classification with Prediction and Real Values

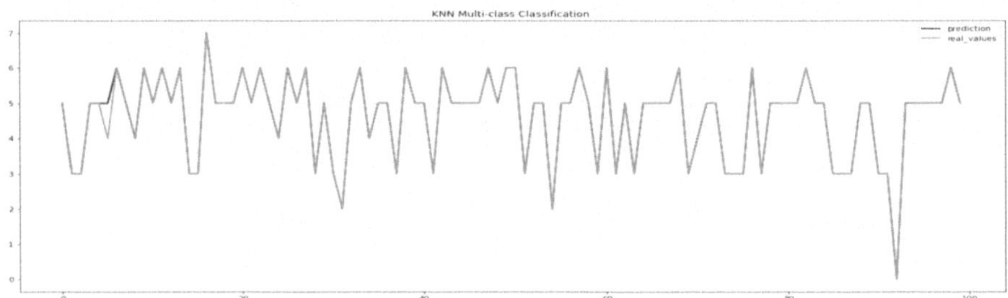


**Fig. 27.** KNN Multi-class Classification with Prediction and Real Values

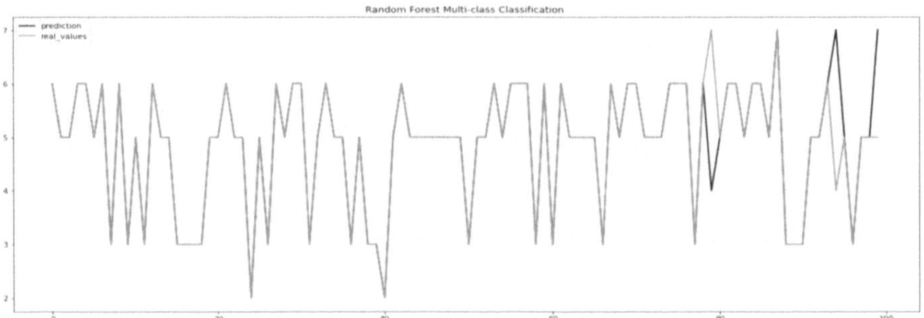


**Fig. 28.** Random Forest Multi-class Classification with Prediction and Real Values

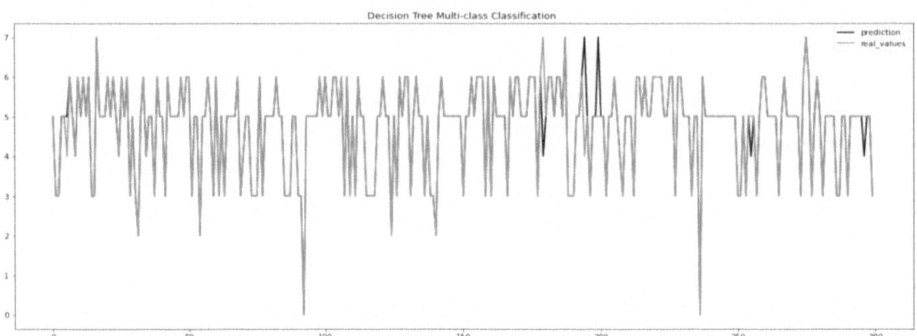


**Fig. 29.** Decision Tree Multi-class Classification with Prediction and Real Values

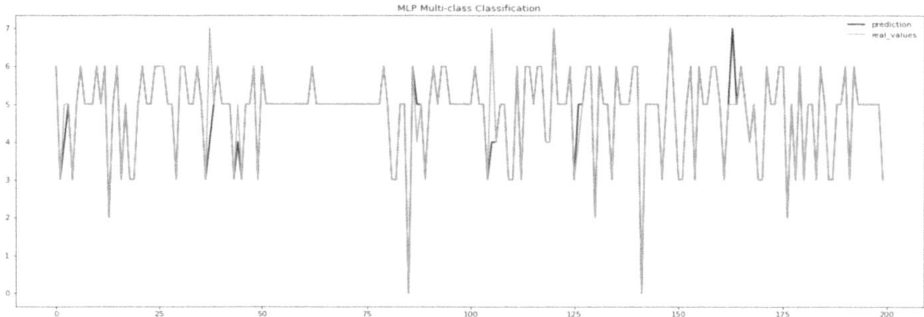


**Fig. 30.** MLP Multi-class Classification with Prediction and Real Values

```
pkl_filename = "/content/mlp_multi.pkl"
if (not path.isfile(pkl_filename)):
  # saving the trained model to disk
  with open(pkl_filename, 'wb') as file:
    pickle.dump(mlp_multi, file)
  print("Saved model to disk")
else:
  print("Model already saved")

Saved model to disk
```

**Fig. 31.** Saving the trained Model to disk

## 9 Conclusions and Future Scope

In conclusion, our research has developed an online network intrusion detection system (NIDS) based on machine learning that is specially designed for IoT architecture. Our NIDS demonstrates the capacity to efficiently protect IoT devices from ever emerging cyberattack techniques by utilizing the power of machine learning algorithms. Our suggested NIDS's potential for online learning is one of its main benefits. The model can adapt and learn from fresh data thanks to this capability, which makes it suitable for dynamic IoT contexts. Despite the limited memory, computation capacity, and electricity that IoT devices have, our model has shown to be effective at identifying intrusions without making performance promises. This efficiency is crucial to prevent obsolescence and ensure long-term applicability in real- world IoT scenarios. Moreover, the incremental upgrade feature of our NIDS stands as a promising avenue for future enhancement. As the threat landscape evolves, our model can be updated and refined to keep up with emerging cyber threats. This adaptability makes our NIDS a sustainable and future-proof solution for protecting IoT networks and devices from potential security breaches.

Looking ahead, we envision further research and development in optimizing the NIDS for even greater resource efficiency, allowing it to run seamlessly on resource-constrained IoT devices. Additionally, exploring the integration of edge computing techniques could alleviate the computational burden on IoT devices while maintaining high-level security. Further- more, conducting large-scale deployment and field tests of our NIDS in diverse IoT environments will provide valuable insights into its real-world effectiveness and robustness. The feedback obtained from practical implementations will pave the way for refining the model and addressing any challenges that may arise in real-world applications. Ultimately, with its proactive threat detection, real-time response capabilities, adaptability, and scalability, our Machine Learning-based NIDS offers a potent solution for safeguarding IoT networks and devices against evolving cyber threats. We believe that this research lays a solid foundation for securing IoT architectures and contributes to the ongoing efforts to ensure the integrity and security of the expanding IoT ecosystem.

## References

1. Morgera, S., Amouri, D., Alaparthy, V.T.: A machine learning based intrusion detection system for mobile internet of things. Sensors **20**(2), 461 (2020)
2. Zhang, W., et al.: An evolutionary study of IoT malware. IEEE Internet of Things Journal (2021)
3. Keserwani, P.K., Govil, M.C., Govil, P., Pilli, E.S.: A smart anomaly-based intrusion detection system for the internet of things (IoT) network using gwo–pso–rf model. J. Reliable Intel. Environ. **7**(1), 3–21 (2021)
4. Muna, A.H., Sitnikova, E., Moustafa, N.: Identification of malicious activities in industrial internet of things based on deep learning models. J. Info. Sec. Appl. **41**, 1–11 (2018)
5. Zhang, D., Xiang, W., Cao, Q., Chen, S.: Application of incremental support vector regression based on optimal training subset and improved particle swarm optimization algorithm in real-time sensor fault diagnosis. Appl. Intell. **51**(6), 3323–3338 (2020). https://doi.org/10.1007/s10489-020-01916-9
6. Gupta, L., Khan, K.M., Zolanvari, M., Teixeira, M.A., Jain, R.: Machine learning-based network vulnerability analysis of industrial internet of things. IEEE Internet of Things Journal **6**(4), 6822–6834 (2019)
7. Moustafa, N., Muna, A.H., Sitnikova, E.: Identification of malicious activities in industrial internet of things based on deep learning models. J. Info. Sec. Appl. **41**, 1–11 (2018)
8. Pandya, S., Mishra, N.: Internet of things applications, security challenges, attacks, intrusion detection, and future visions: a systematic review. IEEE Access (2021)
9. UNSW-NB15Dataset https://www.kaggle.com/search?q=UNSW-NB15

# An Analysis of Machine Learning Tools and Algorithms

R. Sindhuja[1](✉), R. Nanmaran[2], and A. Madhan Kumar[3]

[1] Department of Medical Electronics, Saveetha Engineering College, Chennai, India
Sindhumekna89@gmail.com

[2] Department of ECE, Veltech Rangarajan Dr., Sagunthala R&D Institute of Science and Technology, Chennai, India

[3] Department of Mechanical Engineering, Saveetha Engineering College, Chennai, India
madhankumara@saveetha.ac.in

**Abstract.** Artificial intelligence and Machine learning have become emerging growth in recent years. The concept of Data points, Features, Feature spaces and Similarity measures falls into two main areas of machine learning namely supervised learning and Unsupervised Learning. Machine learning algorithms deals with the data and datasets. The main difference between the supervised and unsupervised learning is to predict the labelled and unlabeled data.

**Keywords:** Artificial Intelligence · Machine learning · Supervised learning · Unsupervised learning · Reinforcement learning

## 1 Introduction

Machine learning uses various algorithms for building mathematical models for making predictions. Using historical data or information. With the help of sample historical data which is known as training data, machine learning algorithms build a mathematical model helps in making predictions or decisions. Machine learning brings computer science and statistics together for creating predictive models. The more the training data, the higher will be the performance. In the below Fig. 1 input data input data is given with the help of machine learning algorithms, by building logical models we can get the desired output data.

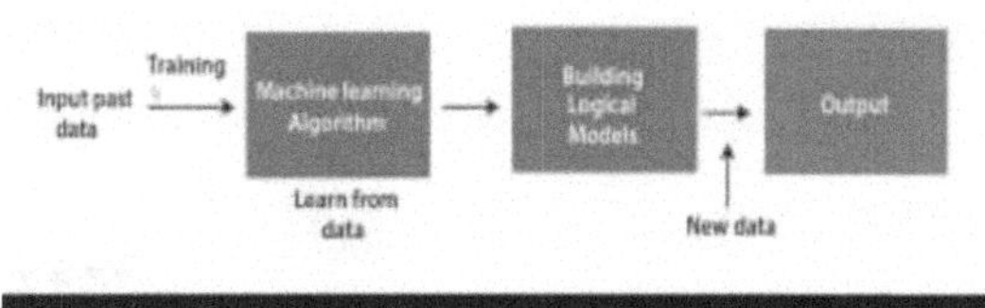


**Fig. 1.** Machine learning Framework


R. Geetha et al. (Eds.): AAIMB 2023, CCIS 2203, pp. 195–204, 2025.
https://doi.org/10.1007/978-3-031-73068-9_16

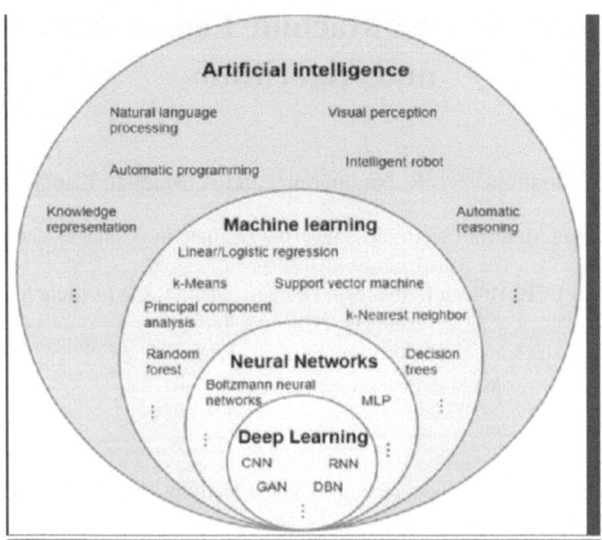


**Fig. 2.** Relationship between AI, ML, Neural Networks and Deep Learning

## 2 Relationship Between Artificial Intelligence, Machine Learning, Neural Network and Deep Learning

Artificial intelligence is the concept of creating smart intelligent machines. In the Fig. 2. Machine learning approach is the subset of Artificial intelligence which helps to build AI driven applications. Neural Networks is a subset of Machine learning which resembles our human brain, and it interconnects nodes or neurons to perform the task. Machine learning Algorithms parse data, learn from that data and make decisions based on what it has learned, whereas Deep learning can learn and make intelligent decisions on its own.

**Some of the Trendy Top 10 Artificial Technologies Are:**

1. Natural language Generation.
2. Natural language Understanding.
3. Speech Recognition.
4. Machine Learning.
5. Virtual Agents.
6. Expert Systems.
7. Decision Management.
8. Deep learning.
9. Robotic process Automation.
10. Text analytics.

## 3 Machine Learning Algorithms

We are using Machine learning in our daily life even without knowing it such as Google maps, Google Assistant, Alexa etc.

1. Image and speech Recognition
2. Medical Diagnosis
3. Statistical Arbitrage
4. Learning Association
5. Classification
6. Prediction
7. Extraction
8. Regression

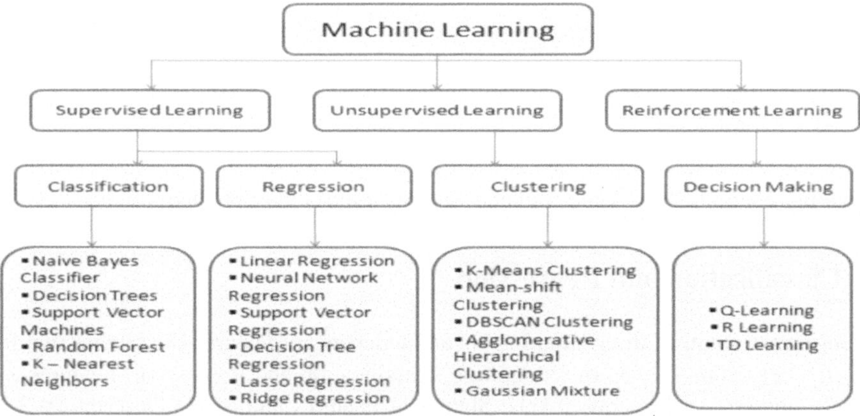


**Fig. 3.** Machine learning algorithm –A Simplifie viwe

Machine learning algorithms [2] are tools, not dogmas (a belief or set of beliefs that people are expected to accept as true without questioning). No free lunch theorem, it implies that there is no single best machine learning algorithm for predictive modelling problems such as classification and regression. We have to try simple classifiers first. Better to have smart features and simple classifiers than simple features and smart classifiers. Use increasingly powerful classifiers with more training data that is bias-variance trade off.

In the above Fig. 3, Machine learning algorithms is classified into

1. Supervised Learning
2. Unsupervised Learning
3. Re-enforcement learning techniques.

Let us elaborately see them in detail.

***Supervised Learning:***

During training the system creates a model using labelled data. Then the model is tested by providing a sample data to check whether it is predicting the exact output or not. Supervised Learning is again classified into

1. Classification
2. Regression.

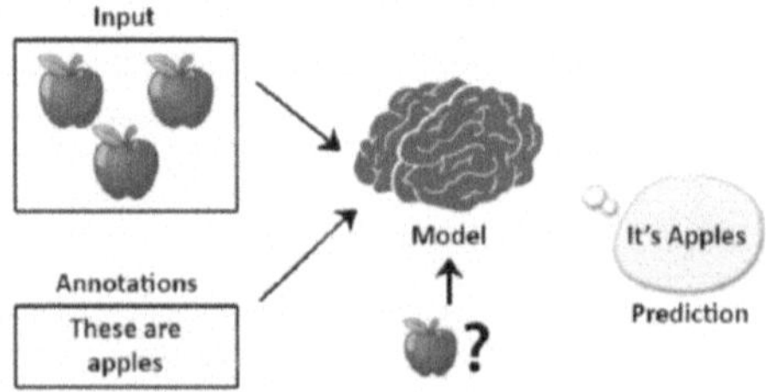


**Fig. 4.** Example of supervised learning in ML

In the above Fig. 4 input data is labelled as apple and then the model is tested by providing a sample data, Apple and checking with the correct prediction.

## 4 Classification and Regression

Choosing informative, discriminating [1] and independent features is a crucial step for effective algorithm in pattern recognition, classification and Regression. Features are usually numeric such as raw pixels, Statistical Features and Shape Descriptors. The classification algorithms are divided into Linear Models and Non-Linear Models.

**Linear Models:** Linear Regression and Support vector machines Non-Linear Models: Some of the Non-linear model algorithms are

1. K-Nearest Neighbor
2. Naive Baiyes
3. Decision Tree Classification
4. Random Forest Classification.

**Binary Classifier:** If the classification problem has only two possible outcomes, then it is called Binary classifier as shown in the Figure. For example, yes or no, male or female.

**Multi-class Classifier:** If the classification problem has more than two outcomes, then it is called multi-class classifier as shown in the figure. For example, Classifications of types of crops, classification of types of Music… In Regression algorithm [4], we predict the output for continuous values, but to predict the categorical values, we need classification algorithms. Classes can be called as target/labels or categories (Fig. 5).

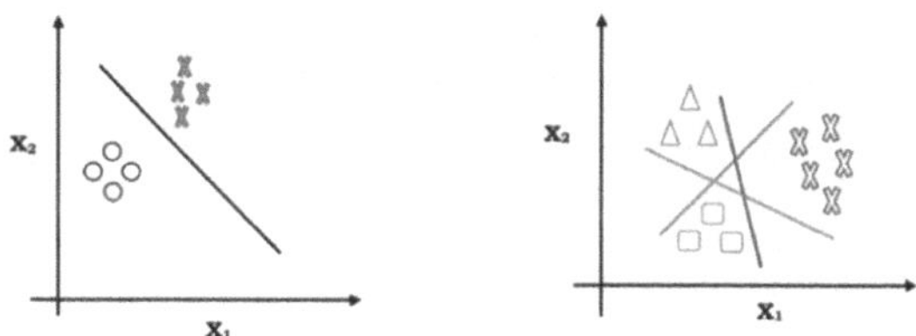


**Fig. 5.** Binary Classifier and Multi-class classifier

***Bias-Variance Trade off Model:***
Once our model is completed, it is necessary to evaluate its performance, either it is a classification or Regression Model. 1. Confusion Matrix: The confusion matrix provides us a matrix/table as output and describes the performance of the model. Accuracy = TP + TN/Total population 2. AUC-ROC Curve: AUC stands for Area under the curve and ROC for Receiver operating characteristics. To predict the multi-class classification model we use AUC-ROC Curve (Table 1).

**Table 1.** Confusion MATRIX

| | Actual positive | Actual Negative |
|---|---|---|
| Predicted positive | True positive | False positive |
| Predicted negative | False Negative | True Negative |

***Some Use Cases of Classification Algorithm Are:***

1. Email spam Detection
2. Speech Recognition
3. Identification of cancer tumor cells.
4. Drugs classification
5. Biometric Identification

***Unsupervised Learning:***
There are two types:

1. Clustering
2. Association

**Clustering:** It is an unsupervised learning method; hence no supervision is provided to the algorithm, and it deals with the unlabeled dataset. The objects [3] with the possible similarities remain in a group that has less or no similarities with another group. It does it by finding some similar patterns in the unlabeled dataset such as shape, size, color, behaviour etc. and divides them as per the presence and absence of those similar patterns.

**Types of clustering methods:**

**Hard Clustering:**
Data point belongs to only one group; example is K- Means Clustering Algorithm.

**Soft Clustering:**
Data point can belong to another group also, example Fuzzy C-means Clustering algorithm.

Fuzzy clustering is a type of soft method in which a data object may belong to more than group or cluster.

Each dataset has a set of membership coefficients, which depend on the degree of membership to be in a cluster.
*Use cases of clustering:*
Some most common uses of this technique are

1. Market segmentation
2. Statistical data analysis
3. Social network analysis
4. *Image segmentation etc.*

**Reinforcement Learning:**
Reinforcement learning is about taking suitable action to maximize reward in a particular situation.

## 5 Example

We have an agent and a reward, with many hurdles in between. The agent is supposed to find the best possible path to reach the reward. Practical applications of Reinforcement Learning: Reinforcement learning can be used in robotics for industrial automation. Reinforcement learning can be used in machine learning and Data processing. Reinforcement learning can be used to create training systems that provide custom instruction and materials according to the requirements of the students.

1. **Machine learning Algorithms in Real-Time Applications:**
   Machine learning Algorithms has become applicable in various real-world application domains, such as Cyber security systems, smart cities, health care, e-commerce, Agriculture and so on.
2. **Cybersecurity and Threat Intelligence:**
   Association rule [5] plays a major role to build a rule- based security system. In cyber security, Machine learning Algorithms plays a crucial role, by analyzes data to identify patterns and it detects malware in encrypted traffic, secures data in the cloud by clustering techniques.
3. **Smart Cities:**

Smart cities should have the following factors: Firstly, Humans should give feedback after interacting with systems. Sensors in smart cities generate data at a high rate. Machine learning Algorithms reviews all these data's and justifies with a high rate of accuracy. Data generated by smart city applications has some degree of Uncertainty and some amount of noise will be there, Machine learning algorithms lowers the noise, and degree of Error is minimized. Re-enforcement learning, and Semi-supervised learning can address these issues and it gives better solutions. In Big data streams, semi-supervised Machine learning approach assures the promising solutions for a noted data in big data streams. The Frame work of Smart cities is categorized by

1. Smart city by means of sensor
2. Smart city by using big data
3. High level Data intelligence & Analytics
4. Smart services.

Machine learning Algorithms plays a role in the above framework through cognitive services, service Discovery, Efficient sampling, Data Recycling and scalable models. All of these factors can be fulfilled by using Machine learning algorithms in smart city applications.

Machine Learning Applications in Some of the top companies, rendering services are: Here are some of the companies which uses Machine learning in an interesting ways. Some of these are listed below.

## 6 IBM'S Watson

IBM'S WATSON is known for Natural language question and answering system, in order to improve organizational performance, advanced oncology, Ameliorate drug discovery, are some of the services which WATSON provides by using Machine learning algorithms, which transforms the health sector.

## 7 GOOGLE and the Machine Learning Algorithms

Google services every day by Machine learning tools and algorithms. Image search, Translation tools, Gmail, Google maps are some of the services in which the machine learning systems are embedded. Tensor Flow is an open-source end to end platform, which is Machine Learning Model is developed. Tensor Flow was released by Google researchers, which uses PYTHON mostly as a Programming language. Tensor Flow has a high level of application programming interface, which is easy to create machine learning models in numerous languages. Machine learning Algorithm is used by GOOGLE translate which translates the content to the user according to the requirements.

4. Baidu-Future Voice Search:

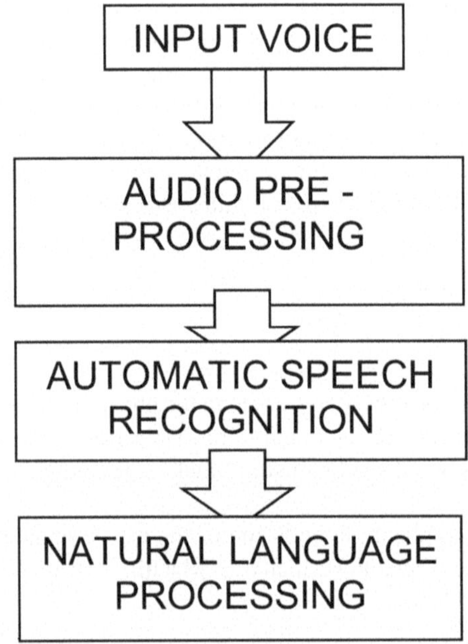


Baidu renders services such as voice search which is integrated into other products and services. Baidu in other terms we will call it as 'Google of china'. In the above mentioned figure the input voice is preprocessed, and then speech is recognized by Machine learning algorithms. MFCC(Mel frequency cepstral coefficient) used for feature extraction in machine learning algorithms, and it employs convolutional neural network, which converts spoken words into text, and it gives the results to the user which is generated by the search engine.

## 8 Microsoft

Machine learning Algorithms is utilized in Microsoft, and in turn it results in various products and services. Azure Machine learning is a service which Microsoft offers, for cloud –based platform for building, deploying and Managing Machine learning models. Microsoft office uses Machine learning algorithms for automatic email sorting, grammar suggestions, and designing power point recommendations.

## 9 Facebook

Facebook utilizes Machine Learning Algorithms for News Rank feeding, Facebook Ads, and it improves the user experience, and Deep Learning Algorithm is used for Face Recognition and Verification.

## 10 Agriculture

5. Machine learning algorithms plays a crucial role in agriculture by monitoring soil parameters such as moisture, minerals etc. By using decision tree approach. By using classification rule it identifies the disease recognition system, through sensor measurements. Farmers can install such systems on their smart devices to classify the anomalies fruits and crops.

## 11 Conclusion

- Machine learning algorithms are the powerful tool for evaluating data. These Algorithms detect patterns in data and learn from them, in order to make their own predictions.
- Machine learning is a branch of Artificial intelligence, processes information fast, takes decision to reaches its threshold. The main factors [6] in Machine Learning Algorithms are Detection, Prediction and Generation.
- The Degree of Error is lowered between the predicted output and the true output in Machine learning Algorithms.
- Some of the limitations Machine learning algorithms facing are inadequate training data, over fitting and Under fitting and lack of skilled resources. Machine learning Algorithms handles large number of complex data. Finally, Machine learning algorithms makes daily life easier and more accessible. It is one of the widely used and adopted language or technology in today's world.

## References

1. Batta Mahesh, J.: Machine Learning Algorithms - A Review. Int. J. Sci. Res. (IJSR) (2018). ISSN: 2319-7064

2. Sarker, I.H.: Machine learning Algorithms, real-world applications and research directions. SN Computer Science **2**(3), 160 (2021)
3. Anzai, Y.: Pattern recognition and machine learning. Elsevier (2012)
4. Mohammed, M., Khan, M.B., Mohammed, B.: Machine learning: algorithms and applications. CRC Press (2016)
5. Balducci, F., Impedovo, D., Pirlo, G.: Machine learning applications on agricultural datasets for smart farm enhancement. Machines **6**(3), 38 (2018)
6. Boukerche, A., Wang, J.: Machine learning-based traffic prediction models for intelligent transportation systems (2020)

# Ensemble Learning-Based Android Malware Detection

V. Priya(✉) and A. Sathya Sofia

PSNA College of Engineering and Technology, Dindigul, India
{vpriyamalar,sathyasofia}@psnacet.edu.in

**Abstract.** Android has retained its global popularity among cell phone users. Simultaneously, there has been a growth of malware targeting the platform, with larger current lines using unexpectedly cutting-edge detection evasion strategies. Options for timely 0-day detection are necessary as standard signature-based approaches become less effective in detecting unknown threats. This study contributes a strategy based on ensemble learning for detecting malicious apps on Android. To improve the accuracy of Android malware detection, it combines the advantages of hybrid analysis with the efficiency and performance of ensemble device research. Machine learning models are constructed using a large collection of malicious and safe software. The built model can predict all real-world test scenarios. The user has complete access to all program information on their devices. The proposed system employs a variety of machine learning algorithms for ensemble learning, including Decision Tree, Random Forest, and Androguard. The testing module is intended to ensure the efficiency and efficacy of the proposed system. The results show that the suggested method is extremely effective at identifying Android malware, with a detection accuracy of up to 100%.

**Keywords:** Random Forest · Kivy · Malware Detection · Andro guard · Android security · zero-day detection

## 1 Introduction

Malware is always developing danger to people, organizations, and governments in the era of improved mobile besides smart connectivity. As a result of the explosive growth of cell-based terminals, Android surpassed iOS as the most popular mobile computing platform. Android is a popular target for malware since it is open source and allows users to download apps from anywhere, including sources with a 30% chance of being malicious. The proliferation of harmful apps for Android devices is the issue that is being tackled by Android malware detection utilising ensemble learning in Python. Due to the increased sophistication of malware, the conventional methods of detection are not always adequate. The task is to create a reliable model that can identify malware with low rates of false positives and false negatives.

Malware designed specifically to infect Android-powered mobile devices is known as Android malware. It operates similarly to other types of malicious software seen on


R. Geetha et al. (Eds.): AAIMB 2023, CCIS 2203, pp. 205–219, 2025.
https://doi.org/10.1007/978-3-031-73068-9_17

personal PCs. Malicious software meant to harm a mobile device illegally is referred to as Android malware or mobile malware. Adware, backdoors, file infectors, PUAs, ransomware, riskware, scareware, spyware, and trojans are all examples of malware. The many forms of malware can be distinguished from one another by their individual properties. Similarly to people, Android malware proliferates. There are numerous malware families in each class of malicious software.

Because of the wide availability of open-source software, Android is projected to account for around 85% of the global smartphone market by 2018. Simultaneously, Cybercriminals are increasingly focusing their attention on Android devices. Recent research by 360 Internet Security Centre found that, during 2018, over 4.342 million unique samples of malware were spotted in malicious apps designed to perform various forms of mobile terminals, or roughly 12,000 unique samples every day. Smartphone users are particularly vulnerable to attacks that can steal personal data, send messages without permission, or lead them to harmful websites.

Organizations shouldn't overlook the risk posed by mobile devices, which might be especially susceptible to the kinds of harmful assaults that cybercriminals employ to steal sensitive information. Various forms of mobile malware exist, such as spyware, ransomware, and Trojan horses. Smishing (short message service phishing) is a relatively new method of attack in which hackers attempt to mislead a smartphone's user into installing malware by sending the victim an infected text message.

Malware of this kind poses a serious threat due to its ability to steal private information, halt operations, corrupt hardware, and expose sensitive data. To be safe, businesses must be aware of the threats posed by mobile malware and take appropriate precautions.

Malware is become more sophisticated and widespread to evade detection, with some malicious programs having more than 50 variants. This is a formidable task for security researchers. To keep up with the evolving sophistication of Android malware, we urgently need better detection methods. The research community has developed a large body of work on the topic of Android malware detection in an effort to address these problems. Due to the necessity of benign programs requiring the same sensitive rights as malware, research based on permissions and intentions of Android apps is prone to false positives.

Adware, backdoors, file infectors, PUAs, ransomware, riskware, scareware, spyware, trojans, trojan-sms, trojan-spies, trojan-bankers, and trojan-droppers are all common types of Android malware. This section describes the roles played by each kind of malware and identifies some prominent malware families within each category.

The goal of Android malware detection using ensemble learning with Python is to develop a reliable and precise model for detecting malicious applications on Android devices. The primary aim is to enhance the model's performance by combining multiple base classifiers in an ensemble model to achieve higher accuracy, precision, recall, and F1-score than a single classifier [24]. Therefore, the purpose of this study is to analyse and present a powerful technique for accurately detecting zero-day Android malware that makes use of static analysis and ensemble machine learning. This research proposes, develops, and explores a feature-based technique that use ensemble learning to tackle the Android malware problem, which differs from the approaches taken in the existing literature.

## 2 Background and Related Work

According to Statista data, Android has the most market share among smartphone operating systems at 71.85% as of July 2022. It's remarkable that mobile phone usage has advanced and innovated. The fear of downloading malware exists because of several free apps and games on Google Play. Android operating system [1] with its open-source nature attracts viruses considering many users unavoidably.

Mobile malware has emerged as a result of these factors. After being installed on a computer, malware begins running in the background, performing its malicious functions invisibly to the user. Given these facts, it is easy to see why Android malware poses such a serious threat to mobile platforms. To exploit the source code besides locating related components when the programme is being performed, there are a variety of Android malware detection methodologies available. Static and dynamic analysis have archaeologically been employed to expose and isolate malicious Android applications. The researchers practiced emulators like GenY motion or Android Studio and platforms like Kali Linux to examine vulnerabilities, ultimately finding the threat and achieving the goal [2]. The authors [3] recommended a dynamic ensemble model centered on fuzzy logic to distinguish malware accessible to the Android OS category. The proposed model incorporates a setup that merges the deliberative ability of a Mamdani-type fuzzy inference system (FIS) with the processing control of ML-based techniques.

AlexNet, VGG16, VGG19, GoogLeNet, and ResNet are just some of the transfer learning algorithms that are being looked into for malware photo classification in this research. In this setup, malware binaries are converted to grayscale photos, after which classification models based on transfer learning are applied. The basic goal of transfer learning is to reduce training time by making use of previously trained models. Furthermore, an efficient model for malware classification will be developed to achieve objective performance [5].

For predicting malware, ensemble approaches outperform conventional machine-learning systems in the work characterized by [4]. They used Light GBM to reduce the number of characteristics from 215 to 100 without sacrificing accuracy (they achieved 99.5%). Attackers are going for smartphones to get their hands on customers' private data for financial gain. In order to detect mobile malware, many methods [25] collect both static and real-time information from apps on various platforms. The authors [6] developed an ensemble learning-based virus detection technique for the Android platform. The results of the experiments showed conclusively that a malware detection strategy based on ensemble learning had a chance of success when applied to the Android system. Risky permissions are used to train a classifier and evaluate a mobile app's maliciousness in the ensemble learning-based malware classification method reported by Guan et al. [7]. Several experiments have been run on real-world Android app datasets, which contain both malicious and secure apps, to further evaluate the efficacy of the proposed malware detection technique. Experiments have proven that the proposed technique for detecting malware is effective.

Mahindru and Sangal [8] supplied us with a framework for selecting useful feature sets to improve malware detection algorithms' practical efficacy.Two additional performance criteria, accuracy and F-measure as a benchmark to detect malware from actual programmes, are considered in the evaluation of the proposed malware detection system. Using a behavioural characteristics dataset, the work by Urmila [9] aims to assess the impact of malware occurrence. LBGM, ENB0, and EEXR are used to include the features into the prediction model intended for malware detection. In that recommended system, there are four phases: collector, cleaner, selector, and detector. Through these four stages, the suggested system carries out the necessary steps for detecting malware. The suggested prediction model EEXR provides results in accordance with performance outcomes.

The technique proposed by the authors [10, 23] uses unsupervised and semi-supervised learning to build anomaly detection models that attribute suspicious behaviour to adversarial samples. The proposed adversarial defensive strategies were integrated into preexisting malware detection frameworks as add-ons. The experimental results demonstrate that their methods can strengthen the malware detection model against attacks. Tests also show that our methods are around 11% more accurate at detecting threats than conventional adversarial training. TC-Droid is an automated technique for identifying Android malware based on a text classification algorithm, created by Zhang et al. [11] due to the frequent failure of feature engineering. Instead of relying on human feature engineers to sift through the report's raw language for insights, they employed a convolutional neural network (CNN). Comparing it to the state-of-the-art Drebin model and many traditional models (NB, LR, KNN, RF), TC-Droid performs better on a variety of real-world data. TC-Droid's many trial configurations and accompanying comparisons contribute to its efficient and adaptable performance in Android malware detection.

### 2.1 State-of-Art-Survey

After referring the literature survey almost in most of the malware detection methods machine learning tools are used widely to perform the malware detection. Techniques like Decision trees, neural network, behavioural analysis etc. have been used. Among all the methods ensemble learning along with random forest technique provides good accuracy in detecting the malware successfully in both static and dynamic analysis.

## 3 Methodology

Classification techniques that have proven their worth include the decision tree classifier [12], the multilayer perceptron [13], and the Naive Bayes classifier [14]. A decision tree classifier produces a binary tree as its output. This is the model for a C4.5 tree. The necessary guidance for these tuples is found by examining a collection of basic tuples known as an education set. A tuple's grandness level can be predicted with the help of the adverb "elegance label." In the first stage, called "learning," we simply pick the training dataset at random. This approach could be quite effective because of the extensive use of typography. The following components can be used to implement the tree's structure:

1. Each tree junction corresponds to an adjective test.
2. A department leaving a node represents a check's useful outputs.
3. A class label is represented by a leaf.

4. A decision tree consists of a collection of rules that can be used to forecast goal functions.
5. The current version of the algorithm employs greedy tactics.

### 3.1 Random Forest Classifier

The supervised learning methodology includes the widely used device learning algorithm random forest [15, 21].

It can be used for categorical and regression ML problems. The concept of ensemble mastery, in which multiple classifiers are combined to solve difficult tasks, was a major inspiration. Just as its name suggests, " To improve the predictive accuracy of a dataset, random forests use a classifier called decision bushes applied to random subsets of the dataset." The random forest considers predictions from all the trees rather than just one to come up with an overall prediction.

### 3.2 Kivy [16]

An open-source framework for creating Graphical User Interfaces (GUIs) for Python is called Kivy.

Expanding mobile apps and multi-touch utility software is now possible thanks to a NUI (natural user interface). Kivy is a popular cross-platform framework for developing software. Using Kivy, Python developers can create cross-platform programmes for Linux, Windows, Android, iOS, and more.Python and Cython form the backbone of the library. It has excellent multi-touch support, making the most of the touchscreen on any given device.. It's a fantastic choice for creating contemporary applications because it supports gestures and other methods of input as well. Kivy is a visual programming language that combines GUIM and node-based approaches. These features contribute to the library's versatility and efficiency.

### 3.3 Androguard

Androguard is an open-source toolkit developed for the express purpose of inspecting and reverse engineering Android applications. Decompiling APK files, extracting information, analysing viruses, and visualising the app's structure are just some of the features and tools it provides for analysing Android apps. It's also used to validate API requests. Androguard is written in Python and supports multiple platforms. Androguard is written in Python and runs on Windows, Linux, and macOS. Because it provides a robust set of tools for developers and security researchers to analyses and comprehend Android applications, it is a popular choice in the domain of Android security.

### 3.4 Database

A database for Android malware detection using ensemble learning would likely have two key components:

- **A training dataset** containing many labeled samples of both benign and malicious Android applications.
- Each example is represented by a set of features that capture different aspects of the application's behavior and attributes.
- Permissions requested, API calls executed, network traffic, file system actions, and so forth are examples of these features.
- The training dataset would be used to train the ensemble learning model to correctly categorize Android applications as benign or malicious.
- **A testing dataset** - This dataset would contain a set of Android applications that had not been seen by the ensemble learning model during training.
- These applications would be used to assess the model's performance and determine its accuracy in identifying applications as benign or malicious.
- The database would also need to be managed and maintained to keep the training and testing datasets up to date and indicative of the most recent Android malware threats.
- This could entail gathering new examples of both benign and malicious programs regularly, as well as upgrading the collection of features used to describe these applications.

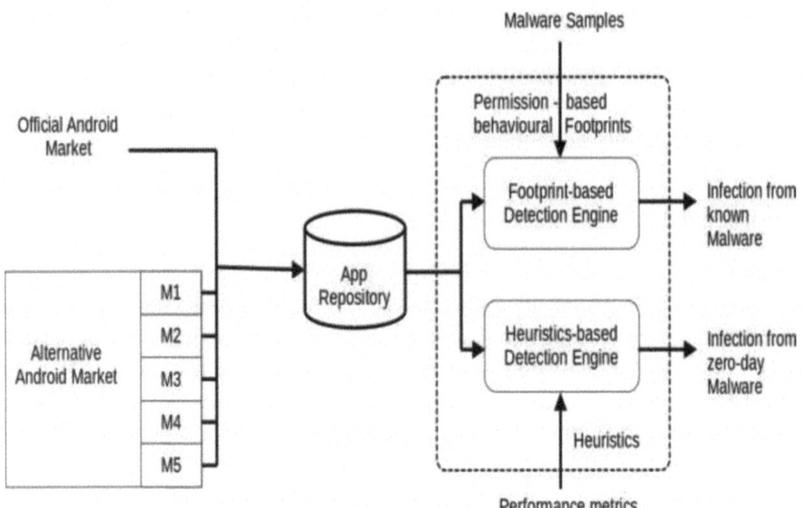


**Fig. 1.** System Architecture

### 3.5 System Architecture

Figure 1 depicts the recommended system architecture for detecting Android malware with ensemble learning. Data selection, preprocessing, data splitting, prediction, scanning, notification, and user interface are only few of the components that make up the system architecture. In order to make an accurate prediction, the ensemble learning model module uses a number of different machine learning algorithms to generate predictions. These algorithms include Random Forest, Decision Trees, SVMs, and others. Using the trained ensemble learning model, the 20-classification module predicts the class label of a new APK file based on the extracted features. The pre-processed data, extracted features, trained models, and classification results are all stored in the database module.

Finally, the system administration module maintains system resources and assures effective system operation.

## 4 Modeling and Analysis

Kaggle's online dataset repository [17] provided the input records that were gathered. Selecting relevant data for Android malware forecasting is known as "facts selection." We should make use of the dataset for Android malware in our project. The dataset that includes details about the unique type of Android malware. We must utilize the pandas' programs in Python to analyze the dataset. Our data set is stored as records with the ".csv" record extension (Fig. 2).

### 4.1 Data Preprocessing

Pre-processing [18] data involves deleting gratuitous information from the dataset. The dataset is converted using pre-processing data metamorphosis ways into a structure applicable for machine literacy. Take out the blanks in this process, 0 is substituted for the null values (also known as missing values or Nan values). Data Ordering Encoding Categorical information refers to variables that can take on only a few distinct values (Fig. 3).

### 4.2 Data Splitting

Data are required all through the machine learning procedure for learning to occur. In order to evaluate the efficacy of the algorithm, it is necessary to collect more data beyond what is needed for training. In our method, we considered 80% of the data to be training data and the remaining 20% to be test data. For pass-validator purposes, information is typically split in half, as described by the information-splitting [19] process. A prediction model is developed using some of the data, and its efficacy is assessed using the other (Fig. 4).

| transact | API call signature |
|---|---|
| 215 unique values | Manifest Permission 53%<br>API call signature 33%<br>Other (30) 14% |
| onServiceConnected | API call signature |
| bindService | API call signature |
| attachInterface | API call signature |
| ServiceConnection | API call signature |
| android.os.Binder | API call signature |
| SEND_SMS | Manifest Permission |
| Ljava.lang.Class.getCanonicalName | API call signature |
| Ljava.lang.Class.getMethods | API call signature |
| Ljava.lang.Class.cast | API call signature |
| Ljava.net.URLDecoder | API call signature |
| android.content.pm.Signature | API call signature |

**Fig. 2.** Data Selection

### 4.3 User Interface

The user interface plays a crucial role in Android's ensemble learning-based malware detection mechanism. The interface allows users to easily interact with the system by selecting the APK files to be scanned. The Kivy module is used to display buttons with all the folder icons so that users can choose the required APK file. Once the file is selected, the prediction module compares the manifest information of the APK file with the prediction model to determine if the file is infected. The outcome of the scan is displayed using the notification module, which shows the combinations of given permissions for infected files and clean results for non-infected files. Overall, the user interface provides an intuitive and easy-to-use experience for users to scan APK (Fig. 5).

### 4.4 Scanning Interface

Users can choose an APK file for scanning using the scanning interface's graphical user interface (GUI). The file is sent to the prediction module for analysis when the user selects it. The prediction module uses the prediction model to determine whether a file is

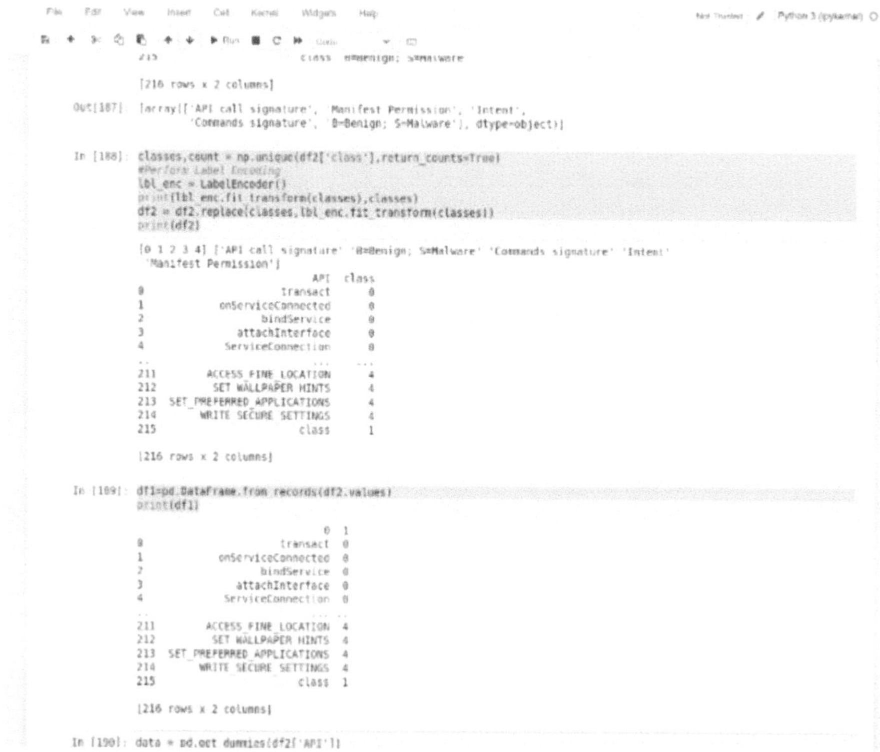

```
215                        class  B=Benign; S=Malware

[216 rows x 2 columns]

Out[187]: [array(['API call signature', 'Manifest Permission', 'Intent',
                 'Commands signature', 'B=Benign; S=Malware'], dtype=object)]

In [188]: classes,count = np.unique(df2['class'],return_counts=True)
          #Perform Label Encoding
          lbl_enc = LabelEncoder()
          print(lbl_enc.fit_transform(classes),classes)
          df2 = df2.replace(classes,lbl_enc.fit_transform(classes))
          print(df2)

          [0 1 2 3 4] ['API call signature' 'B=Benign; S=Malware' 'Commands signature' 'Intent'
           'Manifest Permission']
                                    API  class
          0                    transact      0
          1         onServiceConnected      0
          2                 bindService      0
          3             attachInterface      0
          4           ServiceConnection      0
          ..                        ...    ...
          211        ACCESS_FINE_LOCATION      4
          212          SET_WALLPAPER_HINTS      4
          213  SET_PREFERRED_APPLICATIONS      4
          214        WRITE_SECURE_SETTINGS      4
          215                       class      1

          [216 rows x 2 columns]

In [189]: df1=pd.DataFrame.from_records(df2.values)
          print(df1)

                                      0  1
          0                    transact  0
          1         onServiceConnected  0
          2                 bindService  0
          3             attachInterface  0
          4           ServiceConnection  0
          ..                        ... ..
          211        ACCESS_FINE_LOCATION  4
          212          SET_WALLPAPER_HINTS  4
          213  SET_PREFERRED_APPLICATIONS  4
          214        WRITE_SECURE_SETTINGS  4
          215                       class  1

          [216 rows x 2 columns]

In [190]: data = pd.get_dummies(df2['API'])
```

**Fig. 3.** Data Pre-processing

contaminated with malware. The notification module will show the set of provided permissions that the file requires if it is infected. Users can quickly navigate the application and comprehend the scanning process thanks to the scanning interface's straightforward and user-friendly design. The Kivy module, which enables the display of buttons and folder icons, is used to construct the user interface. The user may quickly grasp the status of the APK file and any associated malware threats thanks to the interface's simple and condensed description of the scanning process.

## 4.5 Feature Scaling

Function scaling [20, 22] is a strategy for standardizing the range of unbiased variables or informational capacities. It is commonly finished at some point during the statistics preparation step and is sometimes referred to as facts normalization in information processing. Characteristic scaling is a technique for putting the information's independent capabilities into a hard-and-fast range. The statistical pre-processing has made it feasible to deal with a diverse set of scales, units of measure, and instruments. Rules learned by machine learning tend to favour larger values and to remember smaller values as the lower values, regardless of the unit of the input if function scaling is not always employed (Fig. 6).

**Fig. 4.** Data Splitting

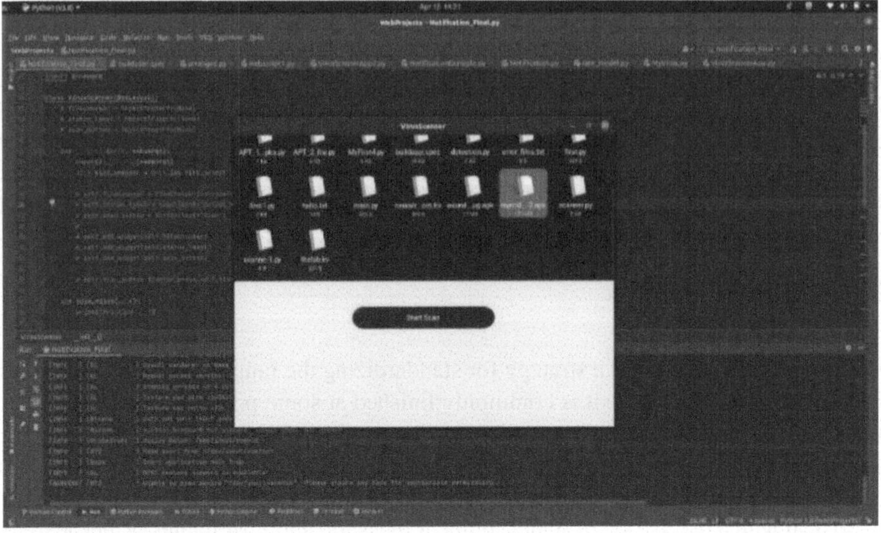

**Fig. 5.** User Interface

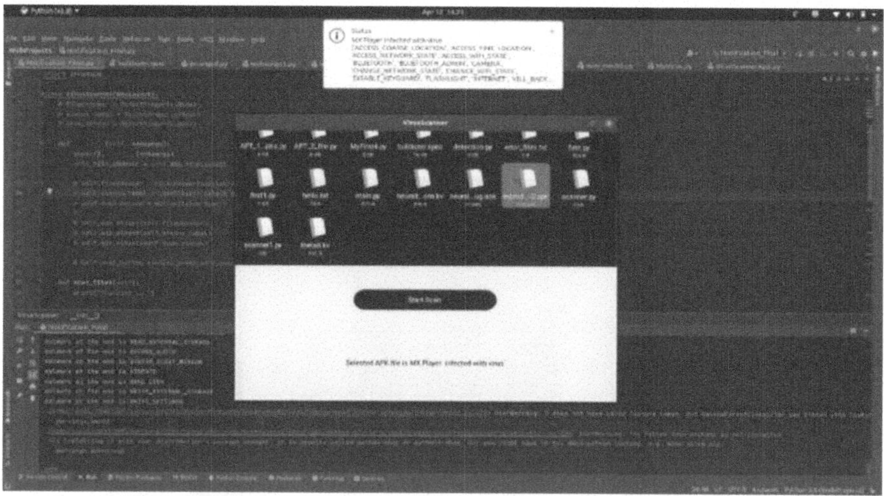

**Fig. 6.** Scanning Interface

## 4.6 Achieved Results

In this study, we used an IoT network dataset consisting of 4,61,043 records (1,61,043 of which were malicious) to examine the efficacy of neural networks in identifying and classifying malware. Malware is discovered with 94.17% accuracy and categorized with 97.08% accuracy using the proposed methodology.

## 4.7 Test Case

A test case is a set of actions and conditions meant to test an individual attribute or component of a software program. It aids in the verification of the software's compliance with the desired requirements and specifications. Test cases are often prepared by software testers and conducted to detect problems and assure the quality of the software product (Fig. 7 and Table 1).

**Table 1.** Table of Test Cases

| Test Case ID | Description | Pre-Condition | Pass/Fail | Expected Result |
|---|---|---|---|---|
| SEMS_001 | Infected file | If the user selected the infected file and starts scanning | Pass | Notification will be popped out and displays the combinations of permissions granted |
| SEMS_002 | Clean File | If the user selected the clean file and starts scanning | Pass | Notification popped out and displays the file is clean |

## 5 Performance Analysis

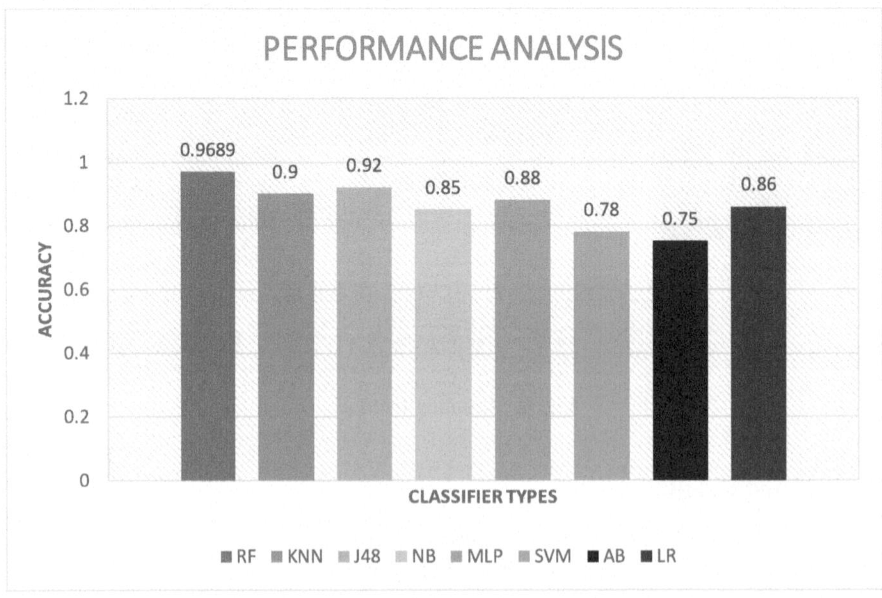


**Fig. 7.** Comparison of different ML Algorithms

The accompanying chart compares the results of various machine learning algorithms considering the new research. The ML algorithms like Random Forest, KNN, J48, Naïve Bias, MLP, SVM, AB, LR about to gives accuracy of 90% and above with little variations. When compared to other ML algorithms, Random Forest gives more accuracy in detecting malwares in Android OS. This can also be used in other operating systems to detect the malware effectively and efficiently.

## 6 Conclusion

In conclusion, employing ensemble learning for Android malware detection is a successful strategy for recognizing and reducing the risks provided by malware. The suggested approach uses an ensemble model that combines many machine learning methods to increase the precision and robustness of malware detection. A clear understanding of the system's design and implementation can be gained from looking at the system architecture and its modules. The suggested approach has the potential to improve the security and dependability of Android-based systems with additional study and improvement.

## 7 Future Work

1. ***Integration of more sophisticated ensemble methods***: Now, the proposed system solely employs bagging and boosting as ensemble methods. To increase the detection accuracy even more, it might be investigated to use more sophisticated techniques such as layered generalization, random subspace, and random oracle.
2. ***Incorporating more various features***: Adding a wider range of features, such as user behavior data, network traffic data, and system call traces, could increase detection accuracy.
3. ***The application of deep learning methods***: Deep learning has displayed outstanding performance in several machine learning tasks. In order to further improve the detection accuracy, the usage of deep learning techniques such as CNNs and RNNs could be investigated.

## References


1. Akhtar, M.S., Feng, T.: Malware analysis detection using machine learning algorithms. Symmetry **14**(11), 2304 (2022)
2. Raymond, V.J., Raj, R., Retna, J.: Investigation of android malware with machine learning classifiers using enhanced pbehca algorithm. Comput. Syst. Sci. Eng. **44**(3), 2147–2163 (2023)
3. Atacak, I.: An ensemble approach based on fuzzy logic using machine learning classifiers for android malware detection. Appl. Sci. **13**(3), 1484 (2023)
4. Al Sarah, N., Rifat, F.Y., Hossain, M.S., Narman, H.S.: An efficient android malware prediction using ensemble machine learning algorithms. Procedia Comput. Sci. **191**, 184–191 (2021)
5. Priya, V., Sathya Sofia, A.: Review on malware classification and malware detection using transfer learning approach. In: 2023 5th International Conference on Smart Systems and Inventive Technology (ICSSIT), Tirunelveli, India, pp. 1042–1049 (2023). https://doi.org/10.1109/ICSSIT55814.2023.10061076
6. Li, W., Cai, J., Wang, Z., Cheng, S.: A robust malware detection approach for android system based on ensemble learning. In: Wang, G., Choo, KK.R., Ko, R.K.L., Xu, Y., Crispo, B. (eds.) Ubiquitous Security, vol. 1557, pp. 309–321. Springer, Singapore (2021). https://doi.org/10.1007/978-981-19-0468-4_23

7. Guan, S., Li, W. Ensemble droid: a malware detection approach for android system based on ensemble learning. In 2022 IEEE MIT Undergraduate Research Technology Conference (URTC), pp. 1–5. IEEE (2022)
8. Mahindru, A., Sangal, A.L.: HybriDroid: an empirical analysis on effective malware detection model developed using ensemble methods. J. Supercomput. **77**, 8209–8251 (2021)
9. Urmila, T.S.: Machine learning-based malware detection on Android devices using behavioural features. Mater. Today: Proc. **62**, 4659–4664 (2022)
10. Chen, J., Yuan, C., Li, J., Tian, D., Ma, R., Jia, X.: ELAMD: an ensemble learning framework for adversarial malware defence. J. Inf. Secur. Appl. **75**, 103508 (2023)
11. Zhang, N., Tan, Y.A., Yang, C., Li, Y.: Deep learning feature exploration for android malware detection. Appl. Soft Comput. **102**, 107069 (2021)
12. Ullah, F., et al.: Modified decision tree technique for ransomware detection at runtime through API calls. Sci. Program. **2020**, 8845833 (2020)
13. Ling, Y.T., Phang, P., Chiew, K.L., Zhang, X.: Malware detection with structural entropy features using multilayer perceptron neural network. In 2022 International Conference on Digital Transformation and Intelligence (ICDI), pp. 01–07. IEEE
14. Mat, S.R.T., Ab Razak, M.F., Kahar, M.N.M., Arif, J.M., Firdaus, A.: A Bayesian probability model for Android malware detection. ICT Express **8**(3), 424–431 (2022)
15. Ravikumar, S., Vijay, K., Pavithra, S., Prithi, S., Kannan, S.: Prediction of rental demands using different machine learning algorithms. In: 2023 International Conference on Computer Communication and Informatics (ICCCI), Coimbatore, India, pp. 1–6 (2023). https://doi.org/10.1109/ICCCI56745.2023.10128331
16. Driscoll, M Build a mobile application with the kivy python framework (2020)
17. Quaranta, L., Calefato, F., Lanubile, F.: KGTorrent: a dataset of python Jupyter notebooks from kaggle. In: 2021 IEEE/ACM 18th International Conference on Mining Software Repositories (MSR), pp. 550–554. IEEE
18. Chemmakha, M., Habibi, O., Lazaar, M.: Improving machine learning models for malware detection using embedded feature selection method. IFAC-Papers OnLine **55**(12), 771–776 (2022)
19. Alhebsi, M. S.: Android Malware Detection using Machine Learning Techniques (2022)
20. Vijay, K., Manikandan, J., Rajendiran, B., Sowmia, K.R., Berna, E.I.: Deep dive on oversampling and under sampling techniques in machine learning. In: Recent Trends in Computational Intelligence and Its Application: Proceedings of the 1st International Conference on Recent Trends in Information Technology and its Application (ICRTITA, 22), p. 423. CRC Press (2023)
21. Khammas, B.M.: Ransomware detection using random forest technique. ICT Express **6**(4), 325–331 (2020)
22. Moon, D., Lee, J., Yoon, M.: Compact feature hashing for machine learning based malware detection. ICT Express **8**(1), 124–129 (2022)
23. Samuel, P., Jayashree, K., Babu, R., Vijay, K.: Artificial intelligence, machine learning, and IoT architecture to support smart governance. In: Saini, K., Mummoorthy, A., Chandrika, R., Gowri Ganesh, N. (eds.), AI, IoT, and Blockchain Breakthroughs in E-Governance, pp. 95–113. IGI Global (2023). https://doi.org/10.4018/978-1-6684-7697-0.ch007

24. Prithi, S., Sumathi, S.: LD2FA-PSO: a novel learning dynamic deterministic finite automata with PSO algorithm for secured energy efficient routing in wireless sensor network. Ad Hoc Netw. **97**, 102024 (2020). https://doi.org/10.1016/j.adhoc.2019.102024
25. Rout, S.S., Vashishtha, L.K., Chatterjee, K., Rout, J.K.: Android malware detection using ensemble feature learning. In: Garg, L., et al. (eds.) SISMS 2021. LNCS, vol. 521, pp. 531–539. Springer, Cham (2023). https://doi.org/10.1007/978-3-031-13150-9_43

# Handling Imbalanced Data for Credit Card Fraudulent Detection: A Machine Learning Approach

E. Sujatha[1(✉)], V. Umarani[1], K. S. Rekha[1], P. V. Gopirajan[2], and V. Manickavasagan[3]

[1] Department of Computer Science and Engineering, Saveetha Engineering College (Autonomous), Sriperumbudur, India
sujathae@saveetha.ac.in

[2] Department of Computational Intelligence, SRM Institute of Science and Technology, Chennai, India

[3] Department of Computer Science and Business Systems, Sri Sairam Engineering College, Chennai, India

**Abstract.** Imbalanced data is crucial for contemporary machine learning models. Evaluating the effectiveness of machine learning classifiers under an unfair dataset is problematic. Several resampling techniques were applied to the unwarranted dataset and pre-processing stages to increase the accuracy of prognosis classifiers. This proposed work discusses the basics of resampling techniques such as Random under-sampling, Random oversampling, SMOTE (Synthetic Minority Over-sampling Technique), SMOTE-ENN (Synthetic minority over-sampling technique and Edited nearest neighbor), ADASYN (Adaptive Synthetic), and SMOTE-Tomek to handle this Weakness with unbalanced data. The proposed system uses an unstable credit card fraudulent dataset collected from Kaggle, which contains 21 columns and 28382 rows, of which 18497 negative samples and 4208 positive samples. The balanced dataset got by applying resampling methods SMOTE, SMOTE-ENN, ADASYN, and SMOTE-Tomek. Afterward, it uses machine learning techniques such as Logistic Regression (LR), Decision tree (DT), and Random Forest (RF) for classification. From this analysis, it selects the optimal scheme and classifier. The grey wolf optimization technique improves the classifier's performance, which optimizes the samples after sampling. This hybrid method's effectiveness was evaluated with standard evaluation parameters. The hybrid Random Forest technique achieved 92% accuracy from this analysis compared to hybrid models.

**Keywords:** Business Intelligence · Data Analytics · Data Reporting

## 1 Introduction

Maintaining credit card-related information is a risky and entire trust process [1]. Due to the advancement of technology, the usage of credit cards has increased rapidly, even during this pandemic [2]. Many machine learning and artificial intelligence-based algorithms were used to safeguard credit card-related information [3–6]. As the usage of


R. Geetha et al. (Eds.): AAIMB 2023, CCIS 2203, pp. 220–233, 2025.
https://doi.org/10.1007/978-3-031-73068-9_18

credit cards increases, fraudulency in credit card-related online payment systems also increases this post-pandemic [7]. Fraudulence in accessing credit cards and other related online financial mechanisms shall be reduced when proper precautionary methods are handled [8–10].

Data analysis is essential for the datasets to extract the required information and interpret necessary results. Proper data analysis provides a complete skeleton of a problem domain and suggests suitable modification of models to be implemented [11]. Data analysis is essential for the applications like medicine, food processing, geography, and many others [12–14]. Sufficient and balanced data provides a technique for perfect analysis. Credit card-related data analysis provides information on the amount spent, outstanding balance, users buying interest, product interests, card holder's characteristics, and purchase time [15].

Data collection is a challenging assignment when done on a large scale. Datasets may be available in dissimilar file types and also in uneven formats. Attempting to learn with skewed statistics is difficult in machine learning [16]. Imbalanced data leads to bias in majority classes. It results in a high error in or entire elimination of minority classes [17]. These imbalanced data will have diversified observations, creating delinquency during data analysis. Trained machine learning models will produce biased results when processed with an imbalanced dataset [18]. An advanced methodology should necessitate retrieving balanced data from the excessive dataset.

Machine learning models provide solutions to classify balanced datasets from the imbalanced dataset. Also, machine learning models shall be applied to any data analysis irrespective of the field of datasets [19]. Machine learning models synchronize effortlessly with cloud repositories and work well with multivariate datasets [20]. Xu Y et al. applied the adaptive subspace optimization ensemble method (ASOEM) [21] to segment slightly elevated unbalanced data and suggested high-dimensional imbalanced data for classifiers. It is applied to 24 elevated, unbalanced datasets from reality using various interpolation techniques. Unbalanced data lower the effectiveness of the different other conventional classifiers. The Deep SMOTE oversampling technique [22] resolves the issues of loss function modifications and instance resampling. It generates high-quality images, improves minority classes, and balances the training set.

Radial-Based Combined Cleaning and Resampling (RB-CCR) [23] used the class possibility to strengthen the energy-based quantization strategy of CCR. Machine learning qualifiers optimize safety, cost, and efficiency [24]. This study proposes a few novel techniques, such as Random under-sampling, Random oversampling, SMOTE, SMOTE-ENN, ADASYN, and SMOTE-Tomek, for handling the imbalanced dataset problem, and the results were compared to demonstrate the viability of said planned research.

## 2 Proposed System

The suggested approach, displayed in Fig. 1, takes the imbalanced dataset and predicts the system's performance using the classifier RF, SVM, DT, and LR. Initially, preprocessing is carried out to create structured data from an unstructured dataset. The resampling technique helps to form a balanced dataset from an unbalanced dataset using SMOTE and its variants. Finally, classifiers are applied to a balanced dataset and find

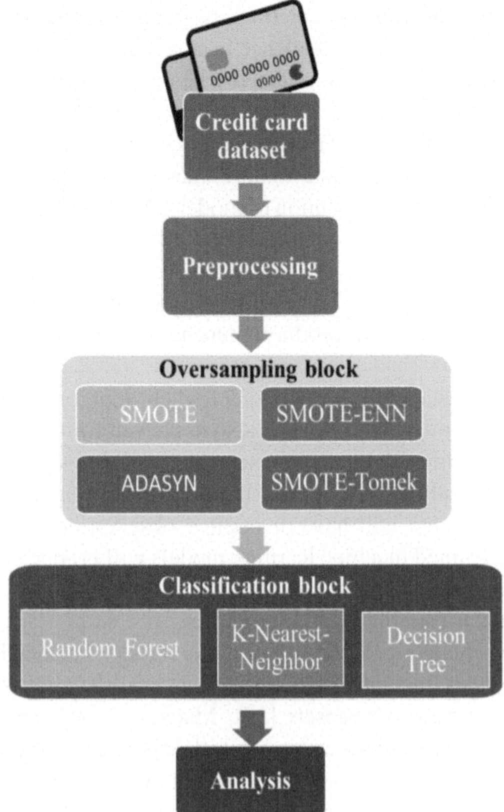


**Fig. 1.** Workflow of Proposed Methodology

the best amongst one. The basics of resampling methods and classifiers are discussed as follows

A. **Smote**

SMOTE [25] is an oversampling approach that produces additional synthesized cases to enhance the number of minority samples in the dataset. This technique is launched in handwritten character recognition successfully. The synthetic pieces are created by interpolation between positive instances. This method produces different data by combining the characteristics of the targeted case with those of its neighbors after sampling the feature set for every target domain and its closest neighbors [26]. The synthetic samples focus on less application-specific and feature space rather than data space [27]. It forces the decision region to be more general in the minority class. The training set consists of several Neural Networks based on the distance metric and random interpolation adapted to obtain new instances.

B. **Smote –Enn**

This SMOTE-ENN enhances the outcomes by combining the SMOTE (for over-sampling) and the ENN (for under-sampling). The assessment and its K-nearest neighbor are eliminated from the dataset rather than only the observed and its 1-nearest neighbor with different classes if there is any predominant class difference between the analysis and its K-nearest correspondent. Hence, ENN provides more in-depth data cleaning. Generally, $k = 3$ where k is the number of nearest-neighbor. Additionally, it eliminates all occurrences of the majority class in a KNN projection that varies from the majority class. It creates a streamlined decision surface by removing the noisy cases as borderline instances. It is beneficial for comprehensive data cleansing.

C. **ADASYN – Adaptive Synthetic Sampling Approach**

This is based on adapting minority data samples according to their distributions using a k-nearest neighbor. It reduces the learning bias and adaptively changes the decision to focus on complicated models. There aren't any presumptions made about the data's density function, and the dispersion is updated. For the KNN algorithm, Euclidean distance is used. By adaptively altering the weights of various minority samples to account for the outlier detection, this uses population distribution as a criterion to dynamically determine the number of synthetic models created for each minority sample.

**Pseudocode: ADASYN – Algorithm**

Step 1: Determine the degree of imbalance between class d from Minority and Majority classes between 0 to 1.

Step 2: Set threshold for imbalance class ratio. If d is less than $d_{th}$, then:

(i) Calculate Syn = (Maj – Min) x A.
Where A is the desired balance level.

(ii) For each Min, determine the ratio $r_i = E_i / k$ where i= 1 to Min, $r_i$ values are between 0 to 1.

(iii) Apply normalization for ri, $R_i = r_i / \sum_{i=1}^{Min} ri$ where $R_i$ is density distribution.

(iv) Calculate $Min_{ex} = R_i \times Syn$, where Syn is calculated in step 2(i).

(v) For i=1 to $M_{ex}$, do

a. Choose a minority example $Min_{ex}$.

b. $Syn_{ex} = Syn_{ex} + (Min_{ex} - Syn_{ex}) \times B$, where b ranges from 0 to 1.

D. **SMOTE-Tomek**

This approach joins the Tomek connections for the under model and the SMOTE as an over-sampling strategy to produce positive outcomes.

It blends the capability of creating artificial minority class data well with the power of removing Tomek linkages from the positive class, which are samples of minority class averages that seem near the mainstream class data.

**Pseudocode – SMOTE-Tomek Algorithm**

Step 1: Start
Step 2: random data and k- nearest neighbor distance calculation
Step 3: Use a random integer between 0 and 1 to determine the difference, then add the result.
Step 4: Repeat Steps 2 and 3 until targeted development.
Step 5: End SMOTE-Tomek

E. **Random Forest**

Both classification and regression may be done with it. It manages values with errors. This predictor won't overfit the dataset if more trees are in the forests. It can model categorical values also. Choose k features from total m features. Create the tree and child nodes and construct the forest.

**Pseudocode: Prediction Pseudo code from created RANDOM FOREST Classifier**

Step 1: Analyzes the test characteristics, predict the result using each randomly generated decision tree's rules, and save the result.
Step 2: Determine how many votes each anticipated target received.
Step 3: Use the objective that received the most votes as the random forest algorithm's final assessment.

F. **Decision Trees**

**Pseudocode: Tree Algorithm**

Input: Training Data D, Attrbs, candidate attributes, Attrbs_Sel_Method.
Output: Decision Tree

G. **K-Nearest Neighbor (kNN)**

**Pseudocode: kNN Algorithm 1**

Input: In is an input matrix, $n_{size}$ is a submatrix, and k is the number of nearest neighbors
Output: kNN graph stores in $G_k$
Step 1: Create $G_k$
Step 2: For i = 0 to $n_{row}$ – 1, do
    $G_k$[i].weight = max
Step 3: Initialize(segments,$n_{GPU}$)
Step 4: For segment = 1 to Seg D,
    Create D and $G_k$
        Host = device (In, D, $G_k$)
        Initialize(splits,segment)
        Create maxk
Step 5: For chunk = 1 to chunks do
        Call Distance(In, $n_{size}$,split,chunk,b)
        Call kNN(D, $n_{size}$,split,chunk,max)
        EndFor
Step 6:EndFor
Step 7: device = host($G_k$`)
Step 8: return $G_k$

**Pseudocode: Distance Algorithm**

Input: In is the input matrix, $n_{size}$ is the dimension of the distance matrix chunk, the split is the split index, the fragment is the chunk index, b is the size of the data block
Output: None, a chunk of the original distance matrix stored in D

```
Step 1: Input X and Y array
Step 2: bx = block. x, by = block.y
Step 3: tx = thread.x, ty = thread.y
Step 4: Xbegin = (bx + n_size / b x chunk) x b x n_col
Step 5: Ybegin = (by + n_size / b x split) x b x n_col
Step 6: Yend = Yend + n_col-1
Step 7: d = 0
 d = d + distance between Y[ty][i] and X[i][tx]
 EndFor
  Step 8: Synchronize. threads()
          EndFor
Step 9:  index = by x b x n_size + ty x b + bx x b + tx
Step 10: D[index] = d
```

**Pseudocode: kNN Algorithm 2**

Input: D is a chunk of the original distance matrix, $n_{size}$ is a dimension of the piece, the split is the split index, the fragment is the chunk index, the mark is an array consisting of farthest neighbors for each row index in the chunks
Output: none, an intermediate kNN graph stored in $G_{K}$.`

## 3 Results and Discussion

For experimental results, the Credit card fraudulent dataset is collected from Kaggle and used to evaluate resampling methods and classifiers. The dataset contains 21 columns and 28382 rows of data. This dataset is not balanced since it includes 18497 negative and 4208 positive samples. The first five rows of the dataset are shown in Fig. 2a, 2b, and 2c. Here churn column specifies the positive and negative samples. The results were taken by Google Collaboratory, which provides a free anaconda python environment and the necessary packages for evaluating the desired model.

The dataset initially has several null values, depicted in Fig. 3. Some matters should replace the null values to avoid the missing value problem.

For example, the missing values in the Gender columns are replaced by –1 if Gender data is unavailable, 0 for dependents and similarly for other columns like "self–employed" for occupation, and 1111 for the city. After replacing missing values, the dataset is divided into a training dataset (80%) and a test dataset (20%). The sampling process is carried out to make the dataset balanced. Here SMOTE, SMOTE + Tomek, SMOTE-ENN, and ADASYN sampling methods are used to create a balanced set. The result of the sampling process is illustrated in Fig. 4a, 4b, 4c and 4d.

After the sampling process, apply the classifier to find their best. Here, machine learning classifiers LR, DT, and RF were used to evaluate the detected credit card fraud dataset. The following parameters are applied to gauge how well the algorithm is doing. [27].

| | customer_id | vintage | age | gender | dependents | occupation | city | customer_nw_category | branch_code | current_balance | previous_month_end_balance |
|---|---|---|---|---|---|---|---|---|---|---|---|
| 0 | 1 | 2101 | 66 | 1.0 | 0.0 | self_employed | 187.0 | 2 | 755 | 1458.71 | 1458.71 |
| 1 | 2 | 2348 | 35 | 1.0 | 0.0 | self_employed | 1020.0 | 2 | 3214 | 5390.37 | 8704.66 |
| 2 | 4 | 2194 | 31 | 1.0 | 0.0 | salaried | 146.0 | 2 | 41 | 3913.16 | 5815.29 |
| 3 | 5 | 2329 | 90 | -1.0 | 0.0 | self_employed | 1020.0 | 2 | 582 | 2291.91 | 2291.91 |
| 4 | 6 | 1579 | 42 | 1.0 | 2.0 | self_employed | 1494.0 | 3 | 388 | 927.72 | 1401.72 |

**a.** First Five rows of the dataset

| average_monthly_balance_prevQ | average_monthly_balance_prevQ2 | current_month_credit | previous_month_credit | current_month_debit | previous_month_debit |
|---|---|---|---|---|---|
| 1458.71 | 1449.07 | 0.20 | 0.20 | 0.20 | 0.20 |
| 7799.26 | 12419.41 | 0.56 | 0.56 | 5486.27 | 100.56 |
| 4910.17 | 2815.94 | 0.61 | 0.61 | 6046.73 | 259.23 |
| 2084.54 | 1006.54 | 0.47 | 0.47 | 0.47 | 2143.33 |
| 1643.31 | 1871.12 | 0.33 | 714.61 | 588.62 | 1538.06 |

**b.** Next Five rows of the dataset

| current_month_balance | previous_month_balance | churn | last_transaction |
|---|---|---|---|
| 1458.71 | 1458.71 | 0 | 2019-05-21 |
| 6496.78 | 8787.61 | 0 | 2019-11-01 |
| 5006.28 | 5070.14 | 0 | NaT |
| 2291.91 | 1669.79 | 1 | 2019-08-06 |
| 1157.15 | 1677.16 | 1 | 2019-11-03 |

**c.** Remaining Five rows of the dataset

**Fig. 2.** First Five rows of the dataset

```
customer_id                      0
vintage                          0
age                              0
gender                         525
dependents                    2463
occupation                      80
city                           803
customer_nw_category             0
branch_code                      0
current_balance                  0
previous_month_end_balance       0
average_monthly_balance_prevQ    0
average_monthly_balance_prevQ2   0
current_month_credit             0
previous_month_credit            0
current_month_debit              0
previous_month_debit             0
current_month_balance            0
previous_month_balance           0
churn                            0
last_transaction                 0
dtype: int64
```

**Fig. 3.** Count of Null values in the dataset

1. Accuracy ($A_{cc}$): The accuracy indicates the proportion of observations from a collection that were accurately identified as suspicious or unusual credit card samples. It is

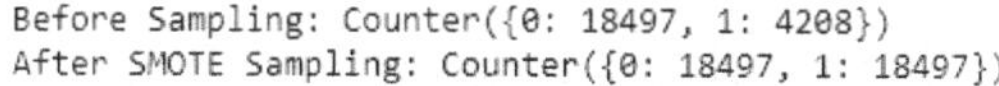


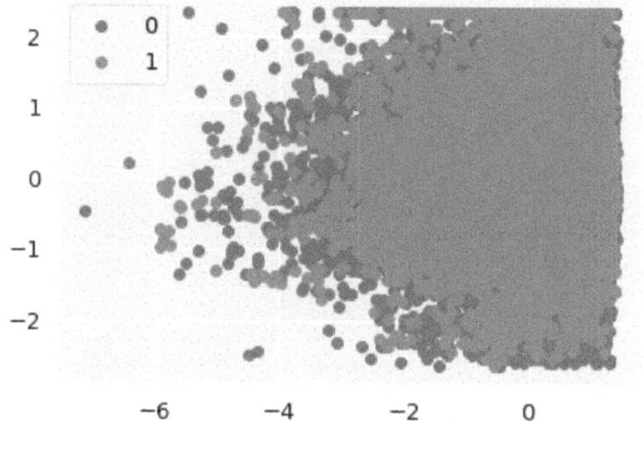


**a.** SMOTE

Before Sampling: Counter({0: 18497, 1: 4208})
After SMOTE+Tomek: Counter({0: 18090, 1: 18090})

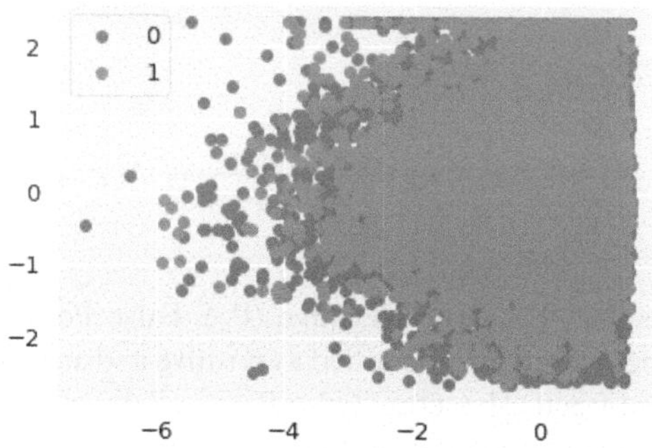


**b.** SMOTE +Tomek

Before Sampling: Counter({0: 18497, 1: 4208})
After SMOTE+ENN Counter({0: 18497, 1: 17388})

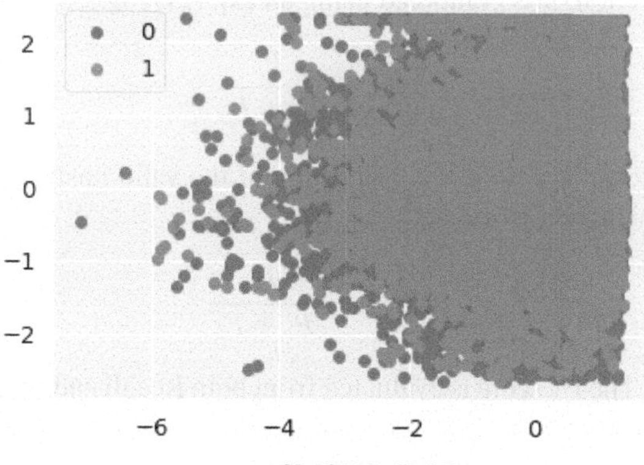


**c.** SMOTE+ENN

**Fig. 4.** SMOTE

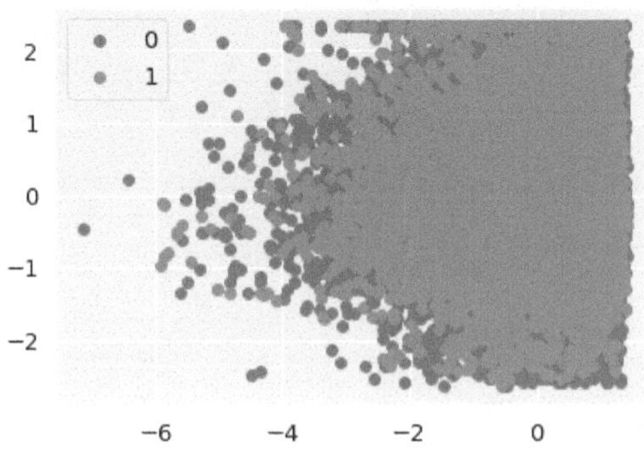


**d.** ADASYN

**Fig. 4.** *(continued)*

computed using an Eq. (1)

$$A_{CC} = \frac{P_t + P_n}{P_t + P_n + F_t + F_n} A_{CC} = \frac{P_t + P_n}{P_t + P_n + F_t + F_n} A_{CC} = \frac{P_t + P_n}{P_t + P_n + F_t + F_n} \tag{1}$$

Finally, True positive ($P_t$), True Negative ($P_n$), False Positive ($F_p$), and False Negative ($F_n$) specimens were first identified as positive and are still anticipated to be positive for credit card fraud. The term "False Positive" describes samples that were first thought to be negative but turned out to be positive. Authentic Negative samples were first classified as unfavorable, and the classifier anticipated to be harmful due to credit card theft. False Negative describes bogus specimens that initially test positive but should be negative.

2. Precision ($P_{rs}$) describes the percentage of actual positive samples vs. all positive pieces in a collection. It is computed using an Eq. (2)

$$P_{rs} = \frac{P_t}{P_t + F_p} \tag{2}$$

3. Recall ($R_{cal}$): Recall specifies the proportion of the valid positive rate of fraudulent samples. It is calculated by Eq. (3).

$$R_{cal} = \frac{P_t}{P_t + F_n} \tag{3}$$

4. F1-Score ($F1_s$): The F1 score is evaluated from both Recall and Precision. It specifies the harmonic mean of Recall and Precision. This F1 score is assessed by Eq. (4)

$$F1_s = \frac{2 * (R_{cal} * P_{rs})}{R_{cal} + P_{rs}} \tag{4}$$

5. ROC AUC score: It specifies the degree of separability between the prediction of positive and negative samples and varies from 0 to 1. The model with a higher ROC-AUC score is better than the lower one.

**Table 1.** Performance report of the classifier with sampling and without sampling

| | Sampling Method | Class | Precision ($P_{rs}$) | Recall ($R_{cal}$) | F1-score ($F1_s$) | support | Accuracy ($A_{cc}$) | AUC score |
|---|---|---|---|---|---|---|---|---|
| Logistic Regression | No sampling | 0 | 0.83 | 0.99 | 0.9 | 4625 | 0.82 | 0.7734 |
| | | 1 | 0.75 | 0.08 | 0.15 | 1052 | | |
| | | 1 | 0.43 | 0.65 | 0.52 | 1052 | | |
| | SMOTE | 0 | 0.92 | 0.81 | 0.86 | 4625 | 0.78 | 0.78 |
| | | 1 | 0.45 | 0.69 | 0.54 | 1052 | | |
| | ADASYN | 0 | 0.92 | 0.85 | 0.88 | 4625 | 0.81 | 0.778 |
| | | 1 | 0.49 | 0.66 | 0.56 | 1052 | | |
| | SMOTE-TOMEK | 0 | 0.92 | 0.81 | 0.86 | 4625 | 0.78 | 0.778 |
| | | 1 | 0.45 | 0.68 | 0.54 | 1052 | | |
| | SMOTE-ENN | 0 | 0.92 | 0.81 | 0.86 | 4625 | 0.81 | 0.787 |
| | | 1 | 0.45 | 0.69 | 0.55 | 1052 | | |
| Decision Tree | No sampling | 0 | 0.88 | 0.95 | 0.91 | 4625 | 0.86 | 0.8 |
| | | 1 | 0.67 | 0.44 | 0.5 | 1052 | | |
| | | 1 | 0.47 | 0.52 | 0.49 | 1052 | | |
| | SMOTE | 0 | 0.91 | 0.78 | 0.84 | 4625 | 0.76 | 0.78 |
| | | 1 | 0.41 | 0.66 | 0.5 | 1052 | | |
| | ADASYN | 0 | 0.91 | 0.81 | 0.86 | 4625 | 0.78 | 0.77 |
| | | 1 | 0.43 | 0.63 | 0.51 | 1052 | | |
| | SMOTE-TOMEK | 0 | 0.91 | 0.8 | 0.85 | 4625 | 0.78 | 0.78 |
| | | 1 | 0.43 | 0.67 | 0.53 | 1052 | | |
| | SMOTE-ENN | 0 | 0.92 | 0.68 | 0.78 | 4625 | 0.7 | 0.769 |
| | | 1 | 0.34 | 0.73 | 0.46 | 1052 | | |
| Random Forest | No sampling | 0 | 0.88 | 0.96 | 0.92 | 4625 | 0.87 | 0.83 |
| | | 1 | 0.74 | 0.44 | 0.55 | 1052 | | |
| | SMOTE | 0 | 0.92 | 0.82 | 0.87 | 4625 | 0.79 | 0.8 |
| | | 1 | 0.46 | 0.67 | 0.55 | 1052 | | |
| | ADASYN | 0 | 0.91 | 0.83 | 0.87 | 4625 | 0.8 | 0.81 |
| | | 1 | 0.46 | 0.66 | 0.54 | 1052 | | |

*(continued)*

**Table 1.** (*continued*)

| | Sampling Method | Class | Precision ($P_{rs}$) | Recall ($R_{cal}$) | F1-score ($F1_s$) | support | Accuracy ($A_{cc}$) | AUC score |
|---|---|---|---|---|---|---|---|---|
| | SMOTE-TOMEK | 0 | 0.92 | 0.83 | 0.87 | 4625 | 0.8 | 0.81 |
| | | 1 | 0.47 | 0.67 | 0.55 | 1052 | | |
| | SMOTE-ENN | 0 | 0.93 | 0.69 | 0.79 | 4625 | 0.8 | 0.81 |
| | | 1 | 0.36 | 0.76 | 0.55 | 1052 | | |

The classification report of machine learning classifiers such as LR, DT, and RF using sampling or without sampling is shown in Table 1. From this table, the highest AUC score obtained for the Random Forest classifier is 0.83. So Random Forest model provides better classification compared with other classifiers. It also shows that Random Forest with SMOTE + ENN provides a better recall percentage is 0.76, which is higher than other sampling methods. The F1-score of Random Forest with sampling methods is higher than another classifier with sampling methods. The accuracy obtained for the Random Forest classifier is high compared with other classifiers. According to these experiments, the Random Forest classifier is a better model than different machine learning Classifiers.

From this analysis, the random forest classifier is the best, and SMOTE + ENN is the best sampling method for the credit card fraud detection system. Finally, we use the Grey wolf optimization technique for optimizing the samples obtained from SMOTE + ENN. The performance of the hybrid grey wolf optimization technique is shown in Table 2.

**Table 2.** Obtained values for sampling and optimization

| Classifier | Sampling and optimization method | Class | Precision | Recall | F1-score | support | Accuracy | AUC score |
|---|---|---|---|---|---|---|---|---|
| Random Forest | No sampling and No optimization | 0 | 0.88 | 0.96 | 0.92 | 4625 | 0.87 | 0.83 |
| | | 1 | 0.74 | 0.44 | 0.55 | 1052 | | |
| | SMOTE + ENN and No optimization | 0 | 0.93 | 0.69 | 0.79 | 4625 | 0.8 | 0.81 |

(*continued*)

**Table 2.** (*continued*)

| Classifier | Sampling and optimization method | Class | Precision | Recall | F1-score | support | Accuracy | AUC score |
|---|---|---|---|---|---|---|---|---|
| | | 1 | 0.36 | 0.76 | 0.55 | 1052 | | |
| | No sampling And Grey wolf optimization | 0 | 0.886 | 0.89 | 0.92 | 4625 | 0.88 | 0.86 |
| | | 1 | 0.73 | 0.43 | 0.56 | 1052 | | |
| | SMOTE ENN + Grey wolf | 0 | 0.91 | 0.92 | 0.9149 | 4625 | 0.92 | 0.91 |
| | | 1 | 0.95 | 0.93 | 0.93 | 1052 | | |

From this table, Random Forest gives better accuracy than hybrid SMOTE ENN + Random Forest. However, hybrid SMOTE ENN + Random Forest gives a better recall score of 0.76 for class 1, which is low in Random Forest without sampling. The grey wolf optimization technique provides better accuracy of 0.88, which is higher than no sampling and hybrid SMOTE ENN + random Forest, but there is no improvement in Recall and F1-Score. So, the proposed combination of SMOTE ENN and Grey wolf optimization with Random Forest classification enhances the overall accuracy of 0.92 and all other parameters. Hence the prediction of credit card fraudulent systems is improved by hybrid SMOTE ENN and Grey wolf optimization with Random Forest.

## 4 Conclusion and Future Work

The unbalanced data problem is addressed in this study using the oversampling techniques SMOTE, ADASYN, SMOTE-Tomek, and SMOTE-ENN. Machine learning classifiers using LR, DT, and RF complete the classification assignment. A fake credit card dataset is used to experiment. Precision, recall, F1-score, AUC score, and accuracy are used to assess the classifier's effectiveness. For the unbalanced dataset, accuracy and prediction are insufficient to determine the classifier model. Hence the AUC score, F1 score, and Recall parameter are used to make this determination. The Sampling method helps to enhance the parameters like precision and recall in an unbalanced credit card fraudulent data set, but it suffers from overfitting. Hence Grey wolf Optimization technique is proposed to optimize the samples obtained from sampling methods. This proposed hybrid Random Forest, SMOTE ENN, and Grey wolf optimization technique provides better accuracy of 92% and AUC scores of 0.91. It offers good precision and recall compared with the non-optimization approach. New ensemble and neural network-based classifiers will be released, and they will lead to an improvement in comparability and improved performance. Additionally, feature selection techniques enhance model outputs and better understand the essential features.

## References


1. Tran, P.H., Heuchenne, C., Tran, K.P., HienTran, P., Huong, T.T., Huong Le, T.M.: Real-time data-Driven approaches for credit card fraud detection. In: ACM International Conference Proceeding Series, pp. 6–9 (2018)
2. Sohony, I., Pratap, R., Nambiar, U.: Ensemble learning for credit card fraud detection. In: ACM International Conference Proceeding Series, pp. 289–294 (2018)
3. Sadgali, I., Sael, N., Benabbou, F.: Fraud detection in credit card transaction using neural networks. PervasiveHealth Pervasive Comput. Technol. Healthc. (2019)
4. Jiang, C., Song, J., Liu, G., Zheng, L., Luan, W.: Credit card fraud detection: a novel approach using aggregation strategy and feedback mechanism. IEEE Internet Things J. **5**(5), 3637–3647 (2018)
5. Rb, A., Kr, S.K.: Credit card fraud detection using artificial neural network. Glob. Transitions Proc. **2**(1), 35–41 (2021)
6. Taha, A.A., Malebary, S.J.: An intelligent approach to credit card fraud detection using an optimized light gradient boosting machine. IEEE Access **8**, 25579–25587 (2020)
7. Zhu, X., et al.: Intelligent financial fraud detection practices in post-pandemic era. Innov. **2**(4), 100176 (2021)
8. Abdallah, A., Maarof, M.A., Zainal, A.: Fraud detection system: a survey. J. Netw. Comput. Appl. **68**, 90–113 (2016)
9. Al-Hashedi, K.G., Magalingam, P.: Financial fraud detection applying data mining techniques: a comprehensive review from 2009 to 2019. Comput. Sci. Rev. **40**, 100402 (2021)
10. Hilal, W., Gadsden, S.A., Yawney, J.: Financial fraud: a review of anomaly detection techniques and recent advances. Expert Syst. Appl. **193**, 116429 (2022)
11. Buvé, C., et al.: Application of multivariate data analysis for food quality investigations: an example-based review. Food Res. Int. **151**, 110878 (2022)
12. Schatz, M.C., et al.: Inverting the model of genomics data sharing with the NHGRI Genomic data science analysis, visualization, and informatics lab-space. Cell Genomics **2**(1), 100085 (2022)
13. Weisser, J., Pohl, T., Heinzinger, M., Ivleva, N.P., Hofmann, T., Glas, K.: The identification of microplastics based on vibrational spectroscopy data – a critical review of data analysis routines. TrAC Trends Anal. Chem. **148**, 116535 (2022)
14. Liu, Z., Chen, J., Wang, K., Wang, B., Zhang, C.: Current status and opportunities in adaptive data analysis for therapeutic cell manufacturing. Curr. Opin. Biomed. Eng. **20**, 100351 (2021)
15. Ho, A.T., Morin, L., Paarsch, H.J., Huynh, K.P.: A flexible framework for intervention analysis applied to credit-card usage during the coronavirus pandemic. Int. J. Forecast. **38**(3), 1129–1157 (2022)
16. Li, J., Liu, Y., Li, Q.: Intelligent fault diagnosis of rolling bearings under imbalanced data conditions using attention-based deep learning method. Measurement **189**, 110500 (2022)
17. Lee, G.K.K., Kasim, H., Sirigina, R.P., How, S.S.Q., King, S., Hung, T.G.G.: Smart robust feature selection (SoFt) for imbalanced and heterogeneous data. Knowl. Based Syst. **236**, 107197 (2022)
18. Huynh, T., Nibali, A., He, Z.: Semi-supervised learning for medical image classification using imbalanced training data. Comput. Methods Programs Biomed. **216**, 106628 (2022)
19. Umarani, V., Julian, A., Deepa, J.: Sentiment analysis using various machine learning and deep learning techniques. J. Niger. Soc. Phys. Sci. **3**(4), 385–394 (2021)
20. Sudharsanan, R., Gopirajan, P., Suresh Kumar, K.: Efficient feature extraction from multispectral images for face recognition applications: a deep learning approach. J. Phys. Conf. Ser. **1767**(1), 012061 (2021). https://doi.org/10.1088/1742-6596/1767/1/012061

21. Xu, Y., Yu, Z., Chen, C.P., Liu, Z.: Adaptive subspace optimization ensemble method for high-dimensional imbalanced data classification. IEEE Trans. Neural Netw. Learn. Syst. **34**(5), 2284–2297 (2021)
22. Dablain, D., Krawczyk, B., Chawla, N.V.: DeepSMOTE: fusing deep learning and SMOTE for imbalanced data
23. Koziarski, M., Bellinger, C., Woźniak, M., Gama, J., Jorge, A., García, S.: RB-CCR: radial-based combined cleaning and resampling algorithm for imbalanced data classification. Mach. Learn. **110**, 3059–3093 (2021)
24. Ben Hajhmida, M., Oueslati, O.: Predicting mobile application breakout using sentiment analysis of Facebook posts. J. Inf. Sci. **47**, 502–516 (2020)
25. Ghorbani, R., Ghousi, R.: Comparing different resampling methods in predicting students' performance using machine learning techniques. IEEE Access **8**, 67899–67911 (2020). https://doi.org/10.1109/ACCESS.2020.2986809
26. Lv, D., Ma, Z., Yang, S., Li, X., Ma, Z., Jiang, F.: The application of SMOTE algorithm for unbalanced data. In: AIVR 2018: Proceedings of the 2018 International Conference on Artificial Intelligence and Virtual Reality, pp. 10–13 (2018). https://doi.org/10.1145/3293663.3293686
27. Ramesh, J., Aburukba, R., Sagahyroon, A.: A remote healthcare monitoring framework for diabetes prediction using machine learning. Healthc. Technol. Lett. **8**(3), 45–57 (2021). https://doi.org/10.1049/htl2.12010

# An Enhanced Learning Model Based on an Improved Random Forest Classifier and an Integrated Attribute Selector for Healthcare Datasets

S. Rajeashwari(✉) and K. Arunesh

Department of Computer Science, Sri S. Ramasamy Naidu Memorial College, Sattur, India
rajeeragavan@gmail.com

**Abstract.** Everywhere, there is an increase in chronic diseases. These diseases affect the immune system and quality of life. Chronic illness treatment can be expensive and difficult to manage. People should focus on healthy lifestyle habits like exercising regularly, eating a well-balanced diet, and getting enough sleep because prevention is the best option. These illnesses can make it hard for doctors to treat. Dimensionality reduction methods allow for the processing of large data sets, allowing for decision support in the area of chronic diseases. A large number of symptoms are included in the records that are important to predict disease. By using feature selection methods, there is a need to remove unnecessary and meaningless symptoms from data sets so as to deliver better classification accuracy. By using feature selection methods, prediction accuracy and model complexity can be reduced. Furthermore, it could improve the interpretation of models and make them easier to understand and provide more insight into the data that lies beneath them. In addition, the option of selecting features may speed up model creation and reduce training time. Moreover, it may reduce the likelihood of overfitting which can have an adverse effect on model performance. The primary contribution of this paper is a comparative analysis of the effects of the analyzed filtering methods, which include the Wrapper methods, which include the Best First Search (BFS) method, the Linear Forward Selection (LFS) method, and the Greedy Step Wise Search (GSS) method, as well as the Information Gain (IG) method, Chi-Square (CS) method, and the Cor-relation Feature Selection (CFS) method. The results show that the Wrapper approach, based on the Best First Search method, has been shown to be more accurate and effective in selecting features. Furthermore, it has been noted that the data gain and correlation element choice strategies have become more persuasive in terms of preciseness. A Random Forest algorithm was used in Python Anaconda as a classifier for this analysis. Attribute significance analysis was performed on the breast cancer, diabetes, kidney disease, and heart disease study datasets. In terms of accuracy and time taken to complete, the CFS approach was better than other filtering techniques. The CFS method, as well as the datasets for heart disease, diabetes, kidney disease, and breast cancer, all had an ac-curacy rate of 96.5 percent. Likewise, similar idle periods were recorded for each of the datasets: 1.07 s, 1.04 s, 1.06 s and 1.01 s. As demonstrated by the accuracy rate and the time taken to complete the analysis, the CFS method was a reliable and efficient method of medical analysis. Furthermore, the CFS method


R. Geetha et al. (Eds.): AAIMB 2023, CCIS 2203, pp. 234–249, 2025.
https://doi.org/10.1007/978-3-031-73068-9_19

has a substantially higher accuracy rate compared to other filter methods. That indicates that the CFS approach can be considered as a viable option for diagnosis. BFS performed exceptionally well in terms of covering strategies compared to other methods. The breast cancer, diabetes, and heart disease datasets achieved maximum accuracy of 94.7%, 89.9%, 92.3%, and 92.8%, respectively. For specific datasets, the same method has been used to detect a latency delay of 1.42 s, 1.44 s, 1.32 s and 1.37 s. An improved random forest classifier has also been used to evaluate the hybrid approach. The classification and clustering are combined by an improved random forest classifier. Its performance has been recorded and verified in 12 separate chronic disease datasets. The results of this classification were very consistent and effective. A mean of 96.7%, 96.5%, 95.6% and 96.2% was achieved for accuracy, specificity, sensitivity and F1-score metrics, respectively.

**Keywords:** Correlation Feature Selection method · Information Gain method · Chi-Square method · Best First Search method · Linear Forward Selection method · Greedy Step Wise Search method · Improved Random Forest

# 1 Introduction

Chronically ill people have weaker immune systems and other bodily functions that can cause long-term health issues and ultimately limit their mobility and independence. The cost of healthcare continues to rise while the standard of living is severely impacted. Additionally, it has an effect on the patient's emotional stability [1]. With the right diagnosis, many chronic diseases can be treated, though therapy can be time-consuming and difficult at times [2]. However, the majority of these conditions remain incurable. The medications available will assist in alleviating symptoms. Patients may live longer as a result of this [3]. Improved public health necessitates accurate diagnosis and treatment. In India, the most common chronic diseases include diabetes, breast cancer, kidney disease, and heart disease [4]. However, it is challenging to treat these diseases because of the rapid expansion of the population. Additionally, the large number of patients has necessitated frequent manual examination of various diseases, which has resulted in ineffective treatment recommendations. Making a diagnosis can take a lot of time and effort for many people who have a chronic illness. Blunders can likewise happen when side effects are physically recorded and analyzed. As a result, effective solutions to this issue necessitate the use of machine learning technologies [4]. The use of machine learning to automatically classify patients based on their symptoms is a topic of ongoing research [4, 5]. Additionally, disease datasets can be analyzed using machine learning to identify the primary symptoms of a particular chronic disease [5, 6]. However, it is difficult to identify the most common symptoms in unstructured raw datasets because this kind of data makes machine learning algorithms less effective.

An unstructured raw data set can be cleaned up using attribute selection, a reliable pre-processing technique [7]. By selecting the relevant properties, it works as an optimization tool to reduce the size of the data [8]. The accuracy and latency of the classification are both reduced as a result [9]. The following are the primary motivations for employing attribute selection approaches: accelerate the execution of algorithms used in data mining; simplify the model and make it easier to understand; a rise in the rate of classification and prediction accuracy; and cut down on overfitting.

The primary objective of this study is to investigate the effects of filter and wrapper approaches on the classification of chronic diseases from datasets and common attribute selection techniques. The effects of filters and packaging on classification performance are tested using data sets on breast cancer, diabetes, heart disease, kidney disease, and diabetes. Using Python software, an experiment demonstrates the benefits and drawbacks of wrappers and filters [9]. Due to its adaptability and simplicity, Random Forest is a classifier utilized in research. Chi-Square, Information Gain, or Correlation Feature Selection (CFS) are the filters used in the analysis. The wrappers, on the other hand, are the approaches known as Best First Search (BFS), Greedy Stepwise Search (GSS), and Linear Forward Selection (LFS). While WrapperSubsetEval is the wrapper methods' predefined search method, the ranking algorithm is the filter technique's predefined search schema. Based on the findings of this study, it was also discovered that BFS outperformed other wrapper methods and that the CFS approach produced superior results. A brand-new hybrid attribute selector with three levels was proposed as a result. BFS was used as a wrapper, and CFS was used as a filter. A classification model employing an enhanced random forest is presented based on the scored attributes. Calculations and additional analysis are performed on the implemented results.

## 2 Literature Review

There are many different ways to collect chronic disease data sets. Healthcare professionals store enormous amounts of patient-related data in unstructured samples [9]. In qualitative research, numerous statistical and machine learning-based analyses are utilized. Medical decisions should benefit from this research. Chronic illness may present with a number of possible signs and symptoms. Important considerations include age, total cholesterol, hypertension, diabetes, intrinsic cardiac risk factors, obesity, and inactivity. A realistic understanding of the risk factors associated with the particular condition is essential for decision-making support. These symptoms are taken from databases using attribute selection methods like wrappers and filters. The selection of features is different for each analysis.

L.A. Simons et al. [10] used decision tables, RBF classifiers, multilayer perceptron's, and support vector machines to analyze coronary artery disease-related datasets and accurately identify Framingham risk disorder in older Australians with heart disease. Surekha S. and others examined the effects of various thyroid disease diagnostic techniques [11]. They compared the efficiency of the fast reduction and Johnson reduction methods of coarse set theory with that of the genetic algorithm and the particle swarm optimization algorithm. In thyroid classification, decision trees and naive Bayes have been utilized as classifiers. The results demonstrated that evolutionary approaches perform better than coarse set theory approaches when it comes to the accuracy of minimal attribute subsets.

Anurag Kumar Verma et al. created a novel method for computer-aided learning [12] for accurately predicting skin conditions. As classification models, AdaBoost, bagging, and gradient boosting were utilized. Using a novel feature selection method that prioritizes the 15 most relevant attributes, an ideal subset of features was chosen. The dermatological test data samples' prediction accuracy improved as a result. Ibrahim M. El-Hasnony compared various approaches to data reduction in the author study [13]. Using classification algorithms like MLP and ANN, the accuracy of these methods for reducing data has been evaluated. The observation indicates that principal component analysis, gain ratio, and correlation feature selection methods performed significantly below the fuzzy rough feature selection method. BM Patil [14] developed a hybrid machine learning model for predicting diabetes risk factors. It used K-means clustering to classify the data using the C4.5 method. With an extremely high accuracy of 92.38 percent, [15] proposed a model that could predict diseases of the knee joints based on VAC signals. The trait assessors were the a priori algorithm and the genetic algorithm, while the classifiers were the LS-SVM. The classification accuracy was 94.31% when the evolutionary method, the a priori algorithm, and LS-SVM were combined.

**Table 1.** Computational Techniques used in the work

| Techniques Description | Methods Description |
|---|---|
| Best First Search Algorithm (BFS) | The attributes that seem to be the best are determined in a priority queue and heuristic, which then analyses them |
| Chi-square Method | Calculating chi-square's value for a class label can demonstrate an attribute's significance |
| Correlation Feature Selection Algorithm (CFS) | An attribute's predictive nature determines how much redundancy exists between two attributes |
| Greedy Step-wise Algorithm | This process finds subsets of attributes greedy from one direction or another. In this process each attribute is assessed on its own basis, and the addition or elimination of attributes leads to a reduction in evaluation. Similarly, reordering and traversing select attributes can generate the rank-based attribute sets |
| Information Gain Method (IG) | The net gain of information associated with the class determines the value of the class attribute **Hence, H (Class) – H (Class\|Attribute) is the InfoGain (Class, Attribute)** |
| Linear Forward Selection (LFS) Method | To determine the impact of attribute selection, an evaluation is based on a limited set of attributes. The restricted set of attributes is defined by a first ordering or ranking process. The direction of the search can be forward or back, with a few steps backwards |

*(continued)*

**Table 1.** (*continued*)

| Techniques Description | Methods Description |
|---|---|
| Ranker Algorithm | When selecting a number of attributes, the ranking attributes are selected from the list of ranked attributes |
| Wrapper Subset Eval Algorithm | For a specific set of attributes using a learning method and a cross validation method, cross validation determines the accuracy of the learning method's classification |

Sushruta Mishra et al. [16] new embedded model applies a wrapper technique based on a random forest classifier and information mining to a variety of disease datasets, including those for diabetes, lung disease, hepatitis, and heart disease. The classification accuracy with the compressed function set was 92.86 percent. A cross breed PCA-LDA model blend for the order of neurodegenerative illnesses was introduced in [17]. An accurate accuracy rate was achieved after ranking the statistical attributes and optimizing them. A 10-fold cross-validation-based sequential forward selection (SFS) method for a pulmonary disease dataset was discussed by the authors of [18]. It conveyed momentous outcomes with the quickest execution times on considerably more information tests while involving SVM for order approval. In order to analyze the existing studies that are relevant to the scope of this research, a comprehensive literature review was carried out. Numerous attribute selection methods for disease diagnosis have been presented by a number of authors. Particle swarm optimization, genetic algorithms, and other common wrapper methods were also discussed.

## 2.1 Evaluation and Comparison of Filter-Wrapper Approach

This research aims to ascertain whether the filter and wrapper strategy has an impact on classification performance. The issue of how to classify disease data has been a key problem in this case. This section provides popular datasets on chronic diseases for research purposes. Public health issues related to chronic diseases around the world are a matter of concern. Diabetes, hepatitis, liver disease, lung cancer, breast cancer, and heart disease are the most prevalent chronic diseases [19]. Due to this, a large number of patients are affected by these prolonged conditions. Therefore, more rapid methods for providing efficient support to patients are essential. Four clinical data sets are used in this study. Diabetes, breast cancer, kidney disease, and heart disease are all included in this. The most prevalent chronic diseases are listed here [20]. The data sets have been collected, processed and compiled from the UCI Repository. In addition, the breast cancer data set contains 10 distinct attributes and 286 separate records while in India's PIMA diabetes dataset there are 8 distinct characteristics and 768 individual entries [21].

There are several computational techniques and methods used in this study, including the Best First Search Algorithm, Chi-Square, Correlation Feature Selection Algorithm, Greedy Stepwise Algorithm, Information Gaining Method, Linear Forward Selection Method, Ranker Algorithm, and Wrapper Subset Evaluation Algorithm.

The overall model of a framework's work process as suggested in this study is shown in Fig. 1. An input file used for preprocessing, which finds and removes features that are not required, is a chronic disease data set. Moreover, information pre-handling manages missing qualities in datasets and the event of dull qualities. This column's mean is used to replace values that are missing or redundant. The data can now be processed with the help of the attribute selection tool.

## 3 Attribute Selection Process Using Filter-Wrapper Approach

In this study, the process was divided into two separate steps. The first component is an attribute selector, which means that the subsets of attributes are evaluated. First of all, the classification algorithm does not apply to attribute selection in a filter approach. These filtering methods are used in order to select the relevant characteristics from a set of chronic disease data that has already been processed. A ranking method is used to create an orderable list of attributes for a selector attribute. This method results in the reduction and optimization of a set of attributes that are applied to classification by an algorithm based on random forest. The second section contains a search schema that allows attributes to look for spaces of possible subsets where the most relevant attribute is located. Combining the wrapper method with a suitable algorithm for searching is necessary to achieve the best set of attributes. The training of the system prototype with a subset of attributes is carried out using wrapper methods. This study covers all filtering techniques used by CFS, IG and Chi-square. The classification process, in which the reduced chronic disease datasets are classified using the random forest algorithm,

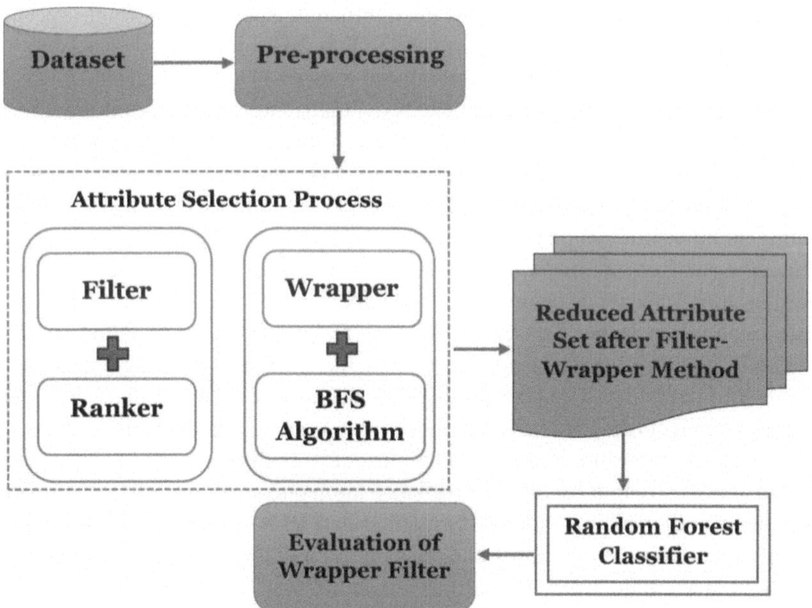


**Fig. 1.** General workflow model of the proposed framework

follows the application of the filter and wrapper method to generate a reduced set of attributes. Diabetes, breast cancer, kidney disease, and heart disease were the conditions on which the random forest algorithm was tested.

In the wrapper approach, the attribute selector was Wrapper Subset Evaluation, and the best first search was linear forward search. The choice strategy and the covetous stepwise hunting method were used as search calculations. In order to demonstrate the significance of each symptom in the use of filters and wrappers, a graph of attribute significance has been presented. The importance of the symptoms was considered and a set of significance charts has been established for determining which symptoms in the dataset are important and which are not. A score of 10 is given to each attribute in the significance Table 1 by an attribute selector. The results of the chart show that only minor symptoms have been removed from the data set.

## 3.1 Process of Attribute Selection Using Filter Approach

The filtering strategy's significance in relation to chronic disease datasets is emphasized in this section. The filtering method was used to narrow the list of characteristics down to just the most important side effects for a variety of persistent disease datasets. From heart disease records, 270 cases with 13 distinct initial characteristics were gathered. The attribute significance plot can be used to determine the individual attribute score. The filtering method deems the attributes ca and exang to be least important. Figure 2 shows highlighted the attribute significance plot for the diabetes data set and the attribute significance analysis for the heart disease data. The diabetes data set contains 768 instances with 8 attributes. The skin attribute received a score of zero, which is quite low. The significance graph for the attribute is displayed below. With just a meaning of 0, the illumination characteristic was the most un-significant. The less important aspects are given less weight. The IG and chi-square methods reduced the heart disease dataset to 10 attributes, whereas the CFS method produced a set of 11 relevant attributes. Priority analysis revealed that the optimized diabetes data attribute set supported the CFS and IG filtering methods with seven attributes. The IG filter method produced an attribute set with 8 attributes, whereas the chi-square method produced a set with 6 attributes and a reduced set with 7 attributes. Table 2 gives a comprehensive analysis of how the separating methods affect a particular set of characteristics and infection records.

The decreased and optimized attribute set shall be classified on the basis of a Random Forest algorithm, once it has been determined which score levels are necessary for each attribute in chronic disease datasets. A filtering technique, using an automatic forest classification method, has been applied to classify the Chronic Disease Dataset and its results are interesting. The CFS method has been more effective than any filtering techniques when it comes to accuracy and execution time. In this work, the accuracy rate was chosen as an algorithm used to classify attributes. In evaluating trait subsets with a significant area of strength for and little reliance upon one an unrelated, CFS performed better than any other method. The CFS method had an accuracy rate of 96.5% and a minimum execution time lag of just one when applied to the breast cancer dataset. The CFS method is accurate at 92.8% and has a delay of less than one second when used in the data on heart disease. Table 3 shows a filtering- wrapper based tabular view of classification accuracy and latency analysis.

**Table 2.** Overall analysis impact of Filter-Wrapper Approach

| Chronic Disease Dataset | Filter Approach | | | Wrapper Approach | | |
|---|---|---|---|---|---|---|
| | CFS | IG | Chi-Square | BFS | LFS | GSS |
| **Heart Disease Dataset** | | | | | | |
| Number of instances | 270 | 270 | 270 | 270 | 270 | 270 |
| Initial Attribute set | 13 | 13 | 13 | 13 | 13 | 13 |
| Reduced Attribute set | 11 | 10 | 10 | 10 | 10 | 10 |
| **Diabetes Dataset** | | | | | | |
| Number of instances | 768 | 768 | 768 | 768 | 768 | 768 |
| Initial Attribute set | 8 | 8 | 8 | 8 | 8 | 8 |
| Reduced Attribute set | 7 | 7 | 6 | 7 | 7 | 7 |
| **Kidney Disease Dataset** | | | | | | |
| Number of instances | 400 | 400 | 400 | 400 | 400 | 400 |
| Initial Attribute set | 25 | 25 | 25 | 25 | 25 | 25 |
| Reduced Attribute set | 22 | 22 | 22 | 22 | 22 | 22 |
| **Breast Cancer Dataset** | | | | | | |
| Number of instances | 286 | 286 | 286 | 286 | 286 | 286 |
| Initial Attribute set | 9 | 9 | 9 | 9 | 9 | 9 |
| Reduced Attribute set | 7 | 8 | 7 | 8 | 8 | 7 |

### 3.2 Process of Attribute Selection Using Wrapper Approach

This section provides a summary of the effects that the wrapper strategy had on the chronic disease dataset. Attribute significance plots were created for each of the three chronic datasets. The wrapper method was tested on the heart disease data instance. The selected attribute in an attribute importance chart. The three characteristics that were found to be least relevant were Ca, Exang, and Oldpeak. Skin characteristics were excluded because they were included in the diabetes data set with a significance level of 0.3. With exposure emerging as the least significant characteristic, each method received a score of approximately 0.3. Three of the attributes are clearly less relevant to the heart disease data set than the others, as the graphical analysis demonstrates. In the samples of the diabetes dataset and the breast cancer dataset, only one attribute was of minor importance. The BFS, LFS, and GSS methods all produced a reduced set of attributes with the same number of attributes: Among the ten that were removed, three were of no use. A wrapper method was tried on the diabetes dataset to find the best set of attributes. It was determined that one of the eight characteristics was irrelevant. As a result, it was excluded from the result sets of all three wrapper methods. The BFS and LFS approaches resulted in a total of eight attributes being removed from the breast cancer data set. Using the GSS method, the reduced set only contained seven attributes. Table 2 demonstrates the application of the wrapper approach through a tabular analysis of the reduced attribute set.

**Table 3.** Classification accuracy and latency analysis using filter-wrapper approaches

| Disease Dataset on Filter Approach | Classifier | Accuracy (Ranker Search Method) | Latency (Ranker Search Method) | Disease Dataset on Wrapper Approach | Accuracy (Wrapper Subset Evaluation Method) | Latency (Wrapper Subset Evaluation Method) |
|---|---|---|---|---|---|---|
| **Heart Disease Dataset** | | | | | | |
| CFS | RF | 92.8% | 1.07 s | **BFS** | 94.7% | **1.32 s** |
| IG | RF | 90.6% | 1.34 s | **LFS** | 92.6% | **1.34 s** |
| Chi-square | RF | 91.3% | 1.16 s | **GSS** | 94.2% | **1.46 s** |
| **Diabetes Dataset** | | | | | | |
| CFS | RF | 89.9% | 1.04 s | **BFS** | 89.9% | **1.42 s** |
| IG | RF | 89.2% | 1.14 s | **LFS** | 90.6% | **1.54 s** |
| Chi-square | RF | 88.9% | 1.05 s | **GSS** | 95.6% | **1.72 s** |
| **Kidney Disease Dataset** | | | | | | |
| CFS | RF | 92.3% | 1.06 s | **BFS** | 92.3% | **1.36 s** |
| IG | RF | 88.8% | 1.22 s | **LFS** | 88.8% | **1.42 s** |
| Chi-square | RF | 91.2% | 1.11 s | **GSS** | 91.2% | **1.61 s** |
| **Breast Cancer Dataset** | | | | | | |
| CFS | RF | 96.5% | 1.01 s | **BFS** | 92.8% | **1.31 s** |
| IG | RF | 89.7% | 1.09 s | **LFS** | 89.7% | **1.59 s** |
| Chi-square | **RF** | **87.8%** | **1.21 s** | GSS | **96.8%** | **1.38 s** |

In order to classify the reduced set of attributes further, a random forest algorithm is used. The best first search method shows the highest accuracy of 94.7% when it comes to detecting heart disease. Linear forward selection, which has an accuracy of 92.6%, appears to be some sort of divergence from the greedy stepwise search and these two methods. When the wrapper technique was applied to the diabetes data, a simple stepwise search accuracy of 95.6% had been achieved and completed in 1.77 s. An accuracy of 90.6% and a latency delay of 1.56 s were achieved by a linear search with forward selection. The best first search method is relatively faster than the linear forward selection method, with an execution time of just 1.56 s and a curiosity rate of 90.6. The best first search method is at least 92.3% accurate for the prediction of kidney disease.

With an accuracy of 88.8%, the linear forward selection method falls some-where in between the best first search and the greedy stepwise search. The greedy stepwise method is remarkably accurate at 96.8% in classifying breast cancer, compared to a low accuracy of 92.8% with the first search method. The greedy stepwise seek latency delay, which is 1.08 s lower than in the other two wrapper methods, is also significantly lower. Overall, it was found that the BFS approach had more favorable results than other approaches. The simplicity of switching between different solutions paths has led to the excellent performance of the BFS method. It's capable of finding a satisfactory solution without unravelling all the knots and getting lost in it. In addition, the BFS approach has considerably reduced time complexity compared to other approaches. A summary of the findings is presented in Table 3.

### 3.3 Proposed Hybrid Classification Model

This section deals with the Hybrid Classification Model of Health Datasets. The proposed machine learning model is accompanied by a enhanced random forest classification and an integrated attribute selection method, shown in Fig. 2. The input of the model is a chronic distress dataset obtained from an appropriate and reliable source. Preprocessing and unstructured filtering are usually required to transform these raw data sets into an appropriate format. Another step is the preprocessing of data from chronic disease samples. These examples are then transferred to a biased data minimizer following the removal of equivocality. The preprocessed data samples are not balanced or even balanced most of the time. Information may not be perceived in a typical manner, which hinders the effective grouping of information. That unequal dissemination of information events. This model uses a transformation technique to effectively deal with the skewed data. The fundamental component of the transformation method is an exponential value between –5 to 5. The highest value is selected when taking into consideration all values within the specified range. The optimal value shall be the values which produce an optimum regular distribution curve.

$$\lambda + \gamma = log(\mathrm{y} + \lambda 2), \lambda 1 = 0 \tag{1}$$

The attribute selection module receives the modified data that has been obtained. In order to provide a more accurate range of features from the COPD dataset, Clustering is used. In this Attribute Selector module K-Means clustering is used together with CFS as a filter and BFS as its wrapper. The similarity of attributes in the attribute sets that serve as inputs is used to create several clusters. "K" characteristics have been selected as cluster centers by K-means Clustering, based on the original disease data set. Using the similarity measure, it is possible to determine distances from a cluster's mean to its attributes. An attribute with the highest similarity shall be given to the appropriate cluster. The Euclidean distance applied to the model we propose reveals attributes shared by similarity, as described in Eq. 2. In the equation, the sum of the squared differences between the coordinates of the two vectors x = (x1, x2) and y = (y1, y2) is referred to as the squared distance between them.

$$v_{xy}^2 = (x1 - y1)^2 + (x2 - y2)^2 \tag{2}$$

$$v_{xy} = \sqrt{(x1 - y1)^2 + (x2 - y2)^2} \tag{3}$$

Integrated Supervised K-Means, a brand-new variant of the K-Means algorithm, is discussed and demonstrated. The regression method is called Integrated Supervised

K-Means, since it consists of unsupervised clustering. In this way the procedure is reconstituted and each sample recalculated as well as its nearest clusters allocated. Integrated Supervised K-Means uses the parameter $p_i$, which represents the sixth sample vector (i = 1,... n), zi represents the cluster's arrangement for the sixth sample, $s_i$ {1,... M}, i = 1...n, I(I,m) returns 1 for $s_i = m$, otherwise 0. $C_m$ is the mean vector corresponding to the $m^{th}$ cluster (m = 1, K), $C_S$ is the cumulative sum of squares distance, β is the convergence factor for the linear difference in the cumulative sum of squared distances, and CMP is the Cluster Member Probability of a Considered Sample. If $s_i = m$, I(I,m) returns 1, otherwise it returns 0.

The Euclidic distance from the measured mean shall be calculated for each sample observation. Each sample is assigned a subsequent cluster. In the case of logistic regression equations from 4 to 7 these steps can be taken:

$$prob(s_i = m) = \frac{1}{1 + \sum_{i=1}^{m} e^{\alpha_m}.p_i} \tag{4}$$

$$prob(s_i = 1) = \frac{e^{\alpha_1}.p_i}{1 + \sum_{i=1}^{m} e^{\alpha_m}.p_i} \tag{5}$$

$$prob(s_i = 2) = \frac{e^{\alpha_2}.p_i}{1 + \sum_{i=1}^{m} e^{\alpha_m}.p_i} \tag{6}$$

$$prob(s_i = m - 1) = \frac{e^{\alpha_{m-1}}.y_i}{1 + \sum_{i=1}^{m} e^{\alpha_m}.p_i} \tag{7}$$

For this purpose, the predicted regression value is used to estimate the cluster member probability of a single sample from each cluster. By adjusting this probability value to a descending order, the proportion of those two highest probabilities is determined. The cumulative sum of the total distance and the cluster mean are recalculated. The convergence factor shall be compared to the variation of the cumulative amount of squared distance measured in the previous round.

The BFS method, which evaluated all attribute values and found a smaller optimal subset of attributes, has also been used to feed in the nonredundant characteristics. In order to classify the attribute set, which determines a patient's risk of developing chronic diseases an improved Random Forest model has then been applied. Finally, by comparing the performance of the developed model with key operating parameters, it has been determined that this classification is effective.

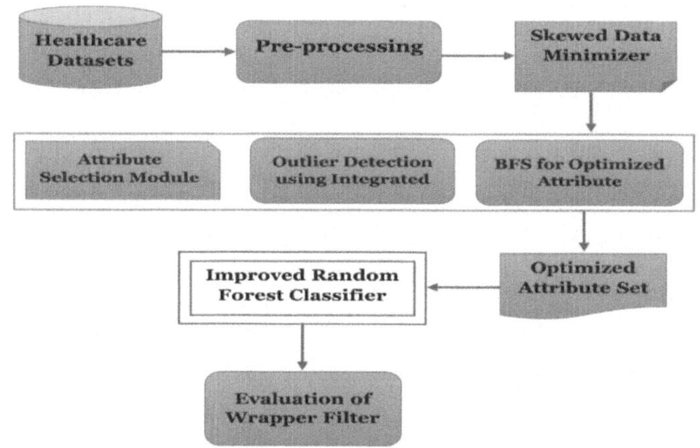


**Fig. 2.** Attribute Selection Method and an Improved Random Forest Classifier

# 4 Results and Analysis

In this section, our proposed model with Attribute Selector was evaluated against a number of Chronic Disease Datasets and its performance in comparison to specific Performance Parameters has been assessed. The BFS method was used to create an optimized attribute subset. It was used for reducing the number of dimensions within 15 sets of disease data generated from UCI's database. It has also been compared to the CFS filter and BFS wrapper method when selecting a reduced attribute set. Table 4 shows the result. It is very clear, in particular for almost all medical records with more attributes, that this difference of performance exists. With a set of somewhat less quality, the proposed model performs significantly better. However, although only eight credits were added to the information on diabetes, it was less clear whether he had been eligible given that his real estate set would be limited to six. Therefore, it can be concluded that if disease data samples analyzed are large and contain a larger relative number of attributes, the recalculation dimension of the proposed "Attribute Selector" is most efficient and beneficial. In contrast, when the model is used to evaluate smaller sets of data, it performs just like a wrapper or filter.

The proposed selector attribute method shall be combined with a random forest in order to classify the dataset. The results of the examination showed that random forest classification has produced a very impressive outcome consistent with almost all disease datasets (see Table 5 for overall performance).

The accuracy of the classification of cervical cancer data was 94.8%, while that of the classification with the highest accuracy was 98.4%. Data on autism spectrum disorders showed a maximum specificity of 93.6%, while the data from Cervical Cancer had a minimum specificity of 98.2%. The F1-score, which is a harmonic mean for specificity and sensitivity, was considered as well. The F score for autism was relatively low, at 93%, while the F score for thyroid and cervical cancer was the highest, at 97.8%. In general, 96.8%, 96.5%, 95.6% and 96.2% were produced by means of the proposed

**Table 4.** Overall Performances of the Proposed Method and Other Wrapper Methods for Chronic Disease Datasets.

| Chronic Disease Dataset | No. of Instances | Attribute Set | CFS | BFS | CFS + BFS | K-means | Proposed work |
|---|---|---|---|---|---|---|---|
| Arrhythmia | 452 | 279 | 271 | 268 | 264 | 274 | 262 |
| Autistic Disorder | 292 | 21 | 20 | 19 | 19 | 20 | 18 |
| Breast cancer | 286 | 9 | 8 | 8 | 8 | 9 | 8 |
| Breast cancer Wisconsin | 699 | 10 | 9 | 9 | 8 | 10 | 8 |
| Cervical Cancer | 858 | 36 | 34 | 33 | 32 | 35 | 30 |
| Chronic Kidney Disease | 400 | 25 | 22 | 22 | 21 | 23 | 20 |
| Hepatitis | 155 | 19 | 18 | 17 | 16 | 19 | 15 |
| Liver Disorder | 345 | 7 | 7 | 6 | 6 | 7 | 6 |
| Lung cancer | 32 | 56 | 53 | 51 | 50 | 54 | 49 |
| Parkinson | 197 | 23 | 21 | 20 | 19 | 22 | 18 |
| Pima Indians Diabetes | 768 | 8 | 7 | 7 | 6 | 8 | 6 |
| Primer Tumor | 339 | 17 | 16 | 15 | 14 | 17 | 13 |
| Statlog (Heart) | 270 | 13 | 11 | 10 | 9 | 12 | 9 |
| Thyroid Disease | 7200 | 21 | 18 | 17 | 16 | 20 | 16 |

characteristic selector approach to exactness, explicitness, responsiveness and F Score (Table 5).

In 14 different datasets of chronic diseases, the CFS filter and BFS wrapper method have been demonstrated and evaluated with a novel integrated selector attribute methodology. The impact of the proposed model on datasets with more attributes was easy to see, given the importance of the attributes and the deletion of outliers. The performance of the classifier was assessed by a range of indicators, and an excellent peak metric score has been obtained. The improvement in the accuracy and error rate of an improved random forest classification model was significantly better than that of a standard Random Forest Classification. In a better random forest, the clustering method makes it more efficient and less likely to make errors because most of the data samples are classified.

**Table 5.** Overall Performances of the proposed work for all Chronic Disease datasets.

| Chronic Disease Dataset | Accuracy | Specificity | Sensitivity | F1-Score |
|---|---|---|---|---|
| Arrhythmia | 96.5 | 96.4 | 95.5 | 96 |
| Autism Disorder | 94.5 | 93.6 | 92.9 | 93 |
| Breast Cancer | 96.2 | 96.1 | 95.6 | 95.7 |
| Breast Cancer Wisconsin | 97.3 | 97.5 | 96.8 | 97.1 |
| Cervical Cancer | 98.4 | 98.2 | 97.4 | 97.8 |
| Chronic Kidney Disease | 96.4 | 96.2 | 95.9 | 96 |
| Hepatitis | 97.9 | 97.5 | 97.3 | 97.4 |
| Liver Disorder | 96.9 | 96.6 | 96.2 | 96.4 |
| Lung Cancer | 95.7 | 95.6 | 95.1 | 95.2 |
| Parkinson's | 94.9 | 94.5 | 94.3 | 94.4 |
| Pima Indians Diabetes | 96.8 | 96.2 | 95.6 | 95.8 |
| Primer Tumour | 97.7 | 97.2 | 97.1 | 97.5 |
| Statlog (Heart) | 97.2 | 96.9 | 96.7 | 96.8 |
| Thyroid Disease | 98.4 | 97.9 | 97.6 | 97.8 |

## 5 Conclusions and Future Scope

The impact of the selection of wrappers and filters on prediction performance from various chronic disease data sets is examined in this paper. In the first part, a comparison was made between wrappers and filters in four chronic disease datasets. Datasets covered diabetes, breast cancer, kidney disease, and heart disease. This study was based on the BFS, LFS, and GSS wrapper techniques, as well as CFS, IIG, Chi Square or Square filters. The CFS method had an optimal accuracy of 92.8%, 92.3%, and 96.5% for heart disease, diabetes, kidney disease, and breast cancer. A similar strategy was utilized to notice dormancy postponements of 1.07s, 1.04s, 1.06s and 1.01s in the separate datasets. The best execution of characterization between covering strategies was carried out by BFS. For heart disease, diabetes, kidney disease and breast cancer, the highest accuracy was 94.7%, 89.9%, 92.3% and 92.8% respectively. A similar method has been used for identifying inactivity in individual datasets with a delay of 1.32 s, 1.42 s, 1.36 s and 1.31 s. Moreover, it has been observed that despite the highest accuracy for classification, wrappers have taken up a great deal of computation. Filter based classifications had a shorter execution period and lower operating costs than wrapper-based classifications. It was argued that many characteristic choice strategies did not give much weight to the design of properties, and that the possibility of exceptions was not taken into account when the quality set was simplified.

In the second section of the paper, an integrated hybrid attribute selection method has been suggested. Integrated Supervised K-Means, a brand-new variant of K-Means cluster analysis, was used in conjunction with the CFS and BFS methods. By participating in the cluster mean finding procedure, this brand-new k-means algorithm generates cluster member probabilities using a regression approach. For the identification of cases which do not have any relevance, these probabilities shall be applied. Improved random forests and a suggested classification method ensure that performance is maximized. The majority of these cases were clustered and classified by the improvable random forest. It was evaluated on 14 distinct chronic disease datasets using various performance metrics. The optimal mean values were 96.2%, 96.5% and 96.2% for accuracy, specificity, sensitivity and f Score respectively. The new integrated attribute evaluator model and the improved random forest classifier had a mean accuracy of 96.85, which was higher than the random forest's 96.1 percent accuracy rate. The mean speed of error is about 0.642 for the mixture order model. Consequently, it could be used by healthcare professionals to provide decision support for the evaluation of their patients' health conditions. Furthermore, this investigation may promote the continuation of disease grouping research efforts in the future. The proposed model can be used with the image related data set in future studies. Furthermore, future research will focus in a major area on the effect of attribute evaluators on ensemble learning and image datasets.

## References

1. Mishra, S., Mallick, P.K., Tripathy, H.K., Bhoi, A.K., González-Briones, A.: Performance evaluation of a proposed machine learning model for chronic disease datasets using an integrated attribute evaluator and an improved decision tree classifier. Appl. Sci. **10**(22), 1–35 (2020). https://doi.org/10.3390/APP10228137
2. Jain, D., Singh, V.: Feature selection and classification systems for chronic disease prediction: a review. Egypt. Inform. J. **19**(3), 179–189 (2018). https://doi.org/10.1016/J.EIJ.2018.03.002
3. Ganiger, S., Rajashekharaiah, K.M.M.: Chronic diseases diagnosis using machine learning. In: 2018 International Conference on Circuits and Systems in Digital Enterprise Technology, ICCSDET 2018 (2018). https://doi.org/10.1109/ICCSDET.2018.8821235
4. Jana, A., Chattopadhyay, A.: Prevalence and potential determinants of chronic disease among elderly in India: rural-urban perspectives. PLoS ONE **17**(3), 0264937 (2022). https://doi.org/10.1371/JOURNAL.PONE.0264937
5. Canlas, R.D.: Data mining in healthcare: current applications and issues (2023). https://evhrdc.herdin.ph/index.php/herdin-home?view=research&cid=55063
6. Chandrashekar, G., Sahin, F.: A survey on feature selection methods. Comput. Electr. Eng. **40**(1), 16–28 (2014). https://doi.org/10.1016/J.COMPELECENG.2013.11.024
7. Balakrishnan, S., Ramaraj, N.: Feature selection using FCBF in type II diabetes databases (2023). https://www.researchgate.net/publication/228975672_Feature_selection_using_FCBF_in_type_II_diabetes_databases
8. Novaković, J., Strbac, P., Bulatović, D.: Toward optimal feature selection using ranking methods and classification algorithms. Yugoslav J. Oper. Res. **21**(1), 119–135 (2011). https://doi.org/10.2298/YJOR1101119N
9. Palaniappan, S., Awang, R.: Web-based heart disease decision support system using data mining classification modeling techniques | Request PDF (2007). https://www.researchgate.net/publication/221237373_Web-based_Heart_Disease_Decision_Support_System_Using_Data_Mining_Classification_Modeling_Techniques
10. Simons, L.A., Simons, J., Friedlander, Y., McCallum, J., Palaniappan, L.: Risk functions for prediction of cardiovascular disease in elderly Australians: the Dubbo Study. Med. J. Aust. **178**(3), 113–116 (2003). https://doi.org/10.5694/J.1326-5377.2003.TB05100.X

11. Surekha, D., JayaSuma.: Comparison of Feature Selection Techniques for Thyroid
12. Verma, K., Pal, S., Kumar, S.: Comparison of skin disease prediction by feature selection using ensemble data mining techniques. Inform Med Unlocked **16**, 100202 (2019). https://doi.org/10.1016/J.IMU.2019.100202
13. El-Hasnony, M., El Bakry, H.M., Saleh, A.A.: Comparative study among data reduction techniques over classification accuracy. Int. J. Comput. Appl. **122**(2), 9–15 (2015). https://doi.org/10.5120/21671-4752
14. Patil, M., Joshi, R.C., Toshniwal, D.: Hybrid prediction model for Type-2 diabetic patients. Expert Syst. Appl. **37**(12), 8102–8108 (2010). https://doi.org/10.1016/J.ESWA.2010.05.078
15. Nalband, S., Sundar, A., Prince, A.A., Agarwal, A.: Feature selection and classification methodology for the detection of knee-joint disorders. Comput. Methods Programs Biomed. **127**, 94–104 (2016). https://doi.org/10.1016/J.CMPB.2016.01.020
16. Mishra, S., Tripathy, H.K., Mishra, B., Sahoo, S.: Implementation of Classification Rule Mining to minimize Liver Disorder risks (2023). https://www.researchgate.net/publication/317179614_Implementation_of_Classification_Rule_Mining_to_minimize_Liver_Disorder_risks_Implementation_of_Classification_Rule_Mining_to_minimize_Liver_Disorder_risks
17. Amin, S., Singhal, A.: Identification and classification of neuro-degenerative diseases using feature selection through PCA-LD. In: 2017 4th IEEE Uttar Pradesh Section International Conference on Electrical, Computer and Electronics (UPCON), vol. 2018, pp. 578–586 (2017). https://doi.org/10.1109/UPCON.2017.8251114
18. Narayanan, B.N., Hardie, R.C., Kebede, T.M.: Performance analysis of feature selection techniques for support vector machine and its application for lung nodule detection. In: Proceedings of the IEEE National Aerospace Electronics Conference, NAECON, vol. 2018-July, pp. 262–266 (2018). https://doi.org/10.1109/NAECON.2018.8556669
19. Ray, C., Kumar Tripathy, H., Mishra, S.: Assessment of autistic disorder using machine learning approach. Adv. Intell. Syst. Comput. **1034**, 209–219 (2020). https://doi.org/10.1007/978-981-15-1084-7_21
20. Sahoo, S., Mishra, S., Mishra, B.K.K., Mishra, M.: Analysis and Implementation of Artificial Bee Colony Optimization in Constrained Optimization Problems, pp. 413–432 (2018) https://services.igi-global.com/resolvedoi/resolve.aspx?doi=10.4018/978-1-5225-2857-9.ch021, https://doi.org/10.4018/978-1-5225-2857-9.CH021
21. Dua, D., Graff, C.: UCI Machine Learning Repository. University of California, School of Information and Computer Science, Irvine, CA. - References - Scientific Research Publishing (2019). https://scirp.org/reference/referencespapers.aspx?referenceid=2607575

# A Systematic Literature Analysis of Scientometric Research in the Field of Cloud-Based Services: A 2023 Update

S. Karthika[1], S. Balachandran[1], and S. Sivankalai[2](✉)

[1] Hindustan Institute of Technology and Science, Chennai, India
[2] Anna University, Chennai, India
skysivan@gmail.com

**Abstract.** This study may access end-user computing services and distant application servers through the internet through cloud-based service. This cloud-based service variant has gained much traction in the computer industry in the past ten years. A crucial aspect of the Modern Library's growth is cloud-based service. Through cloud computing, modern libraries are being quickly revitalised. A scientometric analysis of cloud-based service publications was completed in the presented research article. The data was gathered using the Scopus Database. Eleven thousand three hundred twenty-one articles were obtained from the Scopus database using cloud-based and other services. The obtained data were entered into a Microsoft Excel spreadsheet, where several tables were created to examine the data. 2018: 1,094 articles (9.66%), with Buyya, R., as the top author with 35 papers, followed by Dustdar, S. There were 1139 articles published on cloud-based services in 2010; 1094 papers were published in 2018. The United States is in second place nationally, after China. India has 997 research publications, which puts it in third place. The Top Journal is ACM International Conference Proceeding Series, Followed by IEEE Access and ACM International Conference Proceeding Series.

**Keywords:** Cloud Computing · Cloud Service · scientometrics · Scientific visualization · VOSviewer

## 1 Introduction

Gartner found that global IT leaders ranked cloud computing among their top five technologies in 2013 (McDonald & Aron, 2013). 77% of organisations aim to raise their SaaS investment in the next two years, accounting for 18% of worldwide public cloud service spending of $134 billion in 2017 (Anderson et al., 2013). Cloud computing lets organisations temporarily use software, platforms, and computing infrastructure from the Cloud, given by the cloud provider through the network (Mell & Grance, 2011). Thus, cloud users benefit from resource efficiency, flexibility, risk transference (Armbrust et al., 2010), and energy reduction (Bose & Luo, 2011) due to cloud characteristics like pooled configurable computing resources, on-demand automatic scalability, and


R. Geetha et al. (Eds.): AAIMB 2023, CCIS 2203, pp. 250–262, 2025.
https://doi.org/10.1007/978-3-031-73068-9_20

pay-per-use pricing. Recent calls for papers in relevant journals show that information systems (IS) scholars also pay attention to the new technical, economic, and organisational opportunities and risks of cloud computing services, such as IT security and privacy issues (Benlian and Hess 2010, 2011). At first, scientific contributions focused on accurately defining the cloud computing model (Youseff et al., 2008; Armbrust et al., 2010; Leimeister et al., 2010) to provide a common baseline for discussion and differentiate this new paradigm from pre-existing and well-researched technological concepts like virtualization, multitenancy, web services (Marston et al., 2011), service-oriented architecture (Wieland, 2010), and grid computing. In the context of application service provision (ASP) and web services, services and on-demand software delivery service models have been studied and referred to as SaaS (Currie, W 2004) with varying definitions (Armbrust et al., 2010; Currie & Parikh, 2006), making specific separation, classification, and research of cloud-based services necessary. However, for a profound, complete, and in-depth body of knowledge of cloud computing services and to give researchers an orientation among the existing behavioural approaches, ideas, and theoretical foundations, a review of the current literature on the emerging topic of general cloud service adoption and its organisational enablers and inhibitors is needed. Following this approach, we aim to provide a comprehensive and systematic analysis of theoretical and empirical research on organisational cloud service adoption, focusing mainly on enablers and inhibitors of cloud-based services to promote an explicit understanding of both forces, which our review in line with (Hoberg et al., 2012) shows is lacking. Based on our analysis and practical talks, we suggest five new study fields that explain and analyse theoretical and practical phenomena related to organisational cloud service adoption drivers and repercussions. To lead cloud services research, research questions, theoretical lenses, and methodology are established for each sector.

## 2 Research Methods

In this study, scientometric analysis was carried out to evaluate the various aspects of the bibliographic data impartially. A reliable search engine is essential since there are many articles on the subject. Due to their accuracy, Scopus' highly developed search engines are ideal for this search. Since Scopus comes highly recommended by academics, it was utilised to generate references for this analysis of cloud-based services. The term "cloud-based and services" is used in 11321 articles, according to Scopus data from June 2023. The research included conference papers, journal articles, book chapters, book reviews, books, brief surveys, retracted editorials, notes, errata, letters, and undefined documents. The publication period for this research was chosen from 2001 to 2023. After placing certain limitations. Similar research queries have been made multiple times using the same methodology. Researchers created a method called scientific mapping to manage bibliometric data for scientometric investigations. Scopus data were exported as Comma Separated Values (CSV) files to enable proper programme evaluation. Both the scientific visualisation and the quantitative analysis of the gathered data were performed using VOSviewer (version 1.6.19). VOSviewer is a widely used mapping application that is freely distributable and based on open-source code. Experts strongly advise utilising this software. Therefore, including VOSviewer in this study may assist in accomplishing its objectives. The generated CSV files were imported into VOSviewer, where further analysis was carried out without jeopardising the consistency or integrity of the data. Throughout the bibliographic assessment, we examined factors including publishing locations, well-liked search phrases, highly cited authors, and participating countries. The many features and how they relate were shown statistically and graphically.

## 3 Data Analysis and Interpretation

The provided data must include specific scientometric measures, such as citation counts or impact factors, limiting the scope for in-depth scientometric analysis. However, the data can still be used to identify funding sources, document types, and affiliations associated with Cloud-based Services, providing a general overview of the research landscape. Additional data, including citation data and co-authorship networks, would be required to conduct a comprehensive scientometric analysis to assess research impact, collaboration patterns, and other quantitative measures.

Based on the Table 1 provided, here is the interpretation of the data. The highest number of articles published in a single year is 1,139, which occurred in 2019. The lowest number of articles published in a single year is 1, which occurred in both 2005 and 2001. Top Ten Years: The top ten years with the highest number of research articles published in Cloud-based Services are as follows: 2019: 1,139 articles (10.06%), 2018: 1,094 articles (9.66%), 2017: 1,120 articles (9.89%), 2016: 1,002 articles (8.85%), 2015: 997 articles (8.81%), 2014: 847 articles (7.48%), 2022: 1,012 articles (8.94%), 2021: 961 articles (8.49%), 2020: 1,004 articles (8.87%), 2013: 700 articles (6.18%), These ten years accounted for a significant percentage (79.13%) of the total number of research articles published in the field. Note that the percentages in the table represent the proportion of articles published in a specific year compared to the total number of articles (11,321) in the dataset.

Figure 1 was provided information regarding authorship in the field of Cloud-based services, here is an analysis of the number of articles published by each author. R Buyya has published the highest number of articles, with 35 publications during the specified period. This suggests a significant contribution to research in the field of Cloud-based services followed by Dustdar has published 24 articles, indicating a substantial research output in the domain, A Puliafito is another prolific author with 24 articles published during the specified period, R Bahsoon has published 22 articles, highlighting a notable contribution to the field, J.J.P.C Rodrigues has published 21 articles, demonstrating a significant research presence in Cloud-based services. L Wang has also published 21 articles, indicating a comparable level of research productivity, R Ranjan has published 20 articles, signifying a substantial research output in the field, Y Demchenko has published 19 articles, contributing to the body of knowledge in Cloud-based services and V.C.M Leung has also published 19 articles, indicating a similar research productivity as Demchenko. This analysis showcases the top authors based on the number of articles published during the specified period. These authors have made notable contributions to the field of Cloud-based services, and their research output can provide valuable insights into the advancements and trends within the domain.

Table 2 presents the distribution of articles in the dataset by document type. Here is the interpretation of the data. The most common document type in the dataset is

**Table 1.** Yearly-wise research articles published in Cloud-based and Services

| S. No. | Year | Number of Articles | percentage |
|---|---|---|---|
| 1 | 2023 | 424 | 3.75 |
| 2 | 2022 | 1012 | 8.94 |
| 3 | 2021 | 961 | 8.49 |
| 4 | 2020 | 1004 | 8.87 |
| 5 | 2019 | 1139 | 10.06 |
| 6 | 2018 | 1094 | 9.66 |
| 7 | 2017 | 1120 | 9.89 |
| 8 | 2016 | 1002 | 8.85 |
| 9 | 2015 | 997 | 8.81 |
| 10 | 2014 | 847 | 7.48 |
| 11 | 2013 | 700 | 6.18 |
| 12 | 2012 | 533 | 4.71 |
| 13 | 2011 | 323 | 2.85 |
| 14 | 2010 | 110 | 0.97 |
| 15 | 2009 | 45 | 0.40 |
| 16 | 2008 | 4 | 0.04 |
| 17 | 2007 | 2 | 0.02 |
| 18 | 2006 | 2 | 0.02 |
| 19 | 2005 | 1 | 0.01 |
| 20 | 2001 | 1 | 0.01 |
| | | 11321 | 100.00 |

conference papers, accounting for 54.77% of the total articles. Conference papers are a popular format for presenting research findings at academic conferences. Articles make up 34.75% of the dataset. This category typically includes research papers published in academic journals. Book chapters represent 4.81% of the dataset. These are sections or chapters contributed to edited books. Conference Review: This category accounts for 3.25% of the articles and likely refers to reviews or summaries of conferences. Reviews make up 1.51% of the dataset. These are critical assessments or evaluations of research or literature. Books represent a smaller portion, with 0.49% of the articles. These are complete book publications. Short surveys contribute 0.12% of the dataset. These are concise surveys or summaries of research. Editorials account for 0.11% of the articles. These are typically opinion pieces or introductory notes by editors.

Retracted This category represents articles that have been retracted, amounting to 0.09% of the dataset. Retractions occur when previously published articles are withdrawn due to errors or ethical concerns. Notes contribute 0.05% of the dataset. These are brief observations or comments on specific topics. Erratum accounts for 0.03% of the

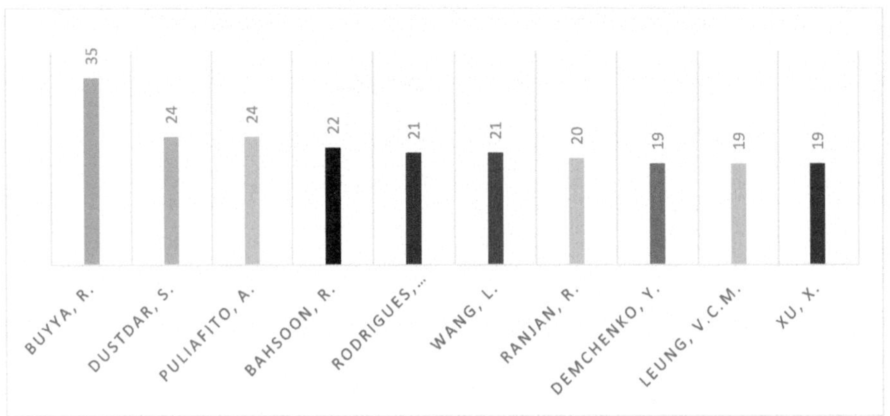


**Fig. 1.** Authorship published articles during the period for Cloud-based Services

**Table 2.** Document Type wise for Cloud-based and Services

| Document Type | No. of Articles | Percentage |
|---|---|---|
| Conference Paper | 6200 | 54.77 |
| Article | 3934 | 34.75 |
| Book Chapter | 544 | 4.81 |
| Conference Review | 368 | 3.25 |
| Review | 171 | 1.51 |
| Book | 56 | 0.49 |
| Short Survey | 14 | 0.12 |
| Editorial | 12 | 0.11 |
| Retracted | 10 | 0.09 |
| Note | 6 | 0.05 |
| Erratum | 3 | 0.03 |
| Letter | 2 | 0.02 |
| Undefined | 1 | 0.01 |
| | 11321 | 100.00 |

articles. These are published corrections or updates to previously published work. Letters represent 0.02% of the dataset. These are short communications or correspondence. This category represents 0.01% of the articles, likely referring to articles with an unspecified or unknown document type. Most articles in the dataset are conference papers, followed by traditional journal articles. However, there is also a notable presence of book chapters and other document types contributing to the diverse scholarly output.

**Table 3.** Top ten Affiliation articles published

| Affiliation | No. of Articles |
|---|---|
| Tsinghua University | 93 |
| Chinese Academy of Sciences | 86 |
| Vellore Institute of Technology | 76 |
| Beijing University of Posts and Telecommunications | 74 |
| King Saud University | 68 |
| The University of Electronic Science and Technology of China | 64 |
| Ministry of Education, China | 59 |
| Nanyang Technological University | 58 |
| Technische Universität Berlin | 51 |
| Huazhong University of Science and Technology | 51 |

Table 3 presents the top ten affiliations based on the number of articles published. Here are the results: Tsinghua University has published 93 articles, making it the top contributor in terms of article count, The Chinese Academy of Sciences follows closely with 86 articles published, Vellore Institute of Technology has published 76 articles, securing the third position, Beijing University of Posts and Telecommunications has published 74 articles, this university ranks fourth in terms of the number of articles published, The King Saud University has contributed 68 articles, placing it in the fifth position, University of Electronic Science and Technology of China has published 64 articles, and the Ministry of Education China is attributed to 59 articles, Nanyang Technological University: With 58 articles, this university is among the top affiliations in terms of article count, Technische Universität Berlin has published 51 articles, Huazhong University of Science and Technology and Along with Technische Universität Berlin, this university also has 51 articles published, These affiliations have shown significant contributions to research based on the number of articles published in the dataset.

Figure 2 provides data on the number of documents, citations, and total link strength for each country in the field of Cloud-based Services; here is the interpretation regarding the highest, lowest, and top ten countries. China has the highest total link strength of 452, indicating a strong co-authorship network with other countries in the field, the United States has the second-highest total link strength of 546, demonstrating a significant collaboration with other countries, and The United Kingdom has a total link strength of 283, reflecting strong co-authorship connections with other countries. Lowest Total Link Strength: Spain has the lowest total link strength of 106, indicating relatively fewer co-authorship connections with other countries. Top Ten Countries by Total Link Strength: China: Germany: 149, Italy: 146, Australia: 193, Saudi Arabia: 122, Canada: 146, India: 190, South Korea: These top ten countries have the highest total link strengths, indicating their significant involvement in co-authorship collaborations in the field of Cloud-based and Services.

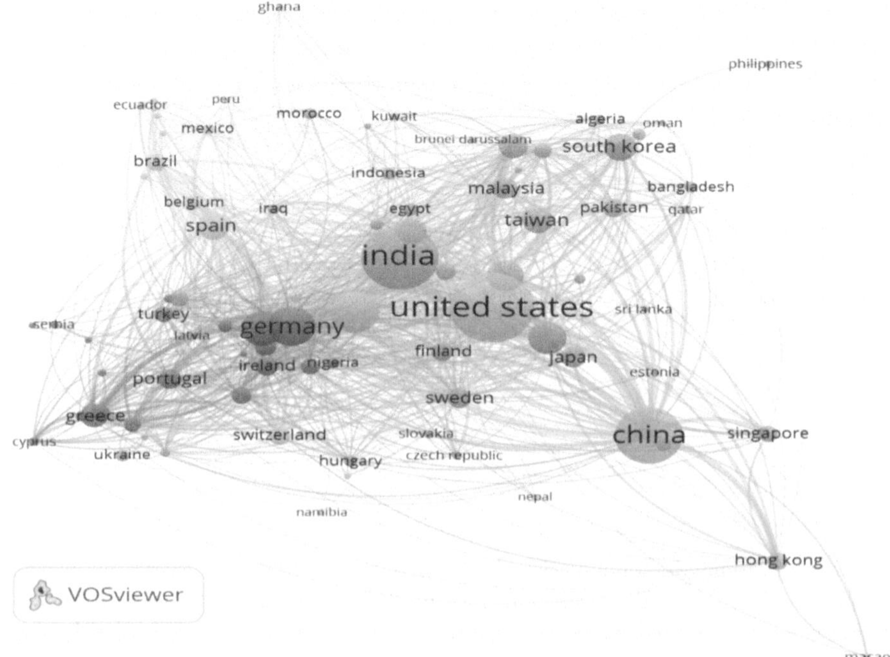


**Fig. 2.** Co-authorship-wise country for Cloud-based and Services

Figure 3 is based on the provided information on keyword occurrences and total link strength in Cloud-based Services. Here are the top ten keywords based on their total link strength; Cloud Computing has a total link strength of 2,208, Web Services have a total link strength of 940, Digital Storage has a total link strength of 817, Internet of Things has a total link strength of 766, Cloud-based has a total link strength of 1,056, Quality of Service has a total link strength of 645, Distributed Computer Systems has a total link strength of 659, Information Management has a total link strength of 494, Cloud Services has a total link strength of 546 and Cloud has a total link strength of 396. These keywords have the highest total link strengths, indicating their significant co-occurrence relationships with other keywords in Cloud-based Services.

Figure 4 provides information on Sources (journals, conference proceedings, etc.) in the field of Cloud-based Services; here is the interpretation regarding the highest, lowest, and top ten Sources based on the number of documents and citations. The highest Number of Citations by IEEE Communications Magazine has the highest number of citations, with 260, indicating its significant impact and influence in the field. IEEE Internet of Things Journal has 116 citations, reflecting its relevance and contribution to the research in the Internet of Things domain. IEEE Access has 154 citations, indicating its wide-ranging scope and popularity among researchers. Future Generation Computer Systems has 17 citations, showcasing its contributions to Cloud-based Services. IEEE Cloud Computing has 37 citations, demonstrating its significance and influence in the cloud computing domain. The lowest Number of Citations by Procedia CIRP has no citations, indicating its limited impact or research activity in the field. Top Ten Sources

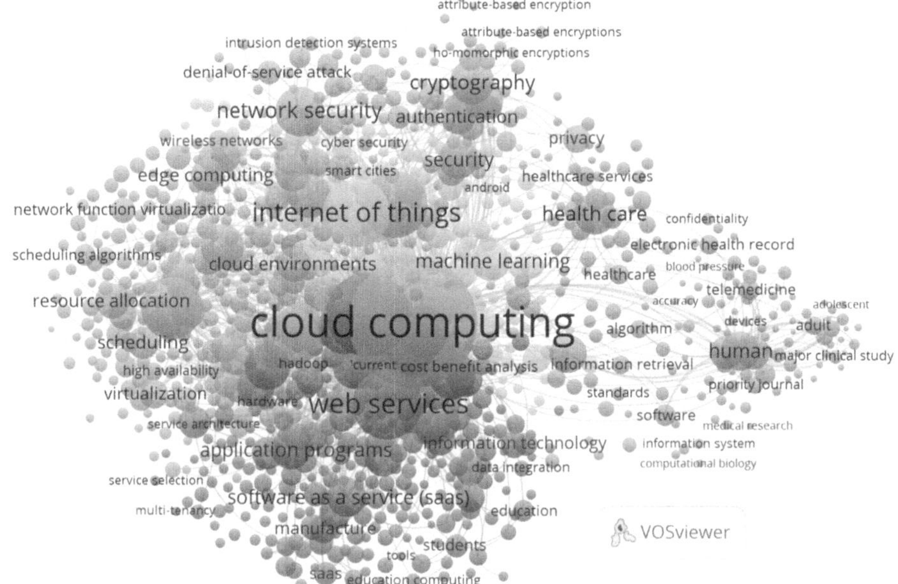


**Fig. 3.** Co-occurrence, wise All keywords for Cloud-based and Services

by Number of Citations by IEEE Communications Magazine: 260 citations, IEEE Internet of Things Journal: 116 citations, IEEE Access: 154 citations, Future Generation Computer Systems: 17 citations, IEEE Cloud Computing: 37 citations, IEEE Transactions on Services Computing: 17 citations, IEEE Transactions on Cloud Computing: 11 citations, Lecture Notes in Computer Science: 13 citations, Computer Networks: 24 citations, IEEE Cloud Computing: Not provided in the data. These top ten Sources have the highest number of citations, indicating their significant contributions and influence in the field of Cloud-based Services.

Figure 5 indicates that 87 countries that meet the minimum thresholds of having at least five documents and five citations. We will calculate the total strength of the citation link for each of these countries and select the countries with the highest total strength. The visualization scale will be 1.0, and we will consider the average annual publication as an additional factor for data interpretation. Here are the selected countries along with their total citation link strengths followed by countries wise Total link strength China 9984, United Kingdom 1794, Germany 1905, Italy 1004, South Korea 795, India 39, United States 19, Canada 18, Spain 22 and Australia 283. Satisfy note that these countries are selected based on the highest total strength of citation links and meet the minimum thresholds for documents and citations. The visualization scale 1.0 indicates that the weights assigned to the citation data are equal to those assigned to the average publication per year data.

Figure 6 described that cited references along with their total co-citation link strengths "armbrust m., fox a., griffith r., joseph a.d., katz r., konwinski a., lee g., patterson d., rabkin a., stoica i., zaharia m., a view of cloud computing, communications

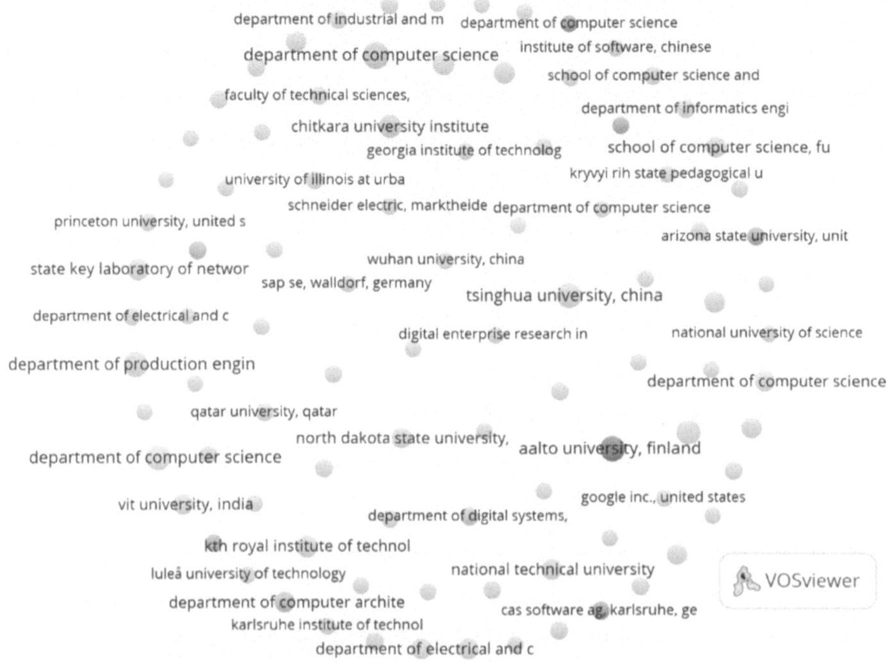


**Fig. 4.** Citation-wise Sources for Cloud-based and Services

of the acm, 53, 4, pp. 50–58, (2010)" - Total link strength: 32 followed by "mell p., grance t., the nist definition of cloud computing, (2011)" - Total link strength: 53, "subashini s., kavitha v., a survey on security issues in service delivery models of cloud computing, journal of network and computer applications, 34, 1, pp. 1–11, (2011)" - Total link strength: 47, "buyya r., yeo c.s., venugopal s., broberg j., brandic i., cloud computing and emerging it platforms: vision, hype, and reality for delivering computing as the 5th utility, future generation computer systems, 25, 6, pp. 599–616, (2009)" - Total link strength: 36, "zhang q., cheng l., boutaba r., cloud computing: state-of-the-art and research challenges, journal of internet services and applications, 1, 1, pp. 7–18, (2010)" - Total link strength: 35, "zissis d., lekkas d., addressing cloud computing security issues, future generation computer systems, 28, 3, pp. 583–592, (2012)" - Total link strength: 37, "oliveira t., thomas m., espadanal m., assessing the determinants of cloud computing adoption: an analysis of the manufacturing and services sectors, information & management, 51, 5, pp. 497–510, (2014)" - Total link strength: 37, "fernando n., loke s.w., rahayu w., mobile cloud computing: a survey, future generation computer systems, 29, 1, pp. 84–106, (2013)" - Total link strength: 22, "bonomi f., milito r., zhu j., addepalli s., fog computing and its role in the internet of things, proceedings of the first edition of the mcc workshop on mobile cloud computing, pp. 13–16, (2012)" - Total link strength: 15, "marston s., li z., bandyopadhyay s., zhang j., ghalsasi a., cloud computing - the business perspective, decision support systems, 51, 1, pp. 176–189, (2011)" - Total link strength: 20.

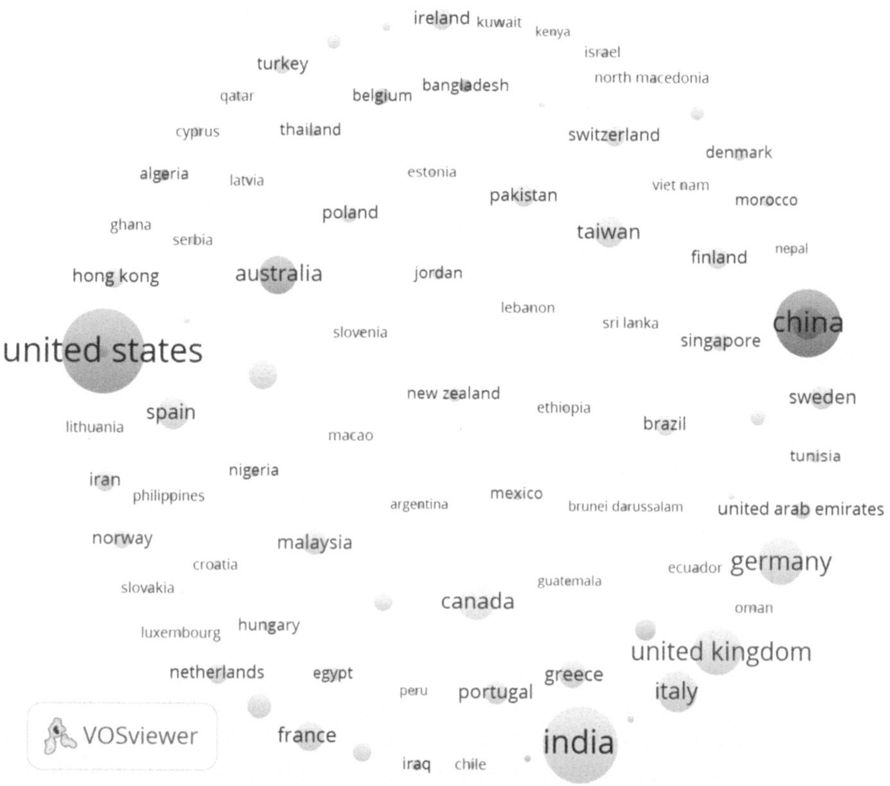

Fig. 5. Citation wise Countries for Cloud-based and Services

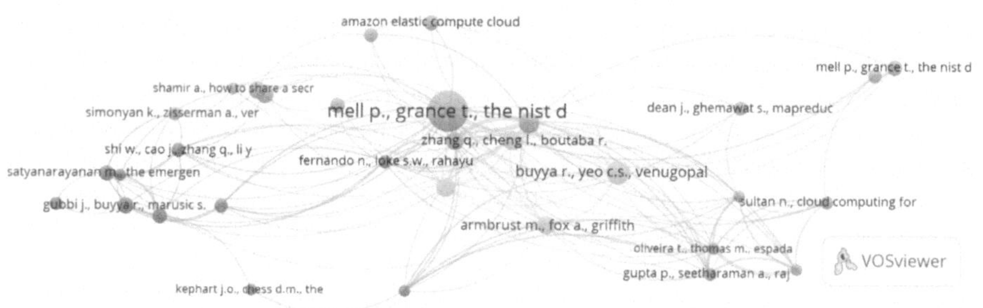


Fig. 6. Co-Citation Wise Cited References for Cloud-based and Services

Table 4 presents the number of articles in the field of Cloud-based Services attributed to various funding sponsors. Here is an interpretation of the data. National Natural Science Foundation of China: This funding sponsor has supported the highest number of articles, with 470 articles in Cloud-based Services. National Science Foundation: The National Science Foundation has funded 304 articles in this field, indicating a significant contribution to research. Horizon 2020 Framework Programme: The Horizon 2020

**Table 4.** Documents by Funding Sponsor for Cloud-Based and Services

| Documents by funding sponsor | No. of Articles |
|---|---|
| National Natural Science Foundation of China | 470 |
| National Science Foundation | 304 |
| Horizon 2020 Framework Programme | 269 |
| European Commission | 213 |
| Seventh Framework Programme | 134 |
| National Key Research and Development Program of China | 107 |
| Fundamental Research Funds for the Central Universities | 92 |
| National Research Foundation of Korea | 76 |
| Engineering and Physical Sciences Research Council | 65 |
| Ministry of Science, ICT and Future Planning | 65 |

Framework Programme, an initiative by the European Union, has sponsored 269 articles related to Cloud-based Services. European Commission: The European Commission has supported 213 articles in this field, showcasing its involvement and investment in research and development. Seventh Framework Programme: This program has funded 134 articles, indicating its contribution to advancing Cloud-based and Services research. National Key Research and Development Program of China: With 107 articles, this Chinese funding program has significantly supported research in this domain. Fundamental Research Funds for the Central Universities: This funding source has sponsored 92 articles, highlighting the support provided by universities for Cloud-based and service research. National Research Foundation of Korea: The National Research Foundation of Korea has funded 76 articles demonstrating Korea's participation and investment in this field. Engineering and Physical Sciences Research Council: This council has supported 65 articles related to Cloud-based and service research. Ministry of Science, ICT and Future Planning: With 65 articles, this ministry has also contributed to advancing Cloud-based and service research. The data suggests a diverse range of funding sponsors actively supporting research in Cloud-based Services, with China's National Natural Science Foundation leading in the number of articles funded.

## 4 Conclusion

We used scientometrics in the current study to examine Cloud-based and service research in the Scopus database. Researchers' access to, exploration of, and analysis of academic literature are all made possible by Scopus. It offers many capabilities and functions for citation tracking, search, and retrieval. Our findings identified the major study areas in cloud-based and service research and the high-impact nations, institutions, journals, and references. The number of publications, authors, and citations indicates a significant interest in studying the extreme effects of cloud computing and services. The best research Year-wise publication, authors, affiliations, document types, funding sponsor,

Citation, co-citation, and co-occurrence have been chosen and examined. The highest number of articles published in a single year is 1,139, which occurred in 2019. R Buyya has published the most reports, with 35 publications during the specified period, while Tsinghua University has published 93 papers. These affiliations have shown significant research contributions based on the dataset's published articles. The most common document type in the dataset is conference papers, accounting for 54.77% of the total articles. Funding sponsors actively support research in Cloud-based Services, with China's National Natural Science Foundation leading in the number of articles funded. Citation-wise Sources The highest Number of Citations by IEEE Communications Magazine has the highest number of citations, with 260 keywords based on their co-occurrence on their total link strength; Cloud Computing has a complete link strength of 2,208, citation-wise Countries with the highest link strength are China (9984), the United Kingdom (1794), and Germany (1905). China has the highest total link strength of 452, indicating a strong co-authorship network with other countries in the field. Our findings identified the important research disciplines in Cloud-based and service research and the high-impact nations and institutions, journals, references, and research orientations.

## References


Anderson Ed, et al.: Forecast: public cloud services, worldwide, 2011–2017, 1Q13 update (2013). http://www.gartner.com/id=2391015

Armbrust, M., et al.: A view of cloud computing. Commun. ACM **53**(4), 50–58 (2010)

Arumugam, P., Sharmila, M., Jeyanthi, R., Virumandi, A., Sivankalai, S., Rega, R.: A scientometric review of global research on the sustainability of paraffin oil. NVEO-NATURAL VOLATILES & ESSENTIAL OILS Journal| NVEO, 9243–9259 (2021)

Benlian, A., Hess, T.: The risks of sourcing software as a service: an empirical analysis of adopters and non-adopters. In: Proceedings of the 18th European Conference on Information Systems, Pretoria (2010)

Benlian, A., Hess, T.: Opportunities and risks of software-as-a-service: findings from a survey of IT executives. Decis. Support. Syst. **52**(1), 232–246 (2011)

Bose, R., Luo, X.: An integrative framework for assessing firms' potential to undertake Green IT initiatives via virtualization–A theoretical perspective. J. Strateg. Inf. Syst. **20**(1), 38–54 (2011)

Currie, W.L.: The organizing vision of application service provision: a process-oriented analysis. Inf Org **14**(4), 237–267 (2004)

Haag, S., Eckhardt, A.: Organizational cloud service adoption: a scientometric and content-based literature analysis. J. Bus. Econ. **84**, 407–440 (2014)

Hoberg, P., Wollersheim, J., Krcmar, H.: The business perspective on cloud computing: a literature review of research on cloud computing. In: Proceedings of the 18th Americas Conference on information systems, Seattle (2012)

Leimeister, S., Böhm, M., Riedl, C., Krcmar, H.: The business perspective of cloud computing: actors, roles and value networks. In: Proceedings of the 18th European Conference on information systems, Pretoria (2010)

Marston, S., Li, Z., Bandyopadhyay, S., Zhang, J., Ghalsasi, A.: Cloud computing: the business perspective. Decis. Support. Syst. **51**(1), 176–189 (2011)

McDonald, M.P., Aron, D.: Research presentation for 'hunting and harvesting in a digital world: the 2013 CIO agenda' (2013). http://www.gartner.com/id=2310018

Mell, P., Grance, T.: The NIST definition of cloud computing Recommendations of the National Institute of Standards and Technology Special Publication 800–145 (2011). http://csrc.nist.gov/publications/PubsSPs.html#800-145

Sivankalai, S.: Authorship Pattern and Collaborative Research in the Field Of Quality Management. Int. J. Innovative Res. Manage. Stud. (IJIRMS) **1**(10), 79–85 (2016)

Sivankalai, S.: Growth of Herpes Research in the 21st Century: A Scientometric Analysis. *GROWTH* (2020)

Sivankalai, S.V., Kaladevi, A., Sharmila, P.: Plagiarism: A scientometric review of the world literature between 2000 and 2019. Ann. Rom. Soc. Cell Biol. 10380–10391 (2021)

Sivankalai, S., Badhusha, K.N.: Bibliometric study on COVID 19 outbreak. Int. J. Library Inf. Stud. **10**(2), 1–19 (2020)

Sivankalai, S., Sivasekaran, K.: Mucormycosis (black fungus) maiming Covid patients: scientometrics analysis through prism of Biblioshiny. Libr. Philos. Pract. 5546 (2021)

Sivankalai, S., Virumandi, A.: Web services in cloud computing research: insights from scientometric. Libr. Philos. Pract. (e-journal), **6**(27), 1–23 (2021)

Sivankalai, S., Yemane, K.: Bibliometric analysis of research output in quality management at African continent from 1990–2016. Int. J. Innovative Res. Manage. Stud. (IJIRMS) **2**(3), 40–48 (2017)

Sivankalai, S., Ashok Kumar, P., Sharmila, M., Rosy Malarvizhi, S.D.: Scientometrics visualization of monkey b virus: status, development, and research directions. J. Pharm. Negative Results, 5325–5336 (2022)

Sivankalai, S., Virumandi, A., Balamurugan, B., Sharmila, M.: Scientometric analysis of the research on the abortion: 2015–2019. Turkish J. Comput. Math. Educ. **4**, 115–124 (2021)

Sivankalai, S., et al.: Scientometric analysis and visualization of astrovirus based on R-packages. Libr. Philos. Pract. (e-journal), 5361 (2021)

Stanleay, P., Nagarajan, M., Sivankalai, S.: Mapping the research productivity on water conservation: a scientometric analysis. In: Proceedings of the First International Conference on Computing, Communication and Control System, I3CAC 2021, Bharath University, Chennai, India (2021)

Usha, S., Sivankalai, S., Sharmila, M., Sharmila, R., Jeyachitra, S., Santhi, J.: Global research trend on yellow fungus: a scientometric analysis. J. Surv. Fish. Sci. **10**(2S), 987–1000 (2023)

Vijayakumar, M., Sivankalai, S., Michael Joseph Stalin, P., Kumaresan, G., Selvakumar, P., Manikandan, V.: A scientometric analysis on development of nanofluids for heat transfer and fluid flow applications. Solid State Technol. **64**(2), 3667–3684 (2021)

Weinhardt, C., et al.: Cloud computing: a classification, business models, and research directions. Bus. Inf. Syst. Eng. **1**(5), 391–399 (2009)

Wieland S SOA in der Cloud. In: Fähnrich, K., Franczyk, B. (eds) INFORMATIK, Service Science: Neue Perspektiven für die Informatik. Band 2. Gesellschaft für Informatik, Bonn, pp 55–60 (2010)

Youseff, L., Butrico, M., Da Silva, D.: Toward a unified ontology of cloud computing. In: Grid Computing Environments Workshop, Austin (2008)

# Optimizing Coronary Illness Prediction Using Hyperparameter Tuning Through Machine Learning

M. G. Vaishnavi[1](✉) and D. Shanthi[2]

[1] Department of CSE, PSNA College of Engineering and Technology, Dindigul, India
vaishnavi.sucila@gmail.com
[2] PSNA College of Engineering and Technology, Dindigul, India

**Abstract.** Numerous illnesses that affect the heart are categorized as heart diseases, commonly referred to as cardiovascular diseases. These can include heart rhythm issues, blood vessel issues, etc., related to congenital cardiac abnormalities. In recent years, it has been the leading cause of death worldwide. Thus, it can save many people's arteries if it can detect this dangerous issue early. Accurate and reliable methods must be found quickly to automate the process of early disease diagnosis and achieve effective disease management. In addition, it has been concentrated on an effective strategy to raise the effectiveness. In the previously existing systems, the heart disease dataset is applied to K-Nearest Neighbors (KNN), Naive Bayes, and Convolutional Neural Network (CNN) for the prediction. The proposed system uses the info-gain feature selection technique to eliminate unneeded features and extract the required features. One of the supervised machine learning algorithms called random forest is used to forecast the presence of cardiac disease in patient bodies. The optimization technique called hyperparameter tuning is used to tune some hyperparameters in the random forest algorithm to eliminate overfitting. The UCI (University of California, Irvine) heart disease dataset obtained from the Kaggle repository is used. Precision, recall, and F1-Score are used to calculate the accuracy. The optimized algorithm is then used to train and test the data, thus predicting the presence of heart disease in people, and increased accuracy is obtained.

**Keywords:** UCI heart disease dataset · Random Forest · Hyperparameter tuning · overfitting · precision · recall · F1-Score

## 1 Introduction

The human coronary heart may be defined as a compound frame organ that collectively incorporates muscle groups with organic nerves Human coronary heart pumps almost 5l of blood within the frame, supplying the human frame with renewed materials. Medical studies literature suggests that there's much hobby from clinical researchers in enforcing Machine Learning, which includes category algorithms in scientific devices. The human coronary heart operation is composite, and any failure is unstable to human lives. Hence,


R. Geetha et al. (Eds.): AAIMB 2023, CCIS 2203, pp. 263–275, 2025.
https://doi.org/10.1007/978-3-031-73068-9_21

coronary heart prognosis structures have been a prime problem for clinical researchers over the decades. Applying linear structures to carry out the coronary heart diagnosis isn't always viable. An advanced coronary heart prognosis system isn't usually available in each clinical center, particularly within regions with much less guidance and care.

Coronary artery disease is a general heart illness where the coronary arteries face problems providing blood, oxygen, and nutrients to the heart's muscles. The prior causes behind this condition are generally cholesterol deposits (plaques) and inflammation within the heart arteries. The human heart is a complicated organ incorporating muscles and nerves that pump almost 5 L of blood throughout the body, delivering necessary nutrients and materials. Medical investigators have shown a significant share in integrating Machine Learning, specifically classification algorithms, into medical devices. Due to the crucial essence of the heart's function, analyzing heart conditions accurately has been a substantial focus for clinical researchers in current decades. Linear systems are inappropriate for heart diagnosis due to the heart's complexity and the potentially life-threatening outcomes of failures. Unfortunately, developed heart diagnosis systems are not universally available in regions with limited medical resources and accessibility. Figure 1 shows the diseases of the coronary artery.

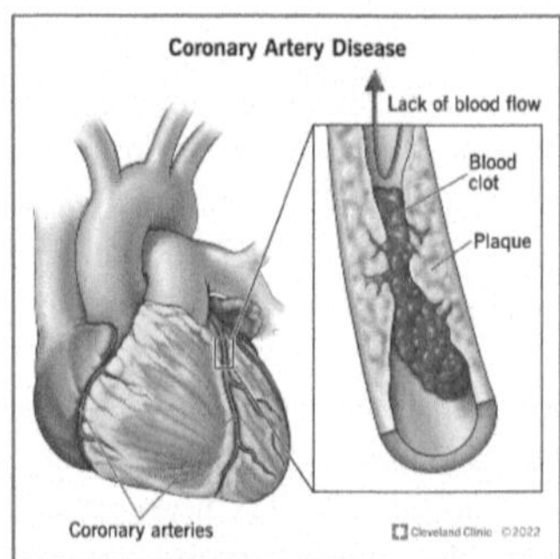


**Fig. 1.** Coronary Artery Diseases

Also, many people need help traveling to faraway medical establishments with exceptional usefulness. Depending solely on physician intuition and knowledge may not yield the best clinical outcomes, leading to medical errors and undesired results. Therefore, there is a growing need for modern computer-based diagnostic systems to reduce fatal mistakes, enhance patient safety, and save lives. This work employed the Random Forest (RF) method to predict and categorize the presence of heart disease. This technique has demonstrated high precision and performance. Similarly, Linear Discriminate Analysis (LDA) for dimensionality deduction eliminates irrelevant data, improving the algorithm's effectiveness.

Moreover, many humans can't tour local scientific facilities where super sanatorium offerings are affordable. Physician instinct and enjoyment aren't constantly sufficient to achieve exceptional clinical results. Consequently, scientific mistakes and unwanted outcomes motivate a want for modern computer-based analysis systems, which in turn lessen clinical deadly errors, growth affected person protection, and store lives. The proposed algorithm uses a Random Forest algorithm to predict and classify whether heart disease is present or not. This use of random forest leads to high accuracy and high performance. The dimensionality reduction called LDA (Linear Discriminant Analysis) has eliminated unwanted data.

## 2 Related work

Senthil Kumar Mohan et al. [1], in their work "Effective Heart Disease Prediction Using Hybrid Machine Learning Approach, proposed. IoT devices are used to generate medical data related to heart diseases and analyzed using machine learning algorithms. The knowledge discovery and extraction are done by a hybrid machine learning model and obtained 88.7% accuracy. Aditi Gavhane et al. [2] proposed work "Prediction of Heart Disease Using Machine Learning" proposed that it is impractical for a not unusual place guy to go through expensive assessments just like the ECG regularly, and as a result, their desires to be a machine in a location that is on hand and on the identical time reliable, in predicting the probabilities of a heart. Pooja Rani et al. [3] have proposed a multivariate imputation using chained equations for identifying and eliminating missing elements in the input data. Then Genetic Algorithm is used for feature selection and extraction, and SMOTE algorithm is used for preprocessing the data. Finally, several machine learning algorithms are used for predicting heart diseases. Dr. M. Kavitha et al. [4] have proposed a Hybrid Machine Learning Model developed by integrating Random Forest and Decision Tree algorithms and experimenting with the Cleveland coronary heart ailment dataset to evaluate the performance. From the experiment, it is identified that the hybrid machine learning model obtained 88.7% accuracy. The benefit is that; the hybrid model provides better results when compared to other algorithms. The future work is to use better algorithms to derive better accuracy. A.H Gonsalves et al. (2019) used historical medical data to predict coronary heart disease (CHD) through the performance of the Machine Learning (ML) method. The investigation concentrates on Naïve Bayes (NB), Support Vector Machine (SVM), and Decision Tree (DT), seeking to determine correlations in CHD data and enhance prediction accuracy. The study utilizes the South African HD dataset containing 462 samples, and intelligent models are formed employing the selected ML processes with 10-fold cross-validation. Finally, NB gives the promising results than the other algorithms. AZSB Habib et al. (2019) This work aims to predict coronary heart disease using an ML algorithm. By examining medical data like BP, hypertension, and diabetes, the ML method seeks to predict the possibility of an individual originating coronary disease. Different ensemble approaches are utilized in this study. The accuracy of each algorithm is calculated, and its performance is estimated using different parameters. Based on the results, the algorithm with the highest accuracy is selected as the model for predicting coronary heart disease. D. Shah et al. (2020) have suggested various parameters related to heart diseases, and the method is based on the different supervised learning algorithms; the data set taken from the UCI repository patient of the HD based on the comparison of previous method K-nearest neighbor has a good performance. Senthil Kumar Mohan et al. (2019) have used the ML method to detect Heart Diseases. This method is employed to improve the accuracy of predicting CVD. The proposed model obtained an improved level of accuracy is 88.7% with reduced complexity. Aditi Gavhane et al. (2018) have utilized ML methods to identify heart diseases. The author has employed the MLP algorithm for better prediction. Compared with the other methods, the proposed ML algorithm obtained an accuracy of 91%. Pooja Rani et al. (2021) have suggested ML through the SVM approach to predicting Heart disease. Using the imputation of multivariate, chained equation methods can be operated to handle the missing values. They have also used different performance metrics, and the SVM has an accuracy of 86.6% [5].

M. Kavitha et al. (2021) have employed a Hybrid Machine Learning method to identify Heart Diseases. The dataset is taken from the Cleveland coronary ailment. They have also utilized the regression algorithm. All the performance metrics are compared and tested. Based on the comparison of all the novel methods, the hybrid approach has a better-predicting accuracy of 88.7% [6].

Dhai Eddine Salhi et al. (2021) have used Machine Learning for Heart Disease Prediction and employed data analytics to identify and forecast HD in patients. They started with a preprocessing stage, selecting the most appropriate functions based on the correlation matrix. Subsequently, they used three data analytics methods (neural networks, SVM, and KNN) to additional data sets to evaluate their accuracy and effectiveness. The findings showed that neural networks were easier to configure and delivered exact results, with an accuracy of 93%. Moreover, they verified the algorithm's stability across various data set sizes. They plan to concentrate on selecting a more optimal attribute sampling method for future work [7].

R. Jane Preetha Princy et al. (2020) employed the Supervised ML method to predict Cardiac Disease. They have also discovered that the Decision Tree classification outperformed and performed better than the other methods in predicting CVD. The above said method achieved the highest accuracy of 73%. Based on the implementation of the algorithms they followed, the dataset size had a positive or negative impact. Finally, their future work strives to study and create a sufficient disease prediction model by estimating ensemble ML methods [8].

Mihir J. Gaikwad et al. (2022) utilized ML to predict Heart Disease Prediction and studied various supervised ML approaches for predicting heart disease. The dataset is taken from the "UCI ML repository for Statlog. After examining the results of these algorithms, they suggested utilizing the algorithm that showed the most increased accuracy in predicting HD. And achieved 82% accuracy. In forthcoming work, they plan to concentrate on eradicating unwanted values and executing dimensionality deduction approaches to improve the prognosis process further [9].

P. Sujatha et al. (2020) used the machine learning method to predict heart diseases and analyze different algorithms to indicate the existence of coronary HD. They estimated the various performance metrics, and based on the metrics results, the outcomes show that the Random Forest shows the highest accuracy, with 83.52%. Also, they seek to investigate combining numerous ML methods to build a more trustworthy and accurate real-time forecast model for HD [10].

Mohammed Nowshad Ruhani Chowdhury et al. (2021) have used ML approaches to detect Heart Disease detection and also employed diverse classification algorithms to create a qualified model for heart disease prognosis. Compared with all algorithms, SVM gives a higher accuracy of 91% and good performance. For future work, they prepare to examine regression studies in supervised ML to select an individual's vulnerability to HD [11].

The image division can be recognized as the fast learning method relying upon the concealing differentiation [13] Deep learning techniques deliver an end-to-end model that classifies medical photos thoroughly. Due to the improved medical picture quality and short dataset size, this approach may have high processing costs and model layer restrictions [14].

## 3 Proposed Methodology

The suggested model is developed to overwhelm the drawbacks of the current system. Thus in this paper, efficient algorithms such as KNN, NB, and DT are applied to predict the presence of heart disease accurately. This method enhances the performance and accuracy of the system or model. As the existing models are time-consuming and may lead to deviations in accuracy and classification, the method of ensemble learning called the random forest algorithm is used. This algorithm method works based on the concept of hierarchy tree structure and follows the process majority voting-based method. Our proposed system uses a dimensionality reduction system called LDA to remove unwanted data from the dataset. Then, the classification and prediction of heart disease is carried out. Finally, the outcome is obtained by predicting system accuracy through precision, recall, and F1 score.

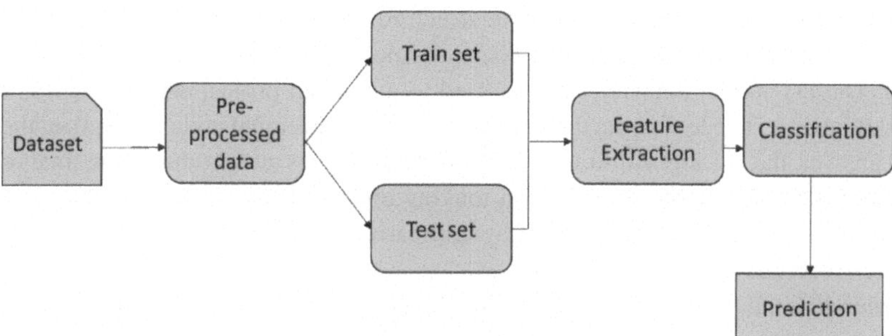


**Fig. 2.** System Architecture

Figure 2 shows the System Architecture and the suggested modules of the models. They are selection, preprocessing, splitting, dimensionality reduction, classification, and performance analysis. All the modules are discussed in the following sections.

### 3.1 K-Nearest Neighbor (KNN)

KNN is a supervised learning algorithm that can be used for classification and regression. KNN algorithm is generally referred to as lazy learning or non-parametric learning algorithm. Because it does not have any special agencies to pre-trained the data, it trains all the input data during classification. The data point is evaluated by matching the similarities between the point in the input datasets. It is achieved by evaluating the distance between the train and test data using Manhattan or Euclidean distance calculation equation. The results are arranged in ascending order based on the resultant distance value. Then based on the majority vote, the final result is produced.

### 3.2 Naive Bayes (NB)

NB is also a type of supervised machine learning algorithm, which is performed based on the Bayes theorem. It is mainly used to solve the classification problems. It is one of the most efficient and simplest classification algorithms, making the model faster to predict the result with less computational time. It is widely used for sentimental analysis, emotion recognition, spam filtration, etc. Based on the probability of the object, it produces the final prediction result.

### 3.3 Decision Tree (DT)

DT is a supervised machine learning algorithm used for classification and regression processes. It is represented in a tree-like structure; the main or root node indicates the dataset, the branch node indicates the decision rule, and the leaf node indicates the final output. The DT model decision node is used to make and project decision rules; it has several branches. The leaf node, the final node, has no additional nodes. Based on the Yes or NO answer, the DT algorithm further splits the branches and produces the final result. The DT algorithm inherits the decision-making ability of human beings and produces the output. It produces the final result by comparing the data in the root node with the attributes recorded in the real datasets.

Generally, the voting-based algorithm helps us to classify the model based on the majority voting given by the sub-decision trees with tuned hyper-parameters. The algorithm to be followed for the classification is as follows;

Step 1: Choose the training set through the random K statistics factors.
Step 2: The algorithm's hyperparameters are tuned based on the available data.
Step 3: The selected information points (Subsets) are constructing the Decision tree
Step 4: Select range N for building the Decision Tree.
Step 5: Continue the step 1 and 2.
Step 6: New data points are created to predict the findings of every decision tree and assign new data points that can categorize that receive the high votes.

**A. Data Selection**

Data choice is an approach to calculating the perfect statistics kind and source and proper devices to collect statistics. Data choice precedes the actual exercise of the statistics series, and it's far the in which information applicable to the evaluation is determined and retrieved from the data collection. After communication is recovered, combined from multiple sources, cleaned, and formatted, it is loaded into a storage system, like a cloud data warehouse. This paper uses the heart disease data collected from the Kaggle datasets to diagnose the disease [13]. The input dataset contains 14 attributes with 10 columns. All the attributes presented in the dataset discussed personal information and health condition data, such as age, BP, sugar, gender, chest pain type, etc., of both male and female patients.

**B. Data Preprocessing**

Data preprocessing is defined as eliminating unneeded data from the dataset, which comprises managing missing data and encoding categorical records. Missing data removal: With the imputer library's help, n eliminates the null values and addresses lost entries. For encoding categorical records, precise data with a finite set of label values are defined as variables. Integer and one-hot encoding methods convert categorical data into a numeric structure to recognize most ML algorithms' numerical input and output variables requirements. In our suggested system, we preprocess the data by handling missing values in the dataset.

**C. Data Splitting**

Splitting data into training and testing sets is essential in assessing data mining measures. Generally, a more substantial part of the dataset is employed for training, while less is reserved for testing. Nevertheless, this approach is crucial for training any ML method of the dataset used. In the point of the UCI heart disease dataset, it was divided into 80% training data and 20% testing data. The sklearn package was employed to achieve this dataset splitting.

**D. Dimensionality Reduction**

Feature reduction is a process that strives to decrease the number of elements in a computationally intensive procedure while maintaining essential data. This study employs Linear Discriminant Analysis (LDA) as a dimensionality reduction approach for feature deduction. The design uses LDA to minimize errors and gain more inferior values within the higher quality range. The elements are converted into minimal values across multiple features by fitting the train and test data rows. The described variance ratio designates the portion of variance attributed to individually specified components.

**E. Classification**

The classification process helps us classify the data available in the dataset into different categories. Our proposed system's main objective is to predict the presence of coronary illness. It helps to predict whether the coronary illness is present or not. The algorithm used for classification in this paper is KNN, NB, and DT, which works based on the hierarchy and voting-based classification algorithm and thus has the tuned hyper-parameters according to the dataset size.

# 4 Result and Discussion

This section depicts various results produced by the proposed model on classifying normal and abnormal heart disease patient records. After analyzing various results of the proposed model, the performance of the model is evaluated using various performance metrics values such as precision, recall, F1-score, and accuracy.

## 4.1 Performance Evaluation

The propped model performs various tasks to predict the classification result. It can classify the new data point from the group or class of data more easily. The following performance metrics are applied and evaluated to evaluate the model's performance.

*Accuracy:* It is one of the simplest forms of metrics, which is determined by the ratio of positively predicted value with a total number of prediction results.

$$Accuracy = \frac{TP + FP}{TP + TN + FP + FN}$$

*Precision:* It is determined by the ratio of correctly predicted positive results with true positive values. It is mainly evaluated to overcome the issues of predicting accuracy.

$$Precision = \frac{TP}{TP + FP}$$

*Recall:* It is similar to precision metrics, which are evaluated to identify incorrectly predicted actual positive values. It is determined by the ratio of the true positive rate with true positive and false negative rate.

$$Recall = \frac{TP}{TP + FN}$$

*F1-Score:* It is analyzed using the result produced by the precision and recall metrics. It is analyzed to evaluate the binary classification efficiency of the proposed model. It is expressed as,

$$F1 - Score = \frac{2 * Precision * Recall}{Precision + Recall}$$

The following section discusses various results analyzed by the proposed ML-based models. In that sense, the Fig. 3 depicts the total number of normal and abnormal heart disease patients in the input dataset.

In Fig. 3, the X-axis represents the target value and the Y-axis represents the count. 0 and 1 indicate the normal and abnormal HD patient records. The analysis shows that, in the input dataset, 54.46% of the patients are affected by HD, and 45.54% are not.

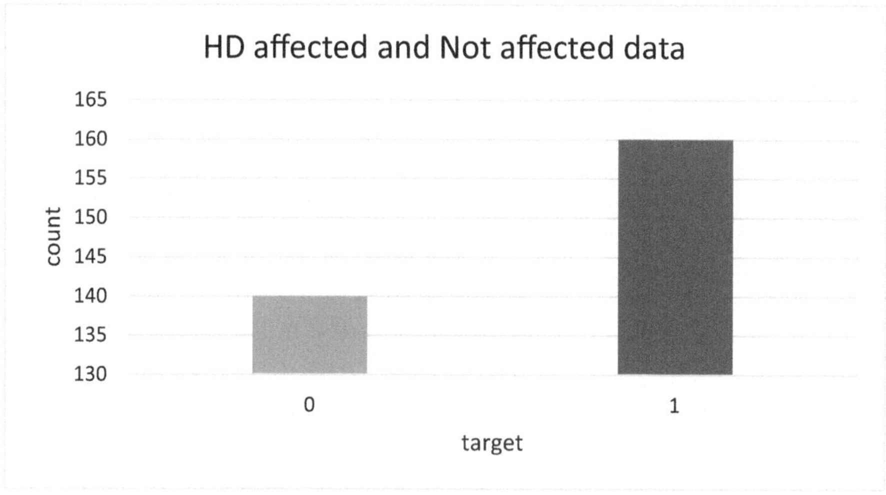


**Fig. 3.** HD Affected and Not Affected Data

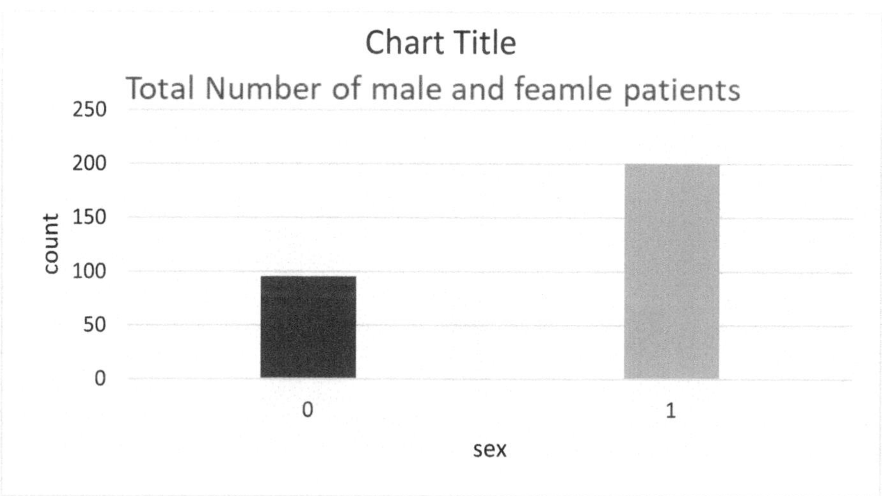


**Fig. 4.** Total Number of Male and Female Patients Records

Figure 4 depicts the total number of male and female patient records in the input datasets. The X-axis represents the gender, and Y-axis represents the count. In figure-4 0 represent female, and 1 represents male records. The analysis shows that more than 90% of the reports are analyzed from males, and 50% of the data are analyzed from female patients.

The histogram representation in Fig. 5 depicts the frequency range of the different attributes used in the proposed model. The X-axis represents the number of data, and Y-axis represents the frequency. The graph clearly shows the frequency distribution range between the attributes in the inputs HD dataset.

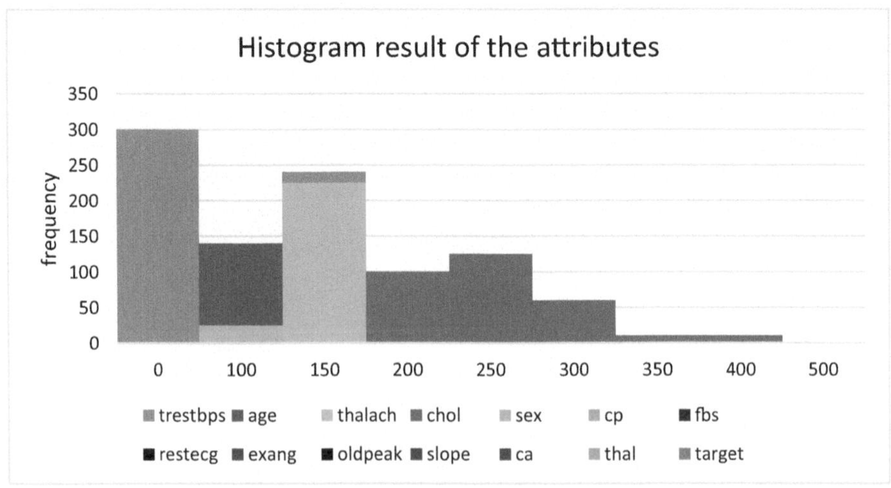


**Fig. 5.** Histogram Result of the Attributes

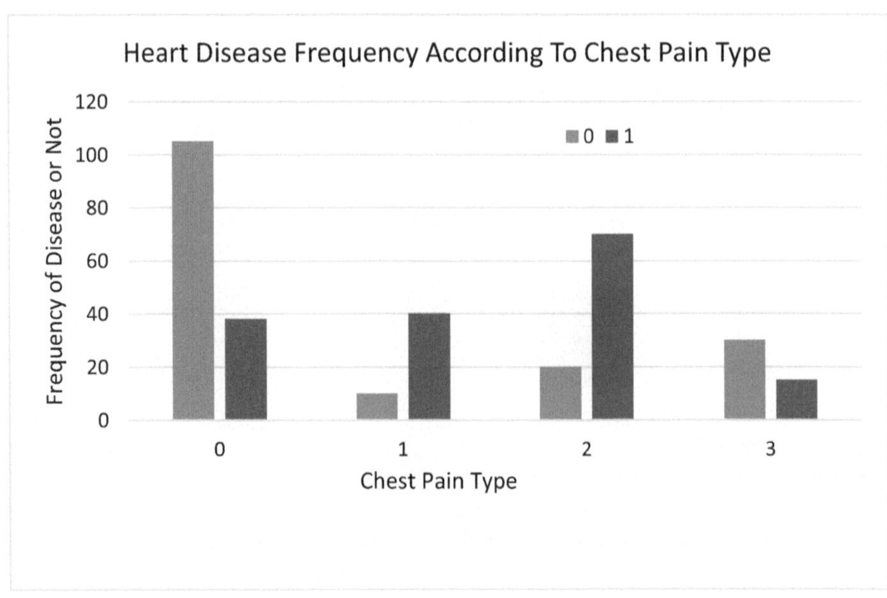


**Fig. 6.** Heart Disease Frequency According to Chest Pain Type

Figure 6 depicts the heart disease frequency according to chest pain type. The X-axis represents the 3 different chest pain types, and the Y-axis represents the disease frequency. In type 0, the frequency of the HD affected ratio is shallow. HD's frequency ratio is higher in types 1, 2, and 3 compared to the non-affected ratio. In Fig. 6, 0 and 1 represent normal and HD-affected data, respectively.

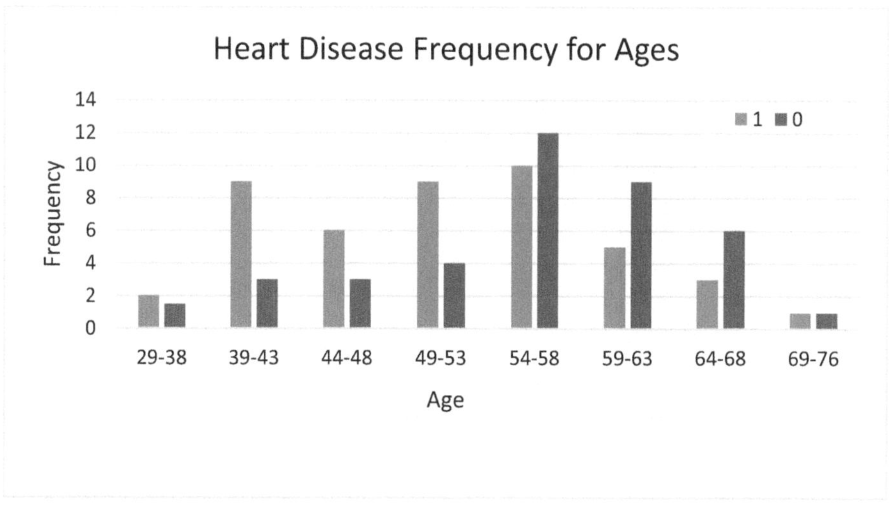


**Fig. 7.** Heart Disease Frequency for Ages

Figure 7 depicts the HD frequency analyses of the various age group patients. The X-axis indicates the different age groups, and Y-axis represents the frequency. The analysis indicates that, between the age group of 39 to 53, most people are affected by HD.

**Table 1.** Performance Metrics

| | Precision | Recall | F1-score | Accuracy |
|---|---|---|---|---|
| KNN | 90.1 | 90 | 92.8 | 90.16 |
| NB | 88.8 | 94.9 | 91.6 | 85.25 |
| DT | 90.5 | 79.8 | 84.5 | 75.41 |

Table 1 depicts the performance metrics value of the proposed model. The experiment result shows that the KNN has achieved 90.1%, 90%, 92.8%, and 90.16% precision, recall, f1-score, and accuracy, respectively. NB has achieved 88.8%, 94.9%, 91.6%, and 85.25% precision, recall, f1-score, and accuracy, respectively. And DT has achieved 90.5%, 79.8%, 84.5%, and 75.41% precision, recall, f1-score, and accuracy, respectively.

Figure 8 depicts the accuracy result of the proposed approaches. The result of the proposed model depicts that while classifying the HD affected and unaffected from the input datasets, the KNN, NB, and DT models have classified the data with 90.16%, 85.25%, and 75.41% accuracy, respectively.

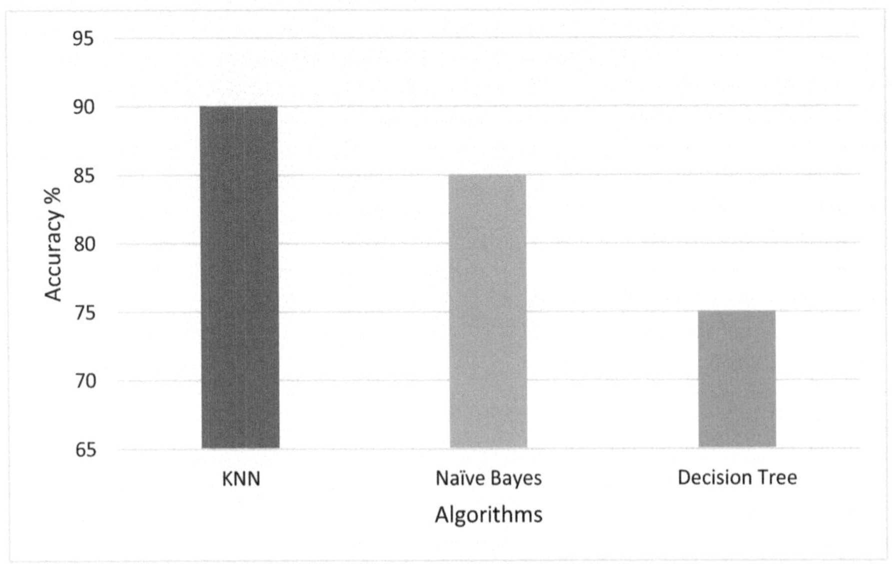


**Fig. 8.** Accuracy Comparison

## 5 Conclusion

Implementing ML algorithms involved various steps to improve the efficiency of the HD prediction model. The input dataset is taken from the Kaggle dataset and experimented using proposed ML-based models KNN, NB, and DT. The obtained are then compared regarding performance metrics (precision, recall, F1-score, and accuracy). The comparison result clearly shows that the KNN model accurately classified the HD affected and not affected patients' records and provides better results. The hyperparameter tuning of the parameters provides improved accuracy in the prediction process and boosts the algorithms' performance for predicting illness. The feature and attribute comparison between the algorithms shows the performance of the KNN algorithm to other algorithms. Thus, by hyper-parameter tuning, unsupervised learning methods can be adopted to predict illness.

## References

1. Gonsalves, A.H., Thabtah, F., Mohammad, R.M.A., Singh, G.: Prediction of coronary heart disease using machine learning: experimental analysis In: Proceedings of the 2019 3rd International Conference on Deep Learning Technologies, pp. 51–56 (2019)
2. Habib, A.Z.S.B., Tasnim, T., Billah, M.M.: A study on coronary disease prediction using boosting-based ensemble machine learning approaches. In: 2019 2nd International Conference on Innovation in Engineering and Technology (ICIET), pp. 1–6. IEEE (2019)
3. Shah, D., Patel, S., Bharti, S.K.: Heart disease prediction using machine learning techniques. SN Comput. Sci. **1**, 1–6 (2020)
4. Mohan, S., Thirumalai, C., Srivastava, G.: Effective heart disease prediction using hybrid machine learning techniques. IEEE Access (2019). https://doi.org/10.1109/ACCESS.2019.2923707

5. Rani, P., Kumar, R., Ahmed, N.M.O.S., Jain, A.: A decision support system for heart disease prediction based upon machine learning. J. Reliable Intell. Environ. **7**(3), 263275 (2021).https://doi.org/10.1007/s40860-021-00133-6
6. Kavitha, M., Gnaneswar, G., Dinesh, R., Sai, Y.R., Suraj, R.S.: Heart disease prediction using hybrid machine learning model. In: 2021 6th International Conference on Inventive Computation Technologies (ICICT). (2021). https://doi.org/10.1109/icict50816.2021.9358597
7. Salhi, D.E., Tari, A., Kechadi, M.-T.: Using machine learning for heart disease prediction international conference on computing systems and applications. In: Proceedings of the 4th Conference on Computing Systems and Applications (2021)
8. Princy, R.J.P., Parthasarathy, S., Hency Jose, P.S., Raj Lakshminarayanan, A., Jeganathan, S.: Prediction of cardiac disease using supervised machine learning algorithms. In: 2020 4th International Conference on Intelligent Computing and Control Systems (ICICCS) (2020). https://doi.org/10.1109/iciccs48265.2020.9121169
9. Gaikwad, M.J., Asole, P.S., Bitla, L.S.: Effective study of machine learning algorithms for heart disease prediction. In: 2022 2nd International Conference on Power Electronics, IoT Applications in Renewable Energy and its Control (PARC) (2022). https://doi.org/10.1109/PARC52418.2022.9726613
10. Sujatha, P., Mahalakshmi, K.: Performance evaluation of supervised machine learning algorithm in prediction of heart disease. In: 2020 IEEE International Conference for Innovation in Technology (INOCON) (2020). https://doi.org/10.1109/incon50539.2020.9298
11. Chowdhury, M.N.R., Ahmed, E., Siddik, M.A.D., Zaman, A.U.: Heart disease prognosis using machine learning classification techniques. In: 2021 6th International Conference for Convergence in Technology (I2CT), (2021). https://doi.org/10.1109/i2ct51068.2021.9418181
12. https://www.kaggle.com/code/charankakaraparthi/prediction-of-heart-disease-by-ml-techniques/notebook
13. Hemanand, D., Bhavani, N.P.G., Ayub, S., Ahmad, M.W., Narayanan, S., Haldorai, A.K.: Multilayer vectorization to develop a deeper image feature learning model. Automatika **64**(2), 355–364 (2023). https://doi.org/10.1080/00051144.2022.2157946
14. Hemanand, D., Selvam, L., Arunachalam, M., Suresh, D.: An image denoising scheme remove unwanted pixel using NLM with sprint deep learning network. Int. J. Intell. Syst. Appl. Eng. **10**(4), 130–137 (2022)

# Detection and Classification of Intracranial Tumor in Machine Learning Using Fuzzy C-Means Algorithm

R. Roobika[1], D. Shanthi[1](✉), and N. Sivakamy[2]

[1] Department of CSE, PSNA College of Engineering and Technology, Dindigul, India
hodcse@psnacet.edu.in
[2] Department of CSE, Panimalar Engineering College, Chennai, India

**Abstract.** The main objective of this paper is to implement various machine-learning algorithms for brain tumor detection and classification to choose the best algorithm. The early stage of brain tumor detection is a crucial task. Detecting and classifying brain tumors can be done by examining the biopsy, which can be done by definitive brain surgery. Initially, all the images are preprocessed, segmented, and processed by machine learning algorithms. Some of the machine learning algorithms, including Logistic Regression, Support Vector Machine, K-Nearest Neighbours, Naïve Bayes, Random Forests, K-Means Clustering, and Neural Networks, are examined with their experimental results. Accuracy, Precision, Recall, and F1-Score are calculated using true positive, false positive, true negative, and false negative values. The comparison shows that the logistic regression and random forest obtained 96% accuracy, and the neural Network obtained 95% accuracy. These models perform well in classifying tumor and non-tumor brain images.

**Keywords:** Brain Tumor · Performance Evaluation · Machine Learning · K-Means · Neural Network · Random Forest

## 1 Introduction

The brain is the organ present in the cranial cavity of the human body. The brain is one the most vital organs among other organs as it is responsible for every action our body makes. The brain's primary function is to supply information to the rest of the body. Without a brain, the human body cannot perceive any information it takes in a sense. A brain not only perceives and shares information but can also perceive feelings such as happiness and fear. Around 40% of the brain is mostly a combination of water, protein, carbohydrates, salts, and the other 60% of the brain is fat. The brain contains nerves, which are responsible for the transfer of information, and blood vessels.

Tumors are abnormally grown cells or a mass of tissue that seems to grow even further but do not die. There are primarily two types of tumors: benign and malignant. A benign tumor grows extensively and rapidly, but it does not metastasize. At the same time, a malignant tumor, also considered cancer grows big and metastasizes. The tumor


R. Geetha et al. (Eds.): AAIMB 2023, CCIS 2203, pp. 276–291, 2025.
https://doi.org/10.1007/978-3-031-73068-9_22

cells can hinder the ability of an organ. In the worst cases, they metastasize and may lead to fatal outcomes. Several helpful medical devices such as CT scans, MRIs, X-Rays, and many more can detect tumors. Treatment of tumors is complicated as it is different for everyone and has types. However, it is not entirely untreatable, and it can be treated with chemotherapy, surgery, Radiation therapy ETC. The tumor in the organ responsible for the human body's function is commonly known as a brain tumor. There are more than 120 types of brain tumors, some benign and some malignant. Meningioma and schwannoma are some of the benign type tumors.

In comparison, chordoma lymphoma is a malignant type of tumor or cancer. Brain tumor does not primarily occur in the brain but also the organs near the brain, such as the pituitary gland, pineal gland, and the membranes around the brain. Brain tumors can be cured by radiotherapy, surgery, and chemotherapy. Brain tumors can be detected with MRI (Magnetic Resonance Imaging). MRI is the method used to diagnose brain tumors. The MRI produces multiple images in which the brain tumor can be found. However, due to the amount of data provided, it is hard for a human to process all the images and find the tumor by himself. Tumors and healthy brain cells look similar, and finding tumors with just one human is even more challenging. If the result provided by the human Is inaccurate, it may lead to fatal consequences. There have been multiple proposals all around the globe for the use of Neural Networks to detect brain tumors from image data. Two types of Neural Networks are used. They are ANN and CNN. The Artificial Neural Network (ANN) works as a human brain and is provided with a set of healthy human brain scans to study and then again provided with data of tumorous brain cells. The ANN learns from the provided data to identify the tumor cell.

One input and output image will be at a time, each containing multiple hidden layers inside. A modal is trained based on the function activation. The ANN works on fully connected layers. The Convolutional Neural Network (CNN) is a linear mathematical operation. The CNN dimension of the image gets reduced as the layers proceed. These dimensions reduce, but they do not lose their information. In general, brain tumor diagnosis needs two different processes, such as segmentation and classification.

Complete segmentation and classification of brain tumors using single modality images is difficult because the brain's structure is complicated. It is possible in MRI to capture multiple images called multi-modality images that provide the whole structure of the brain to classify accurately. Figure 1 depicts the multi-modalities of MRI brain images. Segmentation and classification of brain MRI images getting more attention over the past decades. Uni-modality MRI image processing is explored in various studies. Some of the recent research works have used a multi-modality model of MRI image processing to improve the accuracy of tumor segmentation and classification [1].

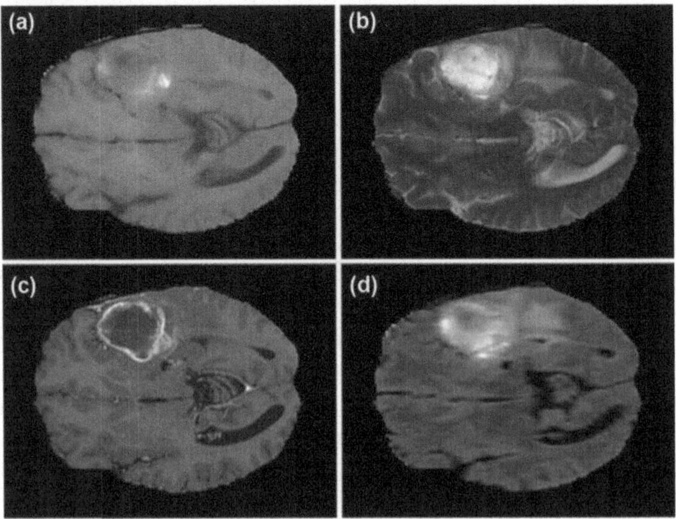


**Fig. 1.** Different Modalities of Brain MRI Images. (**a**) T1, (**b**) T2, (**c**) T1-Contrast (**d**) FLAIR

Edge detection, region detection methods, and machine learning models are used to process multi-modality MRI images [2]. All the machine learning models extract texture features and voxel intensities. Because the input images can be classified using a voxel feature vector [2]. The dataset [3] is taken with essential features, such as intensity, variations in intensity, texture features, and neighborhood values. But, it is well-known that wavelet features are not yet used for classifying multi-modality MRI brain images. This paper uses machine learning models to learn and extract features to improve classification accuracy. This work makes the following contributions:

We are implementing multiple machine learning models by training process using 80% of the MRI image dataset.

The machine learning models are validated and tested using the testing process with 20% of the dataset. And their performance is evaluated by comparing their performance values with one another.

## 2 Literature Review

M.S Majib et al. (2021) focused on creating and exploring different standard and hybrid ML models for the autonomous classification of BT images. They have also discussed the 16 transfer learning methods to select the efficient one for NN-based brain tumor classification. Moreover, a classifier stacked employing advanced technologies was suggested, indicating best performance corresponded to other developed models. In this work, P.R. Kshirsagar et al. (2020) utilizes a software-based algorithm to detect tumors in brain MRI scans. The presented methodology consists of three primary parts: MRI pictures use the preprocessing steps, texture features are extracted using a gray-level matrix co-occurrence, and classification techniques are utilized. Finally, employing these procedures aims to perform precise and automated tumor detection in brain MRI scans.

P. Modiya and S. Vahora (2022) show a specific technique for detecting brain tumors, employing a CNN with transfer learning and dimensionality reduction techniques. An exhaustive investigation is conducted to reach the implementation of various transfer learning models with and without dimensionality reduction approaches, showcasing the usefulness of the proposed model.

M. I Mahmud et al. (2023) this study presented a CNN framework that efficiently determines BT using MR images. They compare our offered architecture with other samples, assessing their implementation using accuracy, recall, loss, and area AUC. They also have results showing that our presented model outperforms the different models. With a dataset of 3264 MR images, our CNN model achieves a 93.3% accuracy. Based on the comparison, the submitted model gives good accuracy. A.E Babu et al. (2018) have examined diverse approaches for detecting and segmenting brain tumors from MRI images. These techniques aim to identify and separate the tumor region from the rest of the brain. Among the methods employed are watershed, bilateral, and others. This examination briefly overviews the distinct approaches used in BT detection and segmentation from MRI images.

A. Keerthana et al. (2021) discussed the different machine-learning algorithms used to predict brain tumor and their effectiveness. The study entailed the preprocessing steps, segmentation, and feature extraction from the images and the related methodologies. M. Sharma et al. also discussed in detail the various models considered for the prediction process and highlighted the usage of the CNN model over others due to the better image processing and detection accuracy. The predictions made through deep learning models are supported by machine learning models like the SVM algorithm to predict and diagnose brain tumors. Though they struggle in image classification, they provide better results in the feature classification obtained from deep learning models. EM Senan et al. (2022) discussed the performance metrics of different deep learning models and hybrid models adopting machine learning algorithms. The improved prediction accuracy with adopting ML models is discussed in detail, with 95.10, 95.25, and 98.25% of accuracy, sensitivity, and specificity. A. Rehman et al. 2019 discussed the effectiveness of transfer learning in improving the prediction accuracy of the detection models. The work compared the performance of different transfer learning models like VGGNet, GoogLeNet, and AlexNet.

## 3 Proposed Methodology

### 3.1 Machine Learning Models

This paper below explains several machine learning models for brain tumor detection and classification, such as LR, SVC, KNN, NB, NN, RF, and K-MC are shown in Fig. 2.

### 3.2 Logistic Regression

One of the linear machine learning models can make predictions based on the probabilities of the input images and their classes. A logistic function matches and maps the input features for classification according to the estimated probabilities.

### 3.3 Support Vector Classifier

One of the classification algorithms that belongs to machine learning is Support Vector Classifier. It uses an optimal hyperplane to cluster the data points in the feature space. It increases the margin among the classes while processing the support vectors.

### 3.4 K-Nearest Neighbors

One of the non-parametric models of machine learning algorithms is K-Nearest Neighbors (kNN), which classify the data concerning the nearest neighbors of a point in the feature space. It creates and assigns labels to each class according to the score of the k nearest neighbors.

### 3.5 Naive Bayes

A probabilistic classification model derived from Bayes' theorem is called Naïve Bayes classifier. It performs well by assuming the features are not depending on any conditions to create and assign classes. Since NB follows a simplistic assumption, it can obtain a successful classification of any data.

### 3.6 Neural Networks

Advanced Artificial Intelligence algorithms are derived from a neural network or artificial neural networks with **N** layers with interconnected nodes. These layers can learn and fetch complex patterns in large datasets by training with assigning appropriate weights.

### 3.7 Random Forest

One ensemble learning model that belongs to machine learning is Random Forest, which comprises a group of decision trees to predict large datasets. Each decision tree in the forest is trained with a subset of the whole dataset, and finally, the average values of individual trees are used for prediction.

### 3.8 K-Means Clustering

An unsupervised learning model of machine learning is K-Means clustering, which splits the data into K clusters according to their similarities. It repeatedly assigns the data points into clusters and updates the centroids until they reach the target value. In order to implement the machine learning models, a common procedure is followed, like data collection or preparation, preprocessing, learning or training data, and validating or testing the model. It is given as a step-wise procedure to experiment with them.

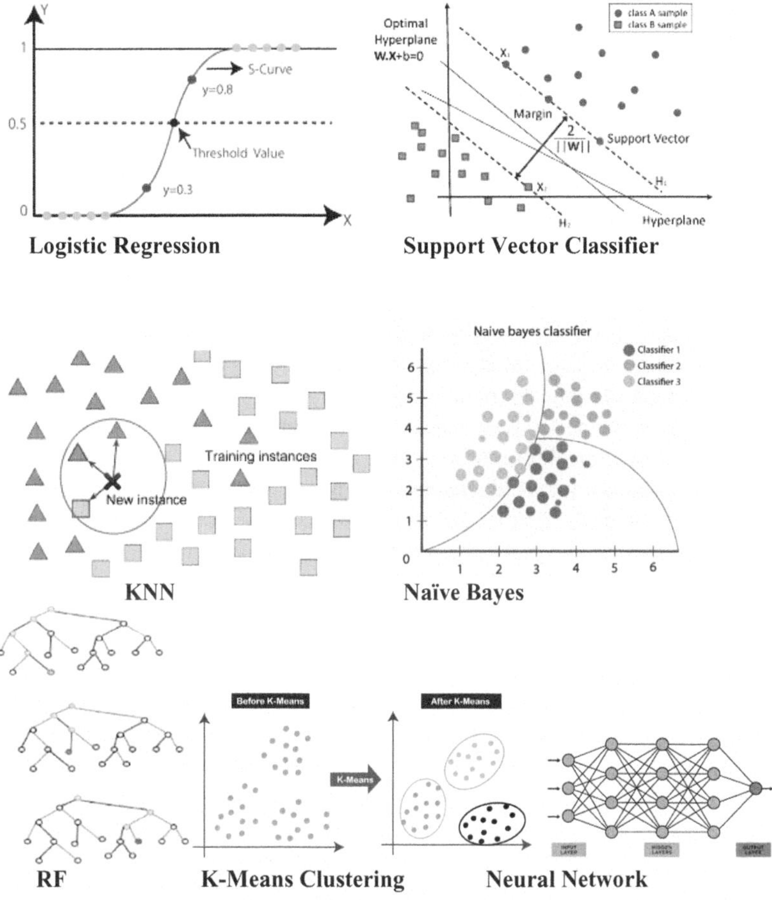


**Fig. 2.** Machine Learning Models

1. *Import packages*

2. *set path to data folder*

3. *Read the images, assign labels persist in Data Frame.*

4. *uniform image size.*

5. *image normalization*

6. *divide data into training, validation, and testing*

7. *Create the model*

8. *Compile the model*

9. *Apply the model on the train set.*

10. *Evaluate the model by applying it on the test set.*

## 4 Experimental Results and Discussion

The above-discussed machine learning models are implemented in Python language and experimented with the Br35H-Mask-RCNN dataset [1] dataset is taken from the website, permits researchers to download the data, and has three folders and one file. The test folder has 101 brain images, the training folder has 501 images, the validation folder has 202 images, and the 245.94 kB sized JSON file has annotations. Totally 3000 MRI images are available with various types, such as glioma, meningioma, and pituitary gland tumor. Initially, the data is prepared by preprocessing and augmenting the MRI images. The augmentation process includes transforming the images geometrically, such as rotation, flip, crop, stretch, and zoom. Then transform the images in terms of color space, such as RGB, to gray, enhancing the brightness or contrast of the images that visually differentiate the objects in the image.

The Figs. 3 show the performance of the different algorithms through which the suitable one for the intracranical tumour detection can be carried out. The results show that neural networks, random forest, K-Nearest Neighbour, logistic regression, and support vector classifier provide better classification results above 90%. The other algorithms like K-means clustering and Naive Bayes algorithms provide very less results as they solely rely on the clustering for prediction.

The Fig. 4a–4g shows the ROC curve for different machine learning algorithms considered. It helps to show the performance of the algorithm by comparing the false positive rate with the true positive rate. The results obtained from the ROC curve also support the findings obtained from Fig. 3. Both K-means clustering and Naive Bayes algorithm struggle to provide better performance, while other algorithm have better prediction results.

The Fig. 5a–5g shows the confusion matrix for different algorithms, which further shows the better performance obtained by the algorithms, and the true and false predictions for the malignant and benign samples are shown in detail. It can be seen that logistic regression, KNN, RF, NN, and SVC algorithms provide better results than other models and provide better true positive rate.

The experimental results show that the LR model performs well in MRI brain tumor detection with 96% accuracy. It obtained 96% precision, 97% recall, and 96% F1-score. It obtained a balanced combination of recall and precision. The AUC shows the difference between tumor and non-tumor images and the classified images in Fig. 6. In the case of SVC, it obtained 94% accuracy, 94% precision, recall, F1-Score, and AUC. The images classified using SVC are shown in Fig. 7. KNN model obtained 90% accuracy, 85% precision, 97% recall, 91% F1-Score, and 89% AUC from the experiment. It obtained a high number of false negatives that show incorrect classification.

| *LR* | precision | recall | f1-score | support |
|---|---|---|---|---|
| *0* | 0.96 | 0.98 | 0.97 | 307 |
| *1* | 0.97 | 0.95 | 0.96 | 289 |
| *accuracy* | | | 0.97 | 595 |
| *macro avg* | 0.97 | 0.97 | 0.98 | 595 |
| *weighted avg* | 0.97 | 0.97 | 0.97 | 595 |

(a).LR

| SVC | precision | recall | f1-score | support |
|---|---|---|---|---|
| *0* | 0.93 | 0.93 | 0.93 | 305 |
| *1* | 0.93 | 0.93 | 0.93 | 287 |
| *accuracy* | | | 0.93 | 593 |
| *macro avg* | 0.93 | 0.93 | 0.93 | 593 |
| *weighted avg* | 0.94 | 0.94 | 0.94 | 594 |

(b). SVC

| NB | precision | recall | f1-score | support |
|---|---|---|---|---|
| *0* | 0.64 | 0.71 | 0.67 | 307 |
| *1* | 0.65 | 0.58 | 0.61 | 289 |
| *accuracy* | | | 0.65 | 595 |
| *macro avg* | 0.65 | 0.64 | 0.64 | 595 |
| *weighted avg* | 0.65 | 0.65 | 0.64 | 595 |

(c). KNN

| NN | precision | recall | f1-score | support |
|---|---|---|---|---|
| *0* | 0.95 | 0.98 | 0.97 | 307 |
| *1* | 0.98 | 0.95 | 0.96 | 289 |
| *accuracy* | | | 0.96 | 595 |
| *macro avg* | 0.97 | 0.96 | 0.96 | 595 |
| *weighted avg* | 0.96 | 0.96 | 0.96 | 595 |

(d).NB

| KNN | precision | recall | f1-score | support |
|---|---|---|---|---|
| *0* | 0.86 | 0.98 | 0.92 | 307 |
| *1* | 0.98 | 0.82 | 0.89 | 289 |
| *accuracy* | | | 0.91 | 595 |
| *macro avg* | 0.92 | 0.90 | 0.90 | 595 |
| *weighted avg* | 0.91 | 0.91 | 0.90 | 595 |

(e).NN

| RF | precision | recall | f1-score | support |
|---|---|---|---|---|
| *0* | 0.99 | 0.95 | 0.97 | 307 |
| *1* | 0.98 | 0.99 | 0.97 | 289 |
| *accuracy* | | | 0.97 | 595 |
| *macro avg* | 0.97 | 0.97 | 0.97 | 595 |
| *weighted avg* | 0.97 | 0.97 | 0.97 | 595 |

(f).RF

| K-MeansC | precision | recall | f1-score | support |
|---|---|---|---|---|
| *0* | 0.65 | 0.77 | 0.70 | 1483 |
| *1* | 0.71 | 0.58 | 0.64 | 1489 |
| *accuracy* | | | 0.67 | 2971 |
| *macro avg* | 0.68 | 0.67 | 0.67 | 2971 |
| *weighted avg* | 0.68 | 0.67 | 0.67 | 2971 |

**(g). KMC**

**Fig. 3.** Performance Comparison

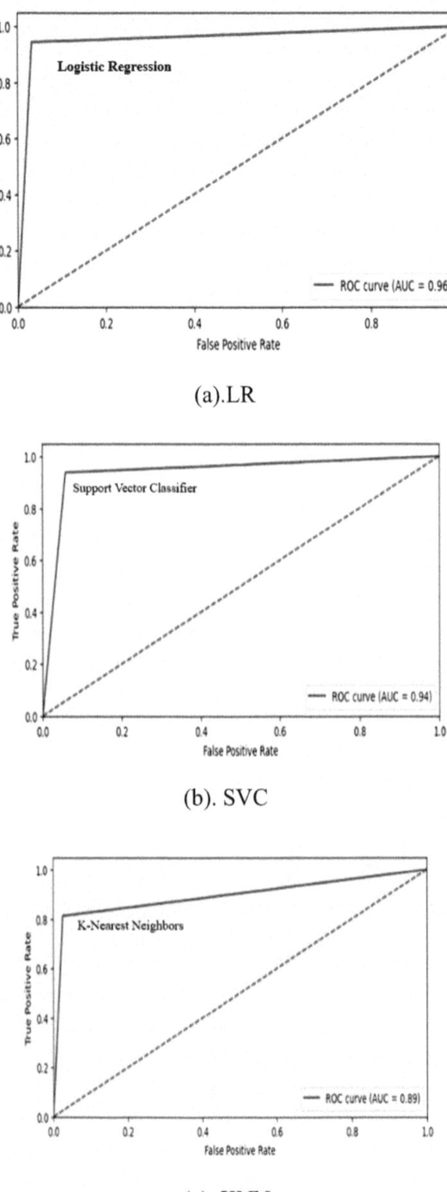


(a).LR

(b). SVC

(c). KNN

**Fig. 4.** ROC Comparison

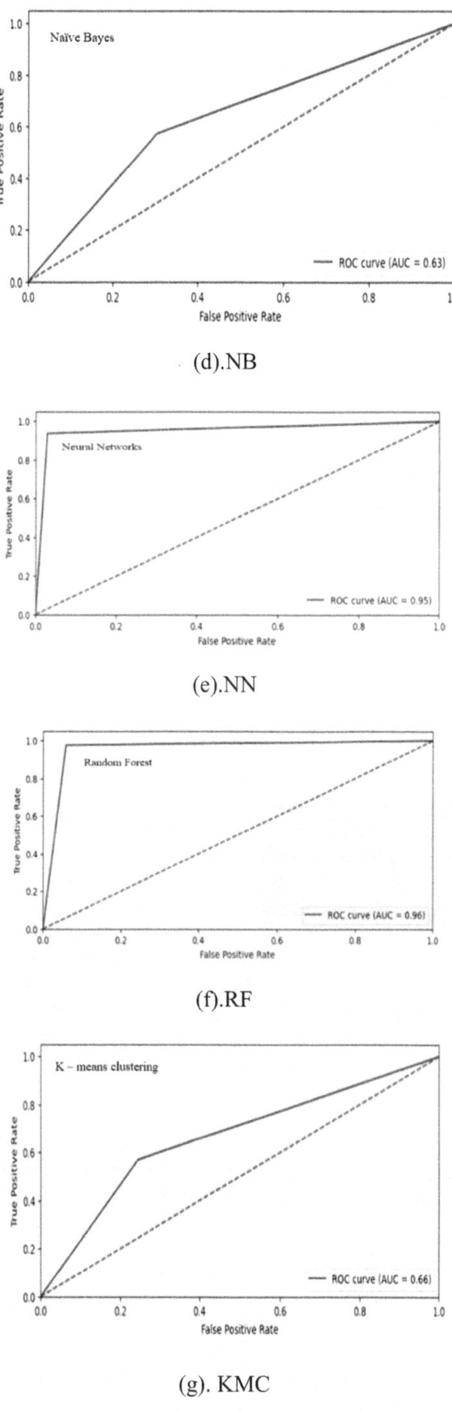


**Fig. 4.** *(continued)*

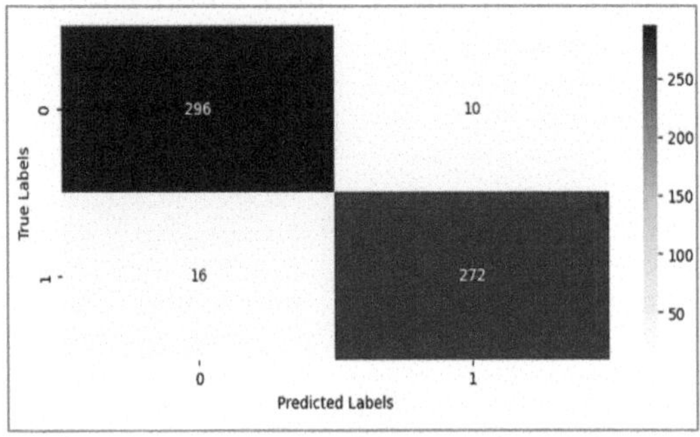


(a). LR

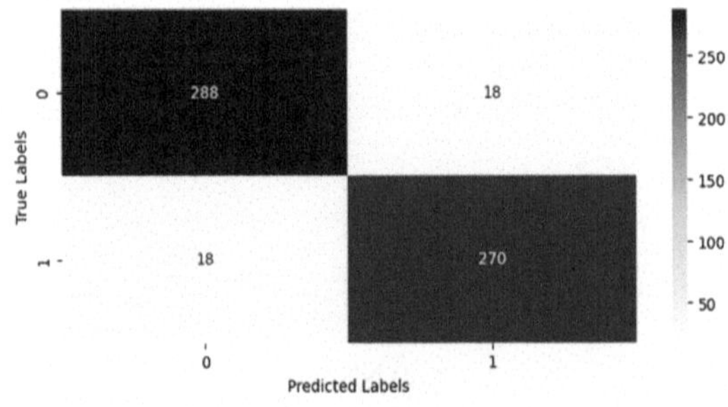


(b).SVC

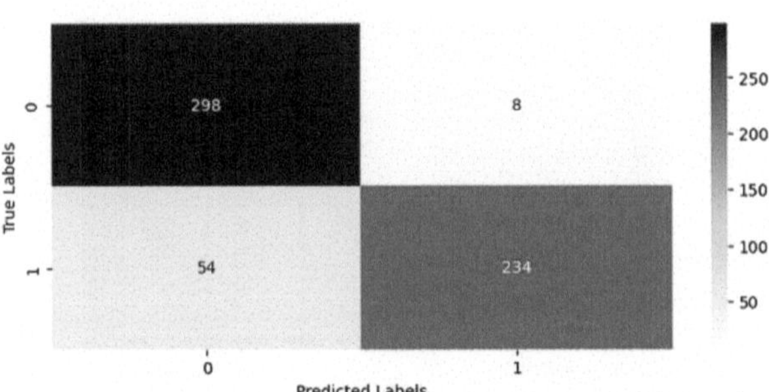


**(c). KNN**

**Fig. 5.** Confusion Matrix Comparison

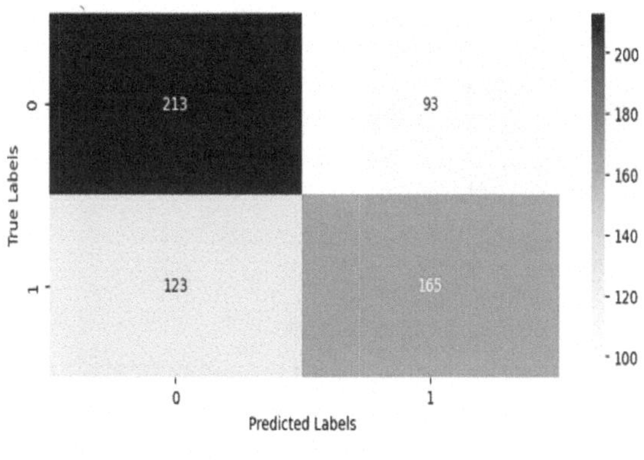

**(d). NB**

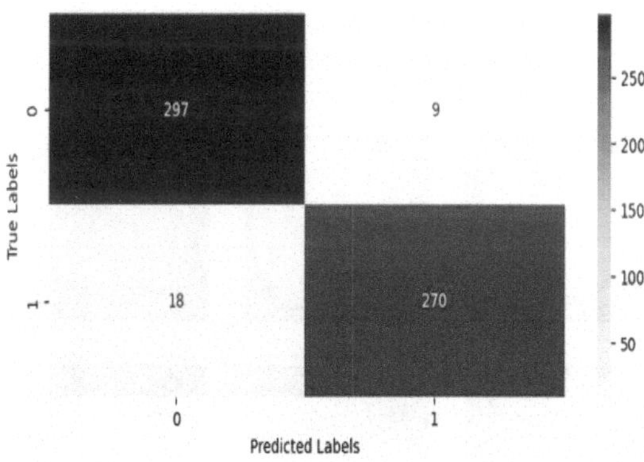


(e). NN

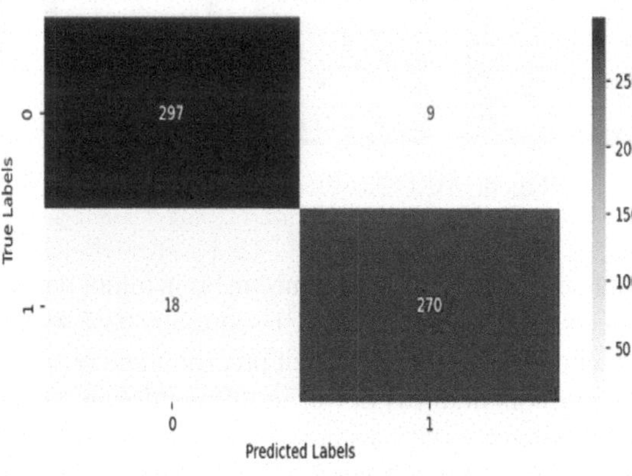


(f). RF

**Fig. 5.** (*continued*)

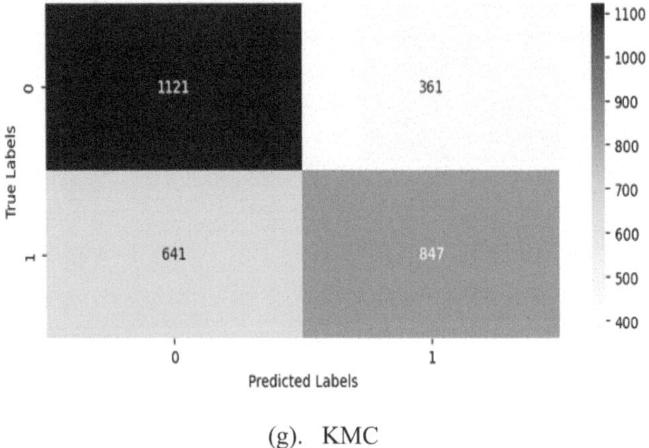


(g). KMC

**Fig. 5.** (*continued*)

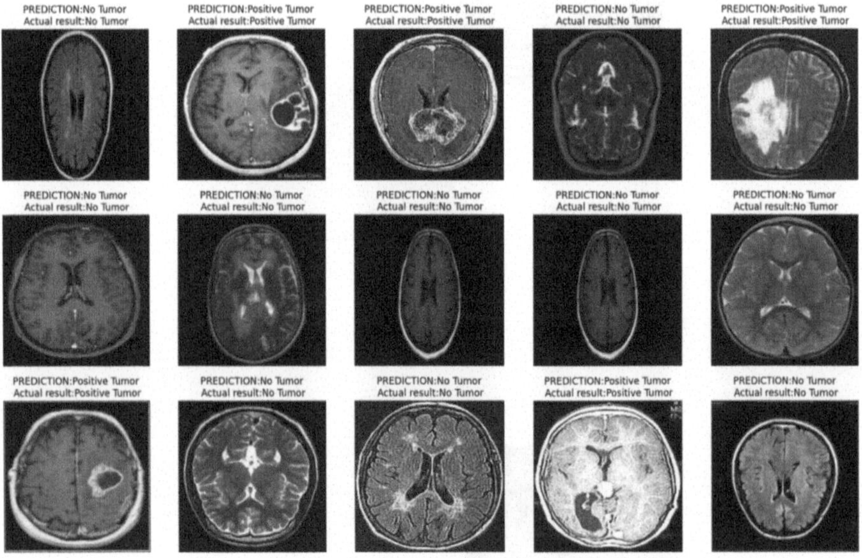


**Fig. 6.** MRI Classification Using LR

The NB model obtained moderate performance in brain tumor prediction. It obtained 64% accuracy, 63% precision, 70% recall, 66% F1-score, and 63% AUC. The experiment shows that NB cannot obtain a balanced value of precision and recall. It has limitations in correctly identifying tumors and less power in differentiating tumor and non-tumor images. Neural Network obtained 95% accuracy, 94% precision, 97% recall, 96% F1-Score, and 95% AUC. It shows that NN can strongly perform well in tumor detection. Random Forest model performs very well in tumor detection. It obtained 96% accuracy, 94% recall, 98% precision, 96% F1-score, and 96% AUC. It shows that it misclassified

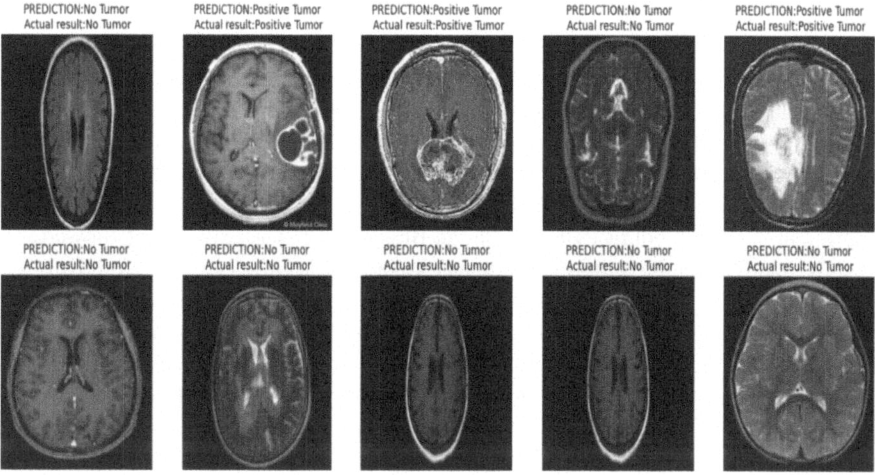


**Fig. 7.** MRI Classification Using SVC

0.7% of the images from the whole dataset and has strong discriminatory power in differentiating tumor and non-tumor images (Fig. 8).

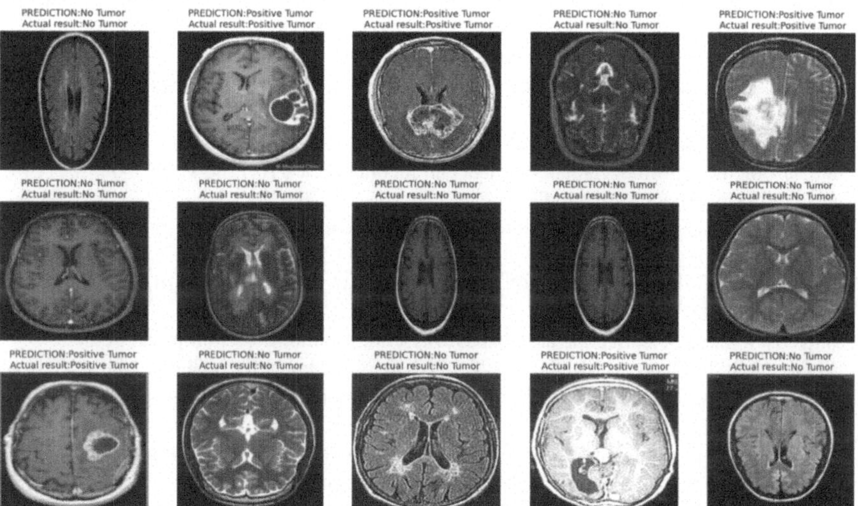


**Fig. 8.** MRI Classification Using RF

Finally, it is noticed that the KMC model performs poorly compared with other models. It obtained 66% accuracy, 64% precision, 64% recall, 76% F1-score, and 66% AUC. It shows that it has limited power to differentiate tumor and non-tumor images. The overall comparison shows that the logistic regression and random forest models obtained 97% accuracy, higher than other models.

## 5 Conclusion

The main objective of this paper is to evaluate the performance of the machine learning models regarding brain tumor detection to know the best model among multiple models. After experimenting with all the models, their performance is evaluated by comparing their performance factors like accuracy, precision, recall, F1-Score, and AUC. A common dataset is taken in the experiment to validate and evaluate the performance. The experiment is carried out in Python with the Bras dataset. The experiment showed that LR, NN, and RF obtained high performance regarding accuracy, precision, recall, and F1-Score. They are also able to differentiate the tumor and non-tumor images accurately. SVC and KNN models perform moderately and can balance precision and recall. They obtained an accuracy of above 90%. Naïve Bayes obtained lower accuracy and lower performance compared with other models. Also, it has limitations in differentiating tumor and non-tumor images. KMC obtained lower accuracy and performance compared to other tumor detection and classification models. From the overall results, the RF model obtained 96% accuracy, 98% precision, 94% recall, 96% F1 score, and 96% AUC.

## References

1. Bauer, S., Wiest, R., Nolte, L.P., Reyes, M.: A survey of MRI-based medical image analysis for brain tumor studies. Phys. Med. Biol. **58**, R97 (2013)
2. Menze, B.H., Jakab, A., Bauer, S., et al.: The multimodal brain tumor image segmentation benchmark (BRATS). IEEE Trans. Med. Imaging **34**(10), 1993–2024 (2015)
3. Festa, J., Pereira, S., Mariz, J.A., Sousa, N., Silva, C.A.: Automatic brain tumor segmentation of multi-sequence Mr images using random decision forests. In: Proceedings MICCAI BRATS (2013)
4. https://www.kaggle.com/datasets/ahmedhamada0/brain-tumor-detection
5. https://bmcmedinformdecismak.biomedcentral.com/articles/10.1186/s12911-023-02114-6#availability-of-data-and-materials
6. https://www.datacamp.com/tutorial/complete-guide-data-augmentation
7. https://link.springer.com/article/10.1007/s10044-017-0597-8#Sec2
8. https://www.academia.edu/36404652/Brain_Tumor_Detection_using_MRI_A_Review_of_Literature
9. https://www.kaggle.com/code/adilababayeva13/brain-tumor-detection-project/edit
10. Majib, M.S., Rahman, M.M., Sazzad, T.S., Khan, N.I., Dey, S.K.: VGG-SCNet: a VGG net-based deep learning framework for brain tumor detection on MRI images. IEEE Access **9**, 116942–116952 (2021)
11. Kshirsagar, P.R., Rakhonde, A.N., Chippalkatti, P.: MRI image based brain tumor detection using machine learning. Test Eng. Manage. **81**, 3672–3680 (2020)
12. Modiya, P., Vahora, S.: Brain tumor detection using transfer learning with dimensionality reduction method. Int. J. Intell. Syst. Appl. Eng. **10**(2), 201–206 (2022)
13. Mahmud, M.I., Mamun, M., Abdelgawad, A.: A deep analysis of brain tumor detection from MR images using deep learning networks. Algorithms **16**(4), 176 (2023)
14. Babu, A.E., Subhash, A., Rajan, D., Jacob, F., Kumar, P.A.: A survey on methods for brain tumor detection. In: 2018 conference on Emerging Devices and Smart Systems (ICEDSS), pp. 213–216. IEEE (2018)
15. Keerthana, A., Kumar, B.K., Akshaya, K.S., Kamalraj, S.: Brain tumour detection using machine learning algorithm. In: Journal of Physics: Conference Series, vol. 1937, no. 1, p. 012008. IOP Publishing (2021)

16. Sharma, M., Sharma, P., Mittal, R., Gupta, K.: Brain tumour detection using machine learning. J. Electron. **3**(4), 298–308 (2021)
17. Senan, E.M., Jadhav, M.E., Rassem, T.H., Aljaloud, A.S., Mohammed, B.A., Al-Mekhlafi, Z.G.: Early diagnosis of brain tumour MRI images using hybrid techniques between deep and machine learning. Comput. Math. Methods Med. (2022)
18. Rehman, A., Naz, S., Razzak, M.I., Akram, F., Imran, M.: A deep learning-based framework for automatic brain tumors classification using transfer learning. Circ. Syst. Signal Process. **39**, 757–775 (2020)
19. Hemanand, D., Selvam, L., Arunachalam, M., Suresh, D.: An image denoising scheme remove unwanted pixel using NLM with sprint deep learning network. Int. J. Intell. Syst. Appl. Eng. **10**(4), 130–137 (2022)
20. Babu, R.G., Hemanand, D., Kumar, K.K., Kanniyappan, N., Vinotha, V.: A survey of satellite images in fast learning method using CNN classification techniques. In: Kumar, A., Ghinea, G., Merugu, S., Hashimoto, T. (eds.) Proceedings of the International Conference on Cognitive and Intelligent Computing. Cognitive Science and Technology, pp. 277–284. Springer, Singapore, (2022), https://doi.org/10.1007/978-981-19-2350-0_27

# A Survey of Anomaly Detection in Video Surveillance

N. Muthurasu(✉) and V. Rajasekar

Department of CSE, SRM Institute of Science and Technology, Chennai, Tamil Nadu, India
muthurasun@gmail.com

**Abstract.** Anomaly detection, also referred to as unusual event detection, plays a critical role in video analytics by identifying abnormal occurrences within a video. However, detecting anomalies in videos presents several challenges, including difficulties in human or object identification, occlusion, clutter, illumination changes, and shape variations. To tackle these obstacles, researchers have developed machine learning and deep learning techniques, with deep learning showing particular promise. With the aid of deep learning, significant advancements have been made in anomaly detection, especially in identifying frames where unusual or abnormal events take place. The main goal of anomaly detection is to pinpoint specific moments in a video where unusual activities occur. This capability proves highly valuable in various settings, including monitoring and detecting mischievous or suspicious activities in places like shopping centers, malls, homes, banks, airports, hospitals, and academic institutions. This survey seeks to comprehensively cover the research areas related to abnormal event detection, shedding light on the progress and potential of this important field.The objective is to review and summarize the existing literature on anomaly detection in video analytics, with a focus on the latest machine learning and deep learning techniques, to identify the key challenges faced in video anomaly detection, including human and object identification, occlusion, clutter, illumination changes, and shape variations, to investigate the advancements made in deep learning-based approaches and their effectiveness in addressing the identified challenges,to explore the applications of video anomaly detection in real-world scenarios, highlighting the significance of these techniques in surveillance and security settings.

**Keywords:** video surveillance · anomaly detection · deep learning

## 1 Introduction

Anomaly detection is the way to identify unusual behaviour or events that differ from the expected behavior. The goal of anomaly detection is to detect and diagnose these anomalies, which can provide important insights into the underlying causes of the abnormal behavior. In this survey, a detailed overview of anomaly detection, including the anomaly types, the techniques used for detecting anomalies, and their applications in various fields.


R. Geetha et al. (Eds.): AAIMB 2023, CCIS 2203, pp. 292–302, 2025.
https://doi.org/10.1007/978-3-031-73068-9_23

### 1.1 Types of Anomalies

The three types of anomalies are explained below

- **Point anomaly:** these are data points that are significantly different from the remaining data.
- **Contextual anomaly:** these are data points that are normal in some contexts but abnormal in others.
- **Collective anomaly:** these are a group of data points that are abnormal when considered together but may appear normal when considered individually.

### 1.2 Anomaly Detection Techniques

Anomaly detection techniques refer to a range of methods used to identify unusual or abnormal behavior within a dataset. These techniques are often used in various applications such as identification of fraudlent activities, intrusion detection, health care, and video analytics. The anomaly detection techniques like statistical methods, Machine learning methods, Clustering methods, Time series analysis, and deep learning methods are discussed.

**Statistical Method:** These methods involve using statistical models to identify anomalies in data. Examples of statistical techniques used in anomaly detection include the Z-score method, the Mahalanobis distance method, and the Gaussian mixture model.

Machine learning methods: Decision trees, SVM classifier, and neural networks can be used to identify the anomalies. It involves training a model on a set of normal data and then using the model to detect anomalies in new data. Clustering methods: Clustering techniques such as k-means clustering can be used to group data points together based

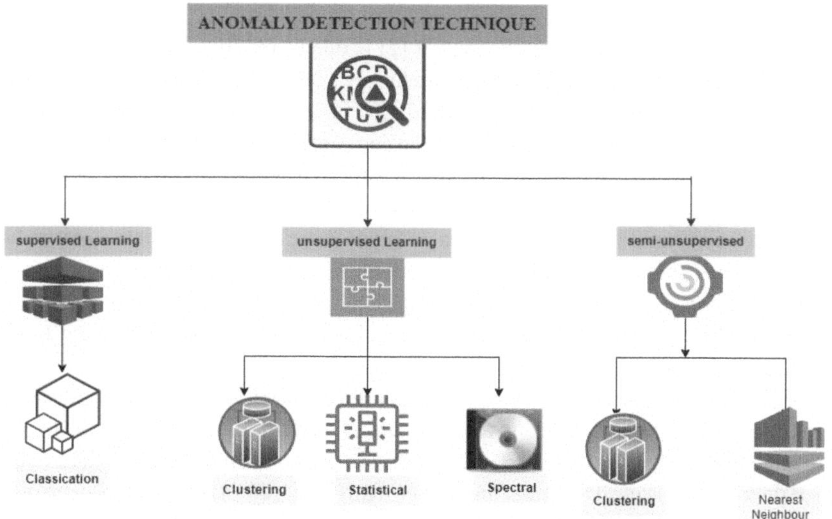


**Fig. 1.** Classification of Anomaly Detection Techniques

on their similarity. Anomalies can then be identified as data points that do not belong to any cluster.

Time series analysis: It analyses data over time to detect anomalies. Autoregressive integrated moving average (ARIMA) and exponential smoothing techniques are used for time series analysis. Deep learning methods: Convolutional neural networks (CNNs) and Recurrent neural networks (RNNs) shown promising results for anomaly detection in various applications, including video analytics. Overall, selecting the appropriate anomaly detection technique relies on the type of application and the type of data used for analysis.

## 2 Literature Review

Anomaly detection has been a topic of active research for several decades, and the literature on the subject is vast. There is an increasing availability of large datasets and the growing need for automated detection of unusual behavior in a wide range of applications, such as intrusion detection, fraud detection, network traffic analysis, medical diagnosis, and fault detection in industrial processes. In this survey, we will review some of the most recent surveys on anomaly detection. Abnormal activity detection is the process of detecting the unexpected behaviour of a person. It has numerous applications, including cybersecurity, fraud detection, predictive maintenance [1], and healthcare. Here is a brief review of abnormal activity detection with references to some notable research papers in the field: One of the earliest and most influential papers in the field of abnormal activity detection is "Statistical Methods for Detecting Malicious Traffic in Networks" by Vern Paxson, published in 1999. The paper proposed a method for detecting network intrusions by analyzing network traffic statistics, such as packet size and interarrival times. The method used a combination of statistical models, such as Poisson and Gaussian, to detect abnormal activity in real-time.

Chandola, V., Banerjee, A., & Kumar, V [2] is one of the most widely cited surveys on anomaly detection. It provides a comprehensive survey for the techniques used in anomaly detection, including statistical methods, machine learning methods, and data mining methods. The survey also discusses the applications of anomaly detection and the evaluation metrics used to measure the performance of anomaly detection algorithms. One-Class SVM is a popular technique for anomaly detection.

Akoglu, L., Chandy, R., &Faloutsos, C [3] Focuses on outlier detection in large graphs, in which size and complexity of graph data is the major challenge to be faced. The survey provides an overview of the various techniques used for outlier detection in graphs, including centrality-based methods, clustering-based methods, and random walk-based methods.

Ruff, L et al. [4] states that deep learning techniques used for anomaly detection have shown remarkable performance in recent years. These methods use neural networks to model the data distribution, and the difference between the observed data and the learned distribution is used to detect anomalies. Ruff et al. (2018) proposed a deep outlier detection (DeepOCC) method that learns an encoder-decoder architecture to reconstruct the input data and uses the difference between the input and output as an anomaly score. The method outperformed other state-of-the-art methods on various benchmark datasets.

Liu, F. T. et al. [5] states that Isolation Forest is a tree-based anomaly detection method that isolates anomalies by recursively partitioning the data into subsets. The method is efficient in high-dimensional data and can handle a large volume of data. Liu et al. (2012) proposed the Isolation Forest method and compared it with other anomaly detection methods on various datasets. The results showed that the Isolation Forest method is robust and outperforms other methods in many cases. Breuniget al. (2000) proposed the LOF method and evaluated it on various datasets. The results showed that the LOF method is effective.

A notable paper in this area is "Unsupervised Anomaly Detection with Generative Adversarial Networks to Guide Marker Discovery" by Schlegl et al., published in 2017 [6]. The paper proposed a method for detecting anomalies in medical images using a combination of a generative adversarial network (GAN) and a marker discovery network. Another recent paper that uses deep learning techniques for abnormal activity detection is "Deep ANT: A Deep Learning Approach for Unsupervised Anomaly Detection in Time Series" by Malhotra et al., published in 2016. Detecting anomalies in time series data using an autoencoder-based deep learning model.

In the field of cybersecurity, abnormal activity detection is used to detect network intrusions and malware attacks[7]. A notable paper in this area is "A Survey of Data Mining Techniques for Malware Detection using File Features" by Sharma et al., published in 2018. The paper provides a comprehensive survey of various data mining techniques.

Many crimes are happening in the world with over a billion people in many countries [8]. Many crimes, related to women harassment, murders, and theft are increasing rapidly during these years with an increased population. So the use of CCTV cameras is also increasing for the purpose of monitoring the daily events from these live data. But it is an inefficient and a time consuming process. Many anomalous events are not properly monitored because of humans [9]. There may be a chance for unnoticing these errors. Therefore automated monitoring systems for analysing the daily events are implemented, analysed and tested. Therefore automatic monitoring systems gain importance in the field of abnormal event detection.

Basically anomaly is referred as the abnormality that deviates from the normal as well as the usual event. Figure 1 shows the general architecture of anomaly detection [10]. There is a lot of data which leads to false alarm and false identification of abnormal events. The false anomaly detection is due to the appearance, variation of shape of the person, false human identification etc. The critical issue in differentiating the normal and abnormal event is the ambiguities present in the video (Fig. 2).

Anomaly detection involves several steps to identify and flag unusual or abnormal patterns or events in data. Here are the typical steps in the process:

**Data Collection:** The first step is to collect and gather the data that will be used for anomaly detection. This can involve gathering data from various sources, such as sensors, logs, databases, or external APIs, [11] depending on the specific application and the type of data being analyzed.

**Data Preprocessing:** The collected data has to be cleaned, formatted, and prepared for analysis. This may involve handling missing values, normalizing data, or transforming data into appropriate formats. Data preprocessing is a critical step to ensure the accuracy and reliability of anomaly detection results.

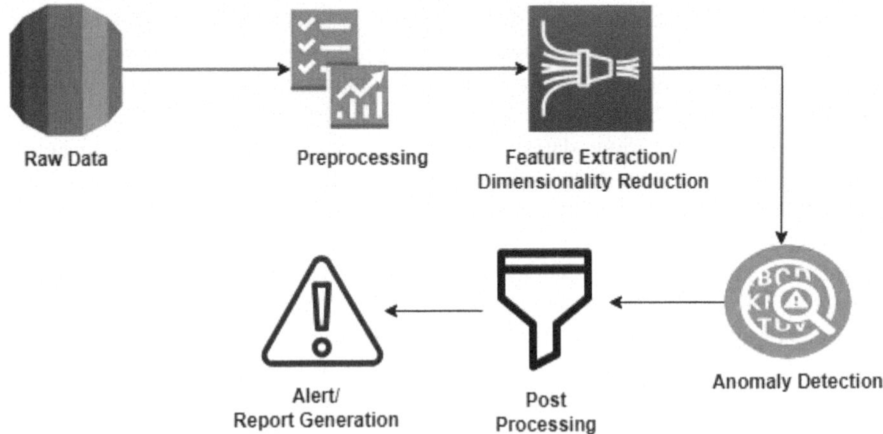


**Fig. 2.** Process for detecting anomalies

**Feature Extraction:** In this step, relevant features or attributes are extracted from the data that are believed to be indicative of normal behaviour [12]. This can involve selecting specific variables or dimensions from the data, or transforming data into meaningful representations, such as time-series data or frequency domain data.

**Model Selection:** The next step is to select an appropriate anomaly detection model or algorithm that is suitable for the specific data and application [13]. There are various types of models, such as statistical-based, machine learning-based, or rule-based, each with their own strengths and weaknesses.

**Model Training:** Once the model is selected, it needs to be trained on the labeled or unlabeled data, depending on the type of algorithm used. Supervised algorithms require labeled data with known anomalies for training [18], while there is no need for labeled in unsupervised learning.

**Anomaly Detection:** After the model is trained, it can be used to detect anomalies in the data [17]. This involves applying the model to the new or unseen data and identifying data points that deviate significantly from the expected or normal behavior based on the established model.

**Evaluation and Validation:** The detected anomalies need to be evaluated and validated to ensure their accuracy and reliability [16]. This can involve comparing the detected anomalies with known anomalies, conducting statistical analysis, or using domain knowledge to validate the findings.

**Alerting and Action:** Once the anomalies are confirmed, appropriate actions can be taken, such as generating alerts, triggering notifications, or initiating corrective measures, depending on the specific application and severity of the anomalies [14].

**Model Update and Maintenance:** Anomaly detection models may need periodic updates and maintenance to adapt to changing data patterns or to improve their accuracy and effectiveness. This may involve retraining the model with new data, tuning model parameters, or updating the rule-based thresholds.

**Monitoring and Continuous Improvement:** Finally, ongoing monitoring of the anomaly detection system is essential to ensure its effectiveness and reliability [15]. Regular review and analysis of the detected anomalies, feedback from users, and continuous improvement of the model and system can help optimize the anomaly detection process and ensure its accuracy and effectiveness over time (Table 1 and Fig. 3).

**Table 1.** Comparison table of Anomaly Detection Techniques

| S.No | Year | Model/Algorithm used | Accuracy Metrics | Dataset Used | Accuracy |
|---|---|---|---|---|---|
| 1 | 2023 | Hierarchical Spatio-Temporal Graph Convolutional Neural Network (HSTGCNN) [24] | AUC - ROC | UCSD Pedestrian2 | 97.73 |
| | | | | ShanghaiTech | 83.40 |
| | | | | CUHK Avenue | 87.50 |
| | | | | IITB-Corridor | 70.40 |
| 2 | 2022 | Temporal-Aware Contrastive Network (TAC-Net) [25] | AUC | UCSD Pedestrian2 | 98.10 |
| | | | | ShanghaiTech | 77.20 |
| | | | | CUHK Avenue | 88.80 |
| 3 | 2022 | Weakly Supervised Adaptive Graph Convolutional Network (WAGCN) [26] | AUC | UCF-Crime | 84.67 |
| | | | | ShanghaiTech | 96.05 |
| 4 | 2021 | video anomaly detection (VAD) [27] | AUC | UCSD Ped2 | 98.20 |
| | | | | UMN data sets, | 99.70 |
| 5 | 2022 | Anomaly Detection with Transformers (ANDTs) [28] | AUC-ROC | AU-AIR | 86.70 |
| 6 | 2019 | Optical Flow and Convolutional Autoencoder (OF-ConvAE-LSTM) [29] | AUC-ROC | UCSD Ped1 | 92.40 |
| | | | | UCSD Ped2 | 96.50 |
| | | | | CUHK Avenue | 89.50 |
| 7 | 2022 | sequential updat-able anomaly detection (SUAD) [30] | AUC-ROC | Metro Anomaly Detection (MAD) | 99.4 |
| 8 | 2021 | particle filtering based anomaly detection [31] | AUC-ROC | UCSD-Peds2 | 96.5 |

(*continued*)

**Table 1.** (*continued*)

| S.No | Year | Model/Algorithm used | Accuracy Metrics | Dataset Used | Accuracy |
|---|---|---|---|---|---|
| 9 | 2022 | Multi-scale and Bidirectional framework using spatio temporal feature learning [32] | AUC-ROC | ShanghaiTech<br>UCSD Ped2<br>CUHK Avenue | 74.50<br>98.10<br>89.0 |
| 10 | 2022 | Collaborative Normality Learning [33] | AUC-ROC | UCF-crime<br>ShanghaiTech | 83.1<br>88.2 |
| 11 | 2023 | Temporal proposal producer (TPP) and online anomaly localizer (OAL) [34] | AUC-ROC | UCF-CRIME<br>XD-VIOLENCE | 85.18<br>78.94 |
| 12 | 2015 | Domain-transformable sparse representations [35] | DIS metric | VDAO-200 | 97.0 |
| 13 | 2021 | LWXception [36] | AUC | UMN dataset | 99.6 |
| 14 | 2021 | Main-Auxiliary Aggregation Strategy [37] | AUC-ROC | UCSD Ped1<br>UCSD Ped2<br>CUHK Avenue<br>Exit | 85.9<br>99.04<br>92.13<br>94.46 |
| 15 | 2022 | Multimodal Supervise-Attention enhanced Fusion [38] | AUC | ShanghaiTech<br>UCF-crime<br>LAD | 93.46<br>81.34<br>82.41 |
| 16 | 2020 | Multipath ConvGRU-based frame prediction network [39] | AUC - ROC | CUHK Avenue | 88.30 |
| 17 | 2019 | Incremental Spatio-Temporal Learner [40] | AUC/ERR | UCSD Ped1<br>UCSD Ped2<br>CUHK Avenue | 75.2/29.8<br>91.9/8.9<br>76.8/29.2 |
| 18 | 2022 | A double low Convolutional LSTM with variational auto encoder [41] | AUC/ERR | UCSD Ped1<br>UCSD Ped2<br>CUHK Avenue | 88.4/16.7<br>88.8/12.2<br>87.2/18.9 |
| 19 | 2022 | Hybrid CNN and Variational auto encoder [42] | ACCURACY | T15<br>QMUL<br>4WAY | 99.0<br>97.3<br>99.5 |

(*continued*)

**Table 1.** (*continued*)

| S.No | Year | Model/Algorithm used | Accuracy Metrics | Dataset Used | Accuracy |
|---|---|---|---|---|---|
| 20 | 2022 | Fusion Net LSTM with optical flow and Spatio temporal feature [43] | AUC | UCSD Ped2<br>CUHK Avenue | 96.9<br>86.1 |
| 21 | 2022 | weakly supervised temporal relation learning [44] | AUC | UCF-CRIME | 83.17 |

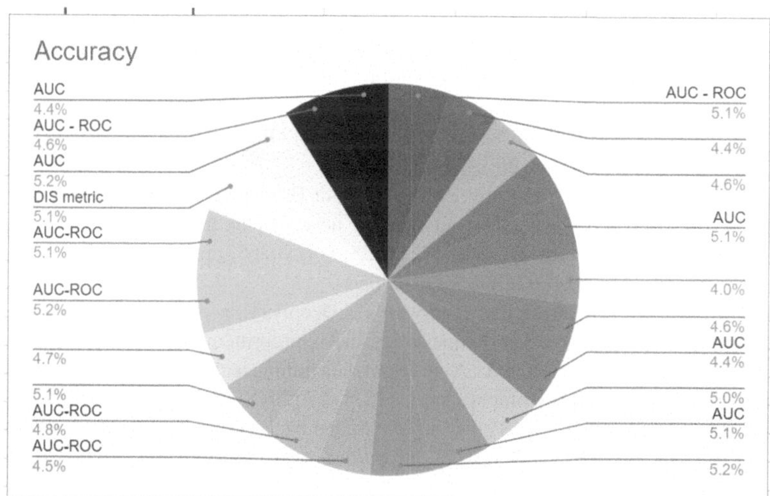


**Fig. 3.** Year wise Accuracy of Anomaly Detection

## 3 Conclusion

In conclusion, anomaly detection is a crucial technique with wide-ranging applications across domains such as cybersecurity, fraud detection, health monitoring, and industrial quality control. It enables organizations to identify unusual behavior, respond in real-time, and mitigate risks, thereby optimizing performance and saving costs. Anomaly detection offers improved accuracy, acting as an early warning system to swiftly detect deviations from normal patterns. Its adaptive and scalable nature makes it suitable for real-time monitoring in big data environments, complementing other data analysis techniques to enhance overall detection capabilities. While considering limitations like false positives and negatives, sensitivity to input parameters, and the need for labeled data, anomaly detection remains a valuable tool to gain insights, protect assets, and achieve business goals. As technology advances, we can anticipate even more sophisticated methods enhancing anomaly detection's effectiveness. Overall, anomaly detection plays a pivotal role in data-driven decision-making, ensuring operational efficiency in today's dynamic business environment. It continues to be an exciting and evolving field with significant potential for future advancements and applications.

## References


1. Paxson, V.: Statistical methods for detecting malicious traffic in networks. ACM SIGCOMM Comput. Commun. Rev. **29**(2), 51–62 (1999)
2. Chandola, V., Banerjee, A., Kumar, V.: Anomaly detection: a survey. ACM Comput. Surv. **41**(3), 1–58 (2009)
3. Akoglu, L., Chandy, R., Faloutsos, C.: Outlier detection in large graphs: a review. ACM SIGKDD Explorations Newsl **15**(2), 1–16 (2015)
4. Ruff, L., Vandermeulen, R.A., Glocker, B., Borgwardt, K.M.: Deep one-class classification. In: ICML, pp. 990–999 (2018)
5. Liu, F.T., Ting, K.M., Zhou, Z.H.: Isolation forest. In: Proceedings of the 2012 11th International Conference on Machine Learning, pp. 1129–1136 (2012)
6. Breunig, M.M., Kriegel, H.P., Ng, R.T., Sander, J.: LOF: identifying density-based local outliers. ACM SIGMOD Rec. **29**(2), 93–104 (2000)
7. Schlegl, T., Seeböck, P., Waldstein, S.M., Schmidt-Erfurth, U., Langs, G.: Unsupervised anomaly detection with generative adversarial networks to guide marker discovery. In: Niethammer, M., et al. (eds.) IPMI 2017. LNCS, vol. 10265, pp. 146–157. Springer, Cham (2017). https://doi.org/10.1007/978-3-319-59050-9_12
8. Malhotra, P., Vig, L., Shroff, G., Agarwal, P.: DeepAnT: a deep learning approach for unsupervised anomaly detection in time series. arXiv preprint arXiv:1609.04476 (2016)
9. Sharma, N., Sharma, V., Sharma, A.: A survey of data mining techniques for malware detection using file features. J. Netw. Comput. Appl. **112**, 1–17 (2018)
10. Bouwmans, T., Maddalena, L., Petrosino, A.: Scene background initialization: a taxonomy. Pattern Recogn. Lett. **96**, 3–11 (2017). https://doi.org/10.1016/j.patrec.2016.12.024. ISSN 0167-8655
11. Bulat, B., Kryjak, T., Gorgon, M.: Implementation of advanced foreground segmentation algorithms GMM, ViBE and PBAS in FPGA and GPU – a comparison. In: Chmielewski, L.J., Kozera, R., Shin, B.S.,Wojciechowski, K. (eds.) ICCVG 2014. LNCS, vol. 8671, pp. 124–131. Springer, Cham (2014). https://doi.org/10.1007/978-3-319-11331-9_16
12. Zhang, G., Jia, J., Xiong, W., Wong, T., Heng, P.-A., Bao, H.: Moving object extraction with a hand-held camera. In: 2007 IEEE 11th International Conference on Computer Vision, pp. 1–8 (2007)
13. Angelov, P., Sadeghi-Tehran, P., Clarke, C.: AURORA: autonomous real-time on-board video analytics. Neural Comput. Appl. **28**(5), 855–865 (2017). https://doi.org/10.1007/s00521-016-2315-7
14. Ramirez-Alonso, G., Ramirez-Quintana, J.A., Chacon-Murguia, M.I.: Temporal weighted learning model for background estimation with an automatic re-initialization stage and adaptive parameters update. Pattern Recognit. Lett. **96**, 34–44 (2017)
15. Déniz, O., Bueno, G., Salido, J., De la Torre, F.: Face recognition using histograms of oriented gradients. Pattern Recognit. Lett. **32**(12), 1598–1603 (2011)
16. Abdel-Hakim, A.E., Farag, A.A.: CSIFT: a SIFT descriptor with color invariant characteristics. In: 2006 IEEE Computer Society Conference on Computer Vision and Pattern Recognition (CVPR 2006), vol. 2, pp. 1978–1983 (2006)
17. Yaseen, M.U., Zafar, M.S., Anjum, A., Hill, R.: High performance video processing in cloud data centres. In: 2016 IEEE Symposium on Service-Oriented System Engineering (SOSE), pp. 152–161 (2016)

18. Yaseen, M.U., Anjum, A., Antonopoulos, N.: Spatial frequency based video stream analysis for object classification and recognition in clouds. In: 2016 IEEE/ACM 3rd International Conference on Big Data Computing Applications and Technologies (BDCAT), pp. 18–26 (2016)
19. Yaseen, M.U., Anjum, A., Rana, O., Hill, R.: Cloud-based scalable object detection and classification in video streams. Futur. Gener. Comput. Syst. **80**, 286–298 (2018)
20. Zhang, X., et al.: Enhancing video event recognition using automatically constructed semantic-visual knowledge base. IEEE Trans. Multimed. **17**(9), 1562–1575 (2015)
21. Escalante, H.J., Guyon, I., Athitsos, V., Jangyodsuk, P., Wan, J.: Principal motion components for one-shot gesture recognition. Pattern Anal. Appl. **20**(1), 167–182 (2017)
22. Gong, L., Yu, M., Gordon, T.: Online code book modeling based background subtraction with a moving camera. In: 2017 3rd International Conference on Frontiers of Signal Processing (ICFSP), pp. 136–140 (2017)
23. Jiang, F., Zhang, S., Wu, S., Gao, Y., Zhao, D.: Multi-layered gesture recognition with kinect. J. Mach. Learn. Res. **16**, 227–254 (2015)
24. Zeng, X., Jiang, Y., Ding, W., Li, H., Hao, Y., Qiu, Z.: A hierarchical spatio-temporal graph convolutional neural network for anomaly detection in videos. IEEE Trans. Circuits Syst. Video Technol. **33**(1), 200–212 (2023). https://doi.org/10.1109/TCSVT.2021.3134410
25. Huang, C., Wu, Z., Wen, J., Xu, Y., Jiang, Q., Wang, Y.: Abnormal event detection using deep contrastive learning for intelligent video surveillance system. IEEE Trans. Industr. Inf. **18**(8), 5171–5179 (2022). https://doi.org/10.1109/TII.2021.3122801
26. Cao, C., Zhang, X., Zhang, S., Wang, P., Zhang, Y.: Adaptive graph convolutional networks for weakly supervised anomaly detection in videos. IEEE Signal Process. Lett. **29**, 2497–2501 (2022). https://doi.org/10.1109/LSP.2022.3226
27. Pillai, G.V., Sen, D.: Anomaly detection in nonstationary videos using time-recursive differencing network-based prediction. IEEE Geosci. Remote Sens. Lett. **19**, 1–5 (2022). https://doi.org/10.1109/LGRS.2021.3072191. Art no. 8010605
28. Jin, P., Mou, L., Xia, G.-S., Zhu, X.X.: Anomaly detection in aerial videos with transformers. IEEE Trans. Geosci. Remote Sens. **60**, 1–13 (2022). https://doi.org/10.1109/TGRS.2022.3198130. Art no. 5628213
29. Duman, E., Erdem, O.A.: Anomaly detection in videos using optical flow and convolutional autoencoder. IEEE Access **7**, 183914–183923 (2019). https://doi.org/10.1109/ACCESS.2019.2960654
30. Zheng, Z., et al.: Anomaly detection of metro station tracks based on sequential updatable anomaly detection framework. IEEE Trans. Circuits Syst. Video Technol. **32**(11), 7677–7691 (2022). https://doi.org/10.1109/TCSVT.2022.3181452
31. Ata-Ur-Rehman, Tariq, S., Farooq, H., Jaleel, A., Wasif, S.M.: Anomaly detection with particle filtering for online video surveillance. IEEE Access **9**, 19457–19468 (2021). https://doi.org/10.1109/ACCESS.2021.3054040
32. Zhong, Y., Chen, X., Hu, Y., Tang, P., Ren, F.: Bidirectional spatio-temporal feature learning with multiscale evaluation for video anomaly detection. IEEE Trans. Circuits Syst. Video Technol. **32**(12), 8285–8296 (2022). https://doi.org/10.1109/TCSVT.2022.3190539
33. Liu, Y., Liu, J., Zhao, M., Li, S., Song, L.: Collaborative normality learning framework for weakly supervised video anomaly detection. IEEE Trans. Circuits Syst. II Express Briefs **69**(5), 2508–2512 (2022). https://doi.org/10.1109/TCSII.2022.3161061
34. Liu, T., Zhang, C., Lam, K.M., Kong, J.: Decouple and resolve: transformer-based models for online anomaly detection from weakly labeled videos. IEEE Trans. Inf. Forensics Secur. **18**, 15–28 (2023). https://doi.org/10.1109/TIFS.2022.3216479
35. Jardim, E., Thomaz, L.A., da Silva, E.A.B., Netto, S.L.: Domain-transformable sparse representation for anomaly detection in moving-camera videos. IEEE Trans. Image Process. **29**, 1329–1343 (2020). https://doi.org/10.1109/TIP.2019.2940686

36. Mehmood, A.: Efficient anomaly detection in crowd videos using pre-trained 2D convolutional neural networks. IEEE Access **9**, 138283–138295 (2021). https://doi.org/10.1109/ACCESS.2021.3118009
37. Wang, Z., Zhang, Y., Wang, G., Xie, P.: Main-auxiliary aggregation strategy for video anomaly detection. IEEE Signal Process. Lett. **28**, 1794–1798 (2021). https://doi.org/10.1109/LSP.2021.3107750
38. Wei, D., Liu, Y., Zhu, X., Liu, J., Zeng, X.: MSAF: multimodal supervise-attention enhanced fusion for video anomaly detection. IEEE Signal Process. Lett. **29**, 2178–2182 (2022). https://doi.org/10.1109/LSP.2022.3216500
39. Wang, X., et al.: Robust unsupervised video anomaly detection by multi-path frame prediction (2020)
40. Nawaratne, R., Alahakoon, D., De Silva, D., Yu, X.: Spatiotemporal anomaly detection using deep learning for real-time video surveillance. IEEE Trans. Industr. Inf. **16**(1), 393–402 (2020). https://doi.org/10.1109/TII.2019.2938527
41. Wang, L., Tan, H., Zhou, F., Zuo, W., Sun, P.: Unsupervised anomaly video detection via a double-flow ConvLSTM variational autoencoder. IEEE Access **10**, 44278–44289 (2022). https://doi.org/10.1109/ACCESS.2022.3165977
42. Kumaran Santhosh, K., Dogra, D.P., Roy, P.P., Mitra, A.: Vehicular trajectory classification and traffic anomaly detection in videos using a hybrid CNN-VAE architecture. IEEE Trans. Intell. Transp. Syst. **23**(8), 11891–11902 (2022). https://doi.org/10.1109/TITS.2021.3108504
43. Song, J.-F., Zhao, H-.L., Wen D.Y., Xu, X.Y.: Video anomaly detection based on optical flow feature enhanced spatio–temporal feature network FusionNet-LSTM-G. IEEE Access **10**, 130314–130325 (2022). https://doi.org/10.1109/ACCESS.2022.3229420
44. Zhang, D., Huang, C., Liu, C., Xu, Y.: Weakly supervised video anomaly detection via transformer-enabled temporal relation learning. IEEE Signal Process. Lett. **29**, 1197–1201 (2022). https://doi.org/10.1109/LSP.2022.3175092

# Prediction and Comparison of ML Algorithm for Heart Disease

G. Belshia Jebamalar[1(✉)], J.A. Adlin Layola[2], J. Rajalakshmi[1], T. Thilagam[1], M. Kausthuban[1], and R. Nareshraj[1]

[1] Department of CSE, S.A. Engineering College, Chennai, India
bel.jeeva@gmail.com
[2] Department of AIML, Rajalakshmi Engineering College, Chennai, India

**Abstract.** Among whole Human body, Heart plays a vital role where it pumps or circulates the blood to all parts of the body. As a result, whole body will function in a proper condition. Heart disease is a major threat globally which takes a human life easily if it is not treated at the proper condition sometimes it can even lead to death. Basically, if the pumping of blood throughout the body is not in proper condition heart disease surely arises. Many risk factors exist which actually stops pumping the blood, those factors are used to get the accurate and early detection of the disease. Researcher generally use data mining and machine learning approaches for dealing large amount of data. For heart or cardiac prediction model that is to handle huge amount of medical data Machine learning approaches can used to produce higher accuracy. In the proposed method three machine Learning approaches are included for predicting and comparing the heart disease. The Models include Logistic Regression, K-Nearest Neighbor and finally Random Forest algorithms are compared for the accuracy and to find the best model which is having greater accuracy.

**Keywords:** Machine learning · Data Mining · Risk Factors · Logistic Regression · KNN Algorithm and Random Forest algorithm

## 1 Introduction

The heart [1] serves as the central organ responsible for pumping blood throughout the entirety of the body and ensuring its proper functioning. Heart disease, a devastating human ailment that is showing an upward trend within both industrialized and developing countries, leading to fatalities. Within this condition, the heart's capacity fails to provide sufficient blood to other regions of the body impairing their ability to perform their regular functions. It links a array of risk factors for heart disease with a urgent demand for accurate, dependable, and feasible ways for making an early diagnosis and managing the condition. Within the healthcare industry, the analysis of extensive datasets commonly employs data mining techniques. To analyze large intricate medical data, researchers use a diverse range of data mining and machine learning methodologies. A machine learning (ML) [2] algorithm, a subset of AI [3], encompasses a variety of precise, probabilistic,


R. Geetha et al. (Eds.): AAIMB 2023, CCIS 2203, pp. 303–313, 2025.
https://doi.org/10.1007/978-3-031-73068-9_24

and enhanced strategies to learn from historical data and discern difficult-to-perceive patterns within large, noisy, or complicate datasets. This data is integral in enabling healthcare providers to predict occurrences of heart ailments. The proposed approach involves developing a machine learning model capable of determining whether or not an individual has cardiac disease.

The software requirements specification is a technical description of a software product's prerequisites. It is the initial stage in the process of requirements analysis. It is a list of a software system's requirements. The ensuing information pertains to specific libraries such as sk-learn, pandas, NumPy, matplotlib, and seaborn. The procedure starts with data collection, using Kaggle to collect prior datasets relevant to heart disease. The data analysis is carried out on the dataset with proper variable identification, i.e., both dependent and independent variables are discovered. The analytical process involves data cleaning and processing, missing value analysis, exploratory analysis, and creation and evaluation of models.

Within the dataset where the data pattern is learned, three distinct techniques, Random Forest, Logistic Regression (LR) and K-Nearest Neighbors (KNN), are deployed. Following the use of two different algorithms, a better algorithm is utilized to predict the outcome. This machine learning model was created with Python's Jupyter Notebook and Flask and trained on Kaggle's standardized dataset. Accuracy, Precision, Sensitivity, Specificity, F Measure, and Error are all metrics used to evaluate the suggested model. The proposed model had a higher level of precision and accuracy. The experimental findings of the proposed model reveal that it is effective in forecasting heart attacks when compared to other current models.

## 2 Related Work

This section points to related research work on the prediction of such heart diseases by various other researchers.

(1) This paper has taken 30-day patients' readmissions which actually does the work with the LVAD. Most important reason for this readmission is due to the heart failure and GIB. Strong Predictor of the readmission is GIB for its Index hospitalization.
(2) in this world there are numerous data, but there exists a process of changing those data into knowledge which is refers to data mining. Health center holds vast amount of data in day today life. This paper explains about various data mining techniques to predict the heart disease at earlier stage. Different techniques have been used and analyzed from which the best and useful one is selected based on priority of efficiency.
(3) The author explains the various reason for CF LVAD where some of the reasons are infections and Gastrointestinal unexpected bleed. There are two types of individuals that is readmitted and non-readmitted but for both the long-term survival holds a little bit difference. The therapy of this disease will be in continuing process for improving its efficacy and cost.
(4) In recent times, death due to heart disease has become a major concern and also a biggest threat of the living being. This disease may occur at any age people since predicting is difficult task for this disease in particular. This paper clearly

explains the various methods, techniques and algorithm engaged to predict has been discussed.

(5) The condition in which the heart and the blood arteries suffers id known as the cardiovascular diseases. Various factors like stress, high blood pressure, obesity, High level of lipids in the blood, diabetes and also certain other factors may contribute in maximum to this cardiovascular disease. Before the start of the diagnosis of this cardiovascular disease, many tests on that disease are taken among which the Electrocardiogram (ECG) is one among that test. Several other test were also discussed in the paper.

(6) In this paper elaboration on heart disease and all the treatments related to the disease were discussed. Additionally, machine learning algorithm for the heart disease exact match model has been explained in detail in the Literature survey part. If the heart suffers to pump the blood naturally the term heart failure arises. Coronary heart disease (CAD) is the term referring to the heart disease where the arteries suffer without the sufficient amount of blood.

(7) Industrial double suction single stage this stage holds various condition on wear. Further analysis on the vibration which has been detected on the pump found to be the good efficiency point. Group of Vibration characteristics has been developed for prediction of heart disease in double suction pumps.

(8) Increase in the knowledge of researcher several prediction techniques are used for the heart disease prediction also now it is highly required to categories and give the overview for the people about the different techniques involved. So many algorithms and techniques are discussed in the paper among all Hybrid Intelligent Algorithm has highest priority since it has good accuracy.

(9) LVAD is so called as Implantable pumps which leads to cardiac several outputs. This LVAD is most commonly and frequently used techniques to predict the hear disease. In this paper they took two patients they have device exchange which leads to lower the acoustic amplitude.

(10) LVAD is highly implemented after care program for the heart failed patients to have implanted the heart surgery. Medolution process actually bind together all the necessary data in single platform, so that all the Doctors shall under their patient stage in a complete picture.

(11) In this paper heart disease has been predicted via machine learning algorithms like KNN, RF and SVM. This Paper analyzed the dataset for the prediction and found good accuracy for the random forest.

# 3 Proposed Enhanced System Model

In the proposed model, a model based on Machine learning has been built for classifying heart disease. Initially process is started by data collecting where all the previously used heart disease data has been taken. The heart disease finding at the right time and proper medication can save the human life. Machine Learning approach generally help the healthcare center in predicting heart disease easily to reduce the human effort and also with highest appropriate results it deals the data. When compared to calculate manually this model has lesser error and also lesser time. First of all, data analysis is initiated where the right and exact variable data are identified and divided the dependent and independent variable. Once the data is identified, Machine Learning algorithm can be applied to predict the disease.

## 3.1 Advantages

- Data Analysis is better.
- Accuracy rate is compared and best model is taken.

## 3.2 Data Processing

Validation process are used in the Machine Learning model to find the error rate. This process clearly displays the accurate wrong values in dataset. Anyways the process of handling the sample data and process of handling real time data is not same scenario. For the real time the data will not be sufficient as of now sample has been taken. All the Variables are necessary in order to find the duplicate, data type and missing values.

After including the Validation dataset into the machine learning model, evaluation will be partial sided. The actual model is frequently checked with validation set. In this place the machine learning experts takes the information and based on the information model hypo parameters are adjusted. The lengthy process includes collection along with data analysis, also data content, its outline and quality. This is time consuming process as a whole.

Data identification is the phase where construction of your model usually takes place also in this process one can understand the data and characteristics in order to construct them. Pythons' panda's library provides various cleaning tasks using data and also it gives priority to large amount of data also missing digits and within fraction of mins it cleans the data very rapidly. It spends very large amount of time in exploring the various data and models but spends very less time in cleaning some of the process make mistakes since the data analyzation is not proper but there are certain more reason for missing data. Based on statistics there are plenty of types in missing data but this situation will influence to fill the missing data and set the process to detect it based on the statistics. Some of the reasons are * Sometimes the user forgets to fill the required field *programming error * in certain case the user wont fill based on the beliefs * during the transfer from the database.

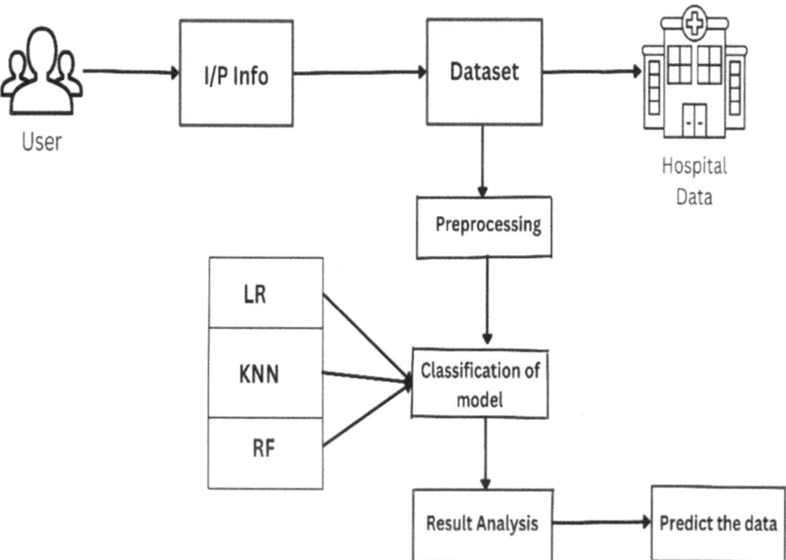


**Fig. 1.** System Architecture

The above Fig. 1 shows that user gives certain values as their inputs in order to predict the heart disease. The data is further processed which include the data from the Hospital and data from Kaggle dataset. once the Data is ready, Preprocessing will take place based on that their classification of algorithms exits. Now LR, KNN and RF model uses the dataset to predict the heart disease. Furthermore, the RF has better accuracy so result has been analyzed. Finally it provides whether the user has heart disease or not.

### 3.3 Data Validation/Cleaning/Preparing Process

First of all, the needed library packages needed to be imported and then need to load the necessary dataset which is required for the prediction process. While data is ready, now its time to check whether all the data are available, any duplicates and variables. when the data is identified 80 percent is taken for training and 20 percent is fixed for testing. Validation dataset is used for measuring competence of the model by the fine tuning. For cleaning different methods and techniques are employed. Data Cleaning process includes certain process like eliminating the mistakes in the data, Missing values and abnormal data. Also this tends in decision making which helps in prediction of the project.

### 3.4 Machine Learning Algorithm

#### 3.4.1 Logistic Regression

Logistic Regression is a statistical technique which is meant for examining the dataset. Dataset holds many independent factors among which more than one factors affect the result. Usually, single variable is not used for measuring the result instead dichotomous variables are used which is one with exactly two outcomes.

To track the Model with the relationship between the predictor and dichotomous features is the main goal of this algorithm.

All the dependent variable has been predicted using Machine learning model with likelihood which is known as Logistic Regression.

The Binary Variable is a dependent Variable where code along with data is done with two values 0 and 1. sometimes 0 is yes or success whereas 1 is no or failure (Table 1).

**Table 1.** Report on Logistic Regression Classification

| | precision | recall | f1-score | support |
|---|---|---|---|---|
| 0 | 0.90 | 0.82 | 0.86 | 45 |
| 1 | 0.80 | 0.89 | 0.84 | 36 |
| accuracy | | | 0.85 | 81 |
| macro avg | 0.85 | 0.86 | 0.85 | 81 |
| weighted avg | 0.86 | 0.85 | 0.85 | 81 |

Confusion Matrix:
[[37 8]
[ 4 32]]
Sensitivity: 0.8222222222222222
Specificity: 0.8888888888888888
Cross validation test results of accuracy:
[0.83333333 0.85185185 0.85185185 0.85185185 0.87037037]
Accuracy: 85.18518518518519.

### 3.4.2 KNN (K-Nearest Neighbor)

Usually, every traditional finds difficulty in categorization model, now to reduce the difficulties this algorithm has been taken. The K-nearest neighbor algorithm also meant to be K-NN, this creates an outline for the data without any certain logics. this functions by adding an additional datapoints to predict the nearby boundary line.

Few Complex models are denoted with Higher k values as they hold smoother curve whereas complex models are denoted by the least k values where the data overflow occurs in this situation.

KNN searches the dataset for correlations between predictors and values. It employs a non-parametric approach because no specific parameters for any given functional form have been identified. It doesn't assume anything about the dataset's properties or results in any way. Lazy Classifier is also known to be KNN because it studies all the training data also keep that in memory but not learn properly so as conclusion it won't fix. While efforts of computer is done in higher priority, This always gives priority to the classification time and don't value the training time.

- Start the process by choosing the k value for computing the KNN algorithm. now distance measurement has to be taken with the help of Euclidean distance formula based on the k neighbors.
- In this all the closer that is nearby to the new point is taken and concluded as the closest to the point which we set. In this location finding K-nearest is pretty difficult.
- out of all highest number of result containing class has been taken into the count then we shall assign a new point to that particular class thus provides the way for using this algorithm (Table 2)

**Table 2.** Report on K-Neighbors Classification

| | precision | Recall | f1-score | Support |
|---|---|---|---|---|
| 0 | 0.69 | 0.82 | 0.75 | 45 |
| 1 | 0.70 | 0.53 | 0.60 | 36 |
| accuracy | | | 0.68 | 81 |
| macro avg | 0.69 | 0.68 | 0.68 | 81 |
| weighted avg | 0.69 | 0.69 | 0.68 | 81 |

Confusion Matrix:
[[37 8]
[17 19]]
Sensitivity: 0.8222222222222222
Specificity: 0.5277777777777778
Accuracy: 68.14814814814815

### 3.4.3 Random Forest

This algorithm mainly addresses the classification where the Regression is average for the different samples.

The data subset is present initially and model is extracts data from it. sometimes the final output will be taken on whole and in some cases, it is based on higher priority Ranking.

1. first of all, the packages and dataset needed will be imported.
2. secondly two variables that is feature variable and target variable will be assigned.
3. training of data followed by testing then proceeded to split.
4. Random forest Classifier will be imported and then include the data in it.
5. Hypo parameter tuning followed by sorting is done in the Algorithm (Table 3).

**Table 3.** Report on Random Forest Classification

| | precision | Recall | f1-score | support |
|---|---|---|---|---|
| 0 | 0.92 | 0.85 | 0.88 | 27 |
| 1 | 0.89 | 0.94 | 0.91 | 34 |
| accuracy | | | 0.90 | 61 |
| macro avg | 0.90 | 0.90 | 0.90 | 61 |
| weighted avg | 0.90 | 0.90 | 0.90 | 61 |

Confusion Matrix:
[[23 4]
[2 32]]
Sensitivity: 0.92
Specificity: 0.888888888888
Accuracy: 90.1616161616

### 3.5 Pseudo Code

---

import library packages

plot Histogram for age distribution

bar plot the ECG and target

variable propagation

plot the Heatmap diagram

preprocessing, split test and dataset, split response variable

add the Data Labels

repeat the process for both the algorithms

Flask deploys

---

## 4 Results and Discussion

The results of three methods accuracy calculations are provided in the table above. The accuracy of the Random Forest, KNN and the logistic regression are computed. By

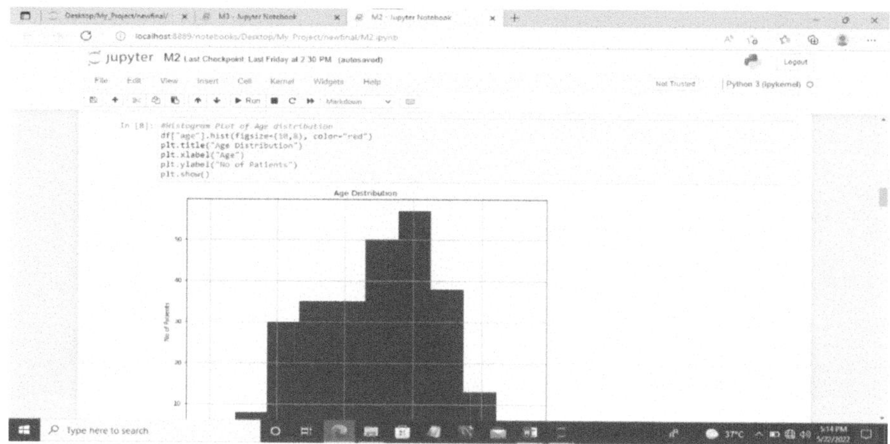


**Fig. 2.** Training the Dataset

comparing the accuracy on all the three models, the Random Forest model predicts the heart disease with the greater accuracy rate (Figs. 2 and 3).

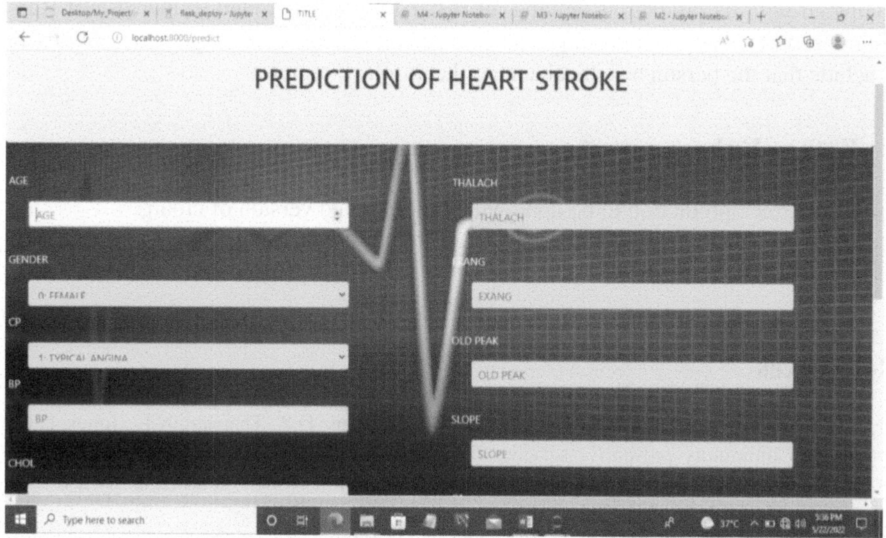


**Fig. 3.** User Interface

The above Fig. 4 displays the better accuracy then other models for heart disease prediction.

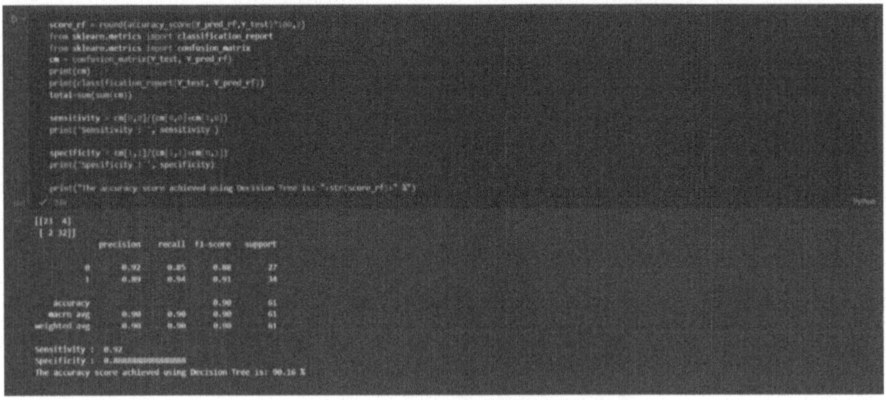

**Fig. 4.** RF algorithm result

## 5 Conclusion

Analytical process consists of all the value analysis which has been missed, analysis on exploratory, processing of data along with its preparation followed by the construction. Above three algorithms of the Machine Learning, Random Forest and KNN has been compared with its predictions among all better accuracy is found in the case of Random Forest. This application can assist in determining heart attack prediction and will conclude that the person has the chance of heart attack or not.

## 6 Future Enhancement

- Heart Attack prediction to be done with an integrated version of cloud.
- All the ML model has been used now it's time to predict in the AI environment.
- Integration with Electronic Health Records (EHR).

## References

1. Jacob, V., Reynolds, J.A., Chattopadhyay, S.K., Hopkins, D.P., Therrien, N.L., Jones, C.D., et al.: Community preventive services task force (CPSTF).: pharmacist interventions for medication adherence: community guide economic reviews for cardiovascular disease. Am. J. Prev. Med. **62**(3), e202–222 (2022)
2. Jin, W.: Research on machine learning and its algorithms and development. In: Journal of Physics: Conference Series, vol. 1544, p. 012003 (2020). https://doi.org/10.1088/1742-6596/1544/1/012003
3. Alsedrah, M.: Artificial intelligence (2017). https://doi.org/10.13140/RG.2.2.18789.65769
4. Akinsola, J.E.T.: Supervised machine learning algorithms: classification and comparison. In: International Journal of Computer Trends and Technology (IJCTT), vol. 48, pp. 128–138 (2017). https://doi.org/10.14445/22312803/IJCTT-V48P126
5. Abdullah, A.S., Rajalaxmi, R.R.: A data mining model for predicting the coronary heart disease using random forest classifier. In: Proceedings of International Conference in Recent Trends in Computational Methods, Communication and Controls, pp. 22–25 (2012)

6. Hazra, A., Mandal, S., Gupta, A., Mukherjee, A., Mukherjee, A.: Heart disease diagnosis and prediction using machine learning and data mining techniques: a review. Adv. Comput. Sci. Technol. **10**, 2137–2159 (2017)
7. Akhter, S.A., Badami, A., Murray, M., Kohmoto, T., Lozonschi, L., Osaki, S., Lushaj, E.B.: Hospital readmissions after continuous-flow left ventricular assist device implantation: incidence, causes, and cost analysis. Ann. Thoracic Surgery **100** (2015). https://doi.org/10.1016/j.athoracsur.2015.03.010
8. Golande, A., Pavan Kumar, T.: Heart disease prediction using effective machine learning techniques. Int. J. Recent Technol. Eng. (IJRTE) **8**(1), 944–950 (2019). ISSN: 2277–3878, Issue-1S4
9. Gutta, C., Mariappan, P., Suganya, G.: Effective prediction of cardiovascular diseases using cluster of machine learning algorithms. J. Critic. Rev. **7**, 2192–2201 (2020)
10. Akhter, S.A., et al.: Hospital readmissions after continuous-flow left ventricular assist device implantation: incidence, causes, and cost analysis. Ann Thoracic Surg. **100**(3), 884–889 (2015). https://doi.org/10.1016/j.athoracsur.2015.03.010. Epub .PMID: 26095106
11. Melinda, H., Norton, M.P.: The effect of change in flow rate on the vibration of double suction centrifugal pumps. In: Proceedings of the Institute of Mechanical Engineers Part E: Process Mechanical Engineering, vol. 216, pp. 47–58 (2002). https://doi.org/10.1243/095440802760075030
12. Chitra, J., Seenivasagam, V.: Review of heart disease prediction system using data mining and hybrid intelligent techniques. Int. J. Soft Comput. **3**, 2229–6956 (2013). https://doi.org/10.21917/ijsc.2013.0087
13. Gardner, Y., Royston, T., Bhat, G., Tatooles, A.: Acoustic characterization of axial flow left ventricular assist device operation in vitro and in vivo. ASAIO J. (Am. Soc. Artif. Internal Organs: 1992). **62** (2015). https://doi.org/10.1097/MAT.0000000000000307
14. Reiss, N., et al.: Telemonitoring and medical care of heart failure patients supported by left ventricular assist devices - the medolution project. Stud. Health Technol. Inf. **236**, 267–274 (2017). PMID: 28508806
15. Praveena Rachel Kamala, S., Gayathri, S., Pillai, N.M., Anto Gracious, L.A., Varun, C.M., Siva Subramanian, R.: In: 4th International Conference _on Electronics and Sustainable Communication Systems (ICESC), Coimbatore, India, pp. 1583–1589 (2023). https://doi.org/10.1109/ICESC57686.2023.10193323.

# Comparative Analysis of Classifiers for Chat Classification: A Study on Random Forest, Support Vector Machine, and Multinomial Naive Bayes with Bag of Words and TF-IDF

K. Hemakirthiga[1] and J. Arunadevi[2](✉)

[1] Raja Doraisingam Govt Arts College, Sivaganga, Affiliated to Alagappa University, Sivaganga, India

[2] PG and Research Department of Computer Science, Raja Doraisingam Govt. Arts College, Sivaganga, India

arunardm2015@gmail.com

**Abstract.** This work focuses on the task of chat classification, aiming to accurately classify chat instances into positive, negative, and neutral classes. The study employs three classifiers: Random Forest, Support Vector Machine (SVM), and Multinomial Naive Bayes (MNB), along with two text representation techniques: Bag of Words (BoW) and TF-IDF (Term Frequency Inverse Document Frequency). The dataset used consists of 584 instances, with 178 positive, 147 negative, and 259 neutral samples. The classifiers are trained and evaluated using metrics such as accuracy, precision, recall, F1-Score, and specificity. The results indicate that SVM with TF-IDF achieved the highest performance across various evaluation metrics, closely followed by Random Forest with TF-IDF. These findings suggest that SVM with TF-IDF is a robust choice for chat classification tasks, while TF-IDF generally outperforms Bag of Words in text representation. The work contributes insights into the effectiveness of different classifiers and text representation techniques for chat classification, providing guidance for future projects in sentiment analysis, customer feedback analysis, and chatbot systems.

**Keywords:** chat classification · text classification · Random Forest · Support Vector Machine (SVM) · Multinomial Naive Bayes (MNB) · Bag of Words (BoW) · TF-IDF (Term Frequency Inverse Document Frequency)

## 1 Introduction

The goal of this work is to build classifiers for a chat dataset consisting of 584 instances, which are classified into three classes: positive, negative, and neutral. The dataset contains 178 instances labeled as positive, 147 as negative, and 259 as neutral. To tackle this classification task, we have employed three popular machine learning algorithms: Random Forest, Support Vector Machine (SVM), and Multinomial Naive.

Bayes. These classifiers have proven to be effective in handling text data and have been widely used in natural language processing tasks.


R. Geetha et al. (Eds.): AAIMB 2023, CCIS 2203, pp. 314–326, 2025.
https://doi.org/10.1007/978-3-031-73068-9_25

In addition to the choice of classifiers, we have also explored two different text representation techniques: Bag of Words (BoW) and TF-IDF (Term Frequency Inverse Document Frequency). These methods transform the raw text data into numerical features that can be processed by the classifiers. Bag of Words (BoW) is a straightforward approach that represents each instance as a vector, where each element corresponds to the frequency of a specific word in the text. This representation disregards the order and context of the words but provides a simple and efficient way to capture the overall vocabulary and word frequencies in the dataset.

TF-IDF (Term Frequency-Inverse Document Frequency) is another widely used text representation technique that takes into account both the term frequency and the inverse document frequency. It assigns higher weights to terms that are frequent in a specific instance but rare in the overall dataset. This approach aims to highlight the importance of distinctive terms in each instance while reducing the influence of common words.

By employing these classifiers and text representation techniques, we aim to build accurate and robust models that can classify new instances of chat data into one of the three predefined classes: positive, negative, or neutral. The evaluation of the models will involve measuring their performance using appropriate metrics, such as accuracy, precision, recall, and F1-score, to assess their effectiveness in handling the imbalanced class distribution and overall predictive capabilities.

Through this work, we hope to contribute to the field of chat classification and provide insights into the effectiveness of different classifiers and text representation techniques in handling this specific task. The outcomes of this study can have practical applications in various domains that rely on chat analysis, such as sentiment analysis, customer support systems, and social media monitoring.

# 2 Literature Review

## 2.1 Literature Review on Text Representations

### 2.1.1 Bag of Words (BoW)

The Bag of Words (BoW) model is a popular and widely used approach for text representation in natural language processing tasks. It treats each document as an unordered collection of words and represents them as a vector of word frequencies. Several studies have explored the effectiveness of BoW in various classification tasks.

One study by Harris et al. [1] applied the BoW model for sentiment analysis on movie reviews. They found that BoW, when combined with machine learning algorithms like Support Vector Machines (SVM) and Naive Bayes, achieved competitive results in sentiment classification tasks. The authors highlighted the simplicity and efficiency of BoW in capturing the overall word frequencies.

Another research by Zhang et al. [2] compared different text representation methods, including BoW, for authorship attribution. The study found that BoW, when combined with SVM, yielded promising results in identifying the authorship of anonymous texts. The authors emphasized the importance of feature selection techniques in enhancing the performance of BoW-based classifiers.

### 2.1.2 TF-IDF (Term Frequency-Inverse Document Frequency)

TF-IDF is another commonly used text representation technique that aims to capture the importance of terms in a document relative to the entire corpus. It assigns higher weights to terms that are frequent in a specific document but rare in the overall dataset. Several studies have investigated the effectiveness of TF-IDF in various text classification tasks.

A study by Sparck Jones [3] introduced the concept of IDF (Inverse Document Frequency) and highlighted its importance in capturing term relevance. The paper demonstrated the effectiveness of IDF in information retrieval tasks and emphasized the need to balance term frequency and inverse document frequency.

In a research work by Wang and Manning [4], the authors compared different text representation methods, including TF-IDF, for sentiment analysis. They found that TF-IDF, when combined with SVM, outperformed other methods in sentiment classification tasks. The study emphasized the ability of TF-IDF to highlight important terms and effectively represent documents.

Another study by Yang et al. [5] explored the use of TF-IDF in document classification for spam filtering. The authors compared TF-IDF with other feature selection methods and found that TF-IDF, when combined with SVM and Random Forest, achieved high accuracy in classifying spam and non-spam emails. The study highlighted the effectiveness of TF-IDF in capturing distinctive terms for classification purposes.

These studies demonstrate the effectiveness of both Bag of Words (BoW) and TF-IDF in various text classification tasks. While BoW provides a simple and efficient representation of word frequencies, TF-IDF captures term importance and relevance in the context of the entire corpus. Both techniques have been successfully applied in sentiment analysis, authorship attribution, information retrieval, and spam filtering tasks, showcasing their versatility and usefulness in text classification.

## 2.2 Random Forest Classifier for Text Classification

Random Forest is a popular machine learning algorithm known for its effectiveness in text classification tasks. It is an ensemble method that combines multiple decision trees to make predictions. Several studies have explored the application of Random Forest in text classification and its performance in comparison to other classifiers.

A study by Breiman [6] introduced the Random Forest algorithm and demonstrated its ability to handle high-dimensional data, such as text, with minimal overfitting. The study highlighted the importance of feature selection and.

randomization in building diverse decision trees, leading to improved generalization and classification accuracy.

In a research work by Joachims [7], the author compared Random Forest with other classifiers, including SVM, in text categorization tasks. The study found that Random Forest achieved competitive results in terms of accuracy and outperformed other classifiers on large-scale datasets. The author emphasized the scalability and robustness of Random Forest for text classification.

Another study by Pang and Lee [8] applied Random Forest for sentiment analysis on movie reviews. They compared Random Forest with other classifiers and found that it consistently achieved high accuracy in sentiment classification tasks. The study highlighted the ability of Random Forest to handle noisy and high-dimensional text data effectively.

In a comparative study by Zhang and Yang [9], the authors evaluated the performance of Random Forest and other classifiers on document classification tasks. They found that Random Forest, when combined with suitable feature selection methods, achieved competitive accuracy and outperformed other classifiers in handling text classification tasks.

A recent study by Chen et al. [10] investigated the performance of Random Forest in multiclass text classification. They compared Random Forest with other classifiers and found that it achieved high accuracy and F1-score in classifying text documents into multiple classes. The study highlighted the robustness and interpretability of Random Forest in multiclass text classification scenarios.

These studies demonstrate the effectiveness of Random Forest in text classification tasks, including sentiment analysis, document classification, and multiclass classification. Random Forest's ability to handle high-dimensional data, its scalability, and its robustness make it a suitable choice for text classification applications.

### 2.3 SVM Classifier for Text Classification

Support Vector Machine (SVM) is a widely used machine learning algorithm known for its effectiveness in text classification tasks. SVM seeks to find an optimal hyperplane that maximally separates instances belonging to different classes. Numerous studies have investigated the application of SVM in text classification and compared its performance with other classifiers.

One of the foundational works by Joachims[7]introduced SVM for text classification and highlighted its effectiveness in handling high-dimensional data, such as text documents. The study demonstrated the ability of SVM to achieve high accuracy and generalization in various text categorization tasks.

In a comparative study by Yang et al. [11], the authors evaluated the performance of SVM and other classifiers on document classification tasks. The study found that SVM consistently outperformed other classifiers, including Naive Bayes and k Nearest Neighbors, in terms of accuracy and F1-score. The authors emphasized the robustness and effectiveness of SVM in handling text classification tasks.

Another study by Cai et al. [12] compared SVM with other classifiers for sentiment analysis. The authors found that SVM achieved competitive results in sentiment classification tasks and outperformed other classifiers, such as Naive Bayes and decision trees. The study highlighted the ability of SVM to handle complex text features and capture nonlinear relationships.

In a research work by Huang et al. [13], the authors explored the effectiveness of SVM in multiclass text classification. They compared SVM with other multiclass classifiers and found that SVM achieved high accuracy and F1-score in classifying text documents into multiple classes. The authors emphasized the ability of SVM to handle imbalanced class distributions and effectively handle large-scale text datasets.

A recent study by Wang et al. [4] investigated the performance of SVM and other classifiers in sentiment analysis tasks. They compared SVM with Random Forest and

Naive Bayes and found that SVM achieved competitive accuracy and F1- score in sentiment classification. The study highlighted the effectiveness of SVM in handling different sentiment analysis scenarios.

These studies demonstrate the effectiveness of SVM in text classification tasks, including document classification, sentiment analysis, and multiclass classification. SVM's ability to handle high-dimensional data, its robustness, and its ability to capture nonlinear relationships make it a popular choice for text classification applications.

### 2.4 Multinomial Naive Bayes Classifier for Text Classification

Multinomial Naive Bayes (MNB) is a popular probabilistic classifier widely used in text classification tasks. It is based on the Naive Bayes principle and assumes independence between features. Numerous studies have explored the application of MNB in text classification and evaluated its performance in comparison to other classifiers.

A seminal work by McCallum and Nigam [14] introduced the use of Naive Bayes classifiers, including MNB, for text classification tasks. The study demonstrated the effectiveness of MNB in document classification and highlighted its simplicity and efficiency in handling text data.

In a comparative study by Rennie et al. [15] the authors evaluated the performance of MNB and other classifiers on document classification tasks. The study found that MNB achieved competitive accuracy and outperformed other classifiers, including SVM and decision trees, on large-scale datasets. The authors emphasized the effectiveness of MNB in handling high-dimensional text data.

Another study by Wang and Manning [4] compared MNB with other classifiers for sentiment analysis tasks. The authors found that MNB achieved competitive results in sentiment classification and outperformed other classifiers, such as SVM and Random Forest, in terms of accuracy and F1-score. The study highlighted the simplicity and effectiveness of MNB in sentiment analysis.

In a research work by Zhang et al. [16] the authors explored the effectiveness of MNB in text categorization tasks. They compared MNB with other classifiers and found that MNB achieved high accuracy in classifying text documents into predefined.

categories. The authors emphasized the ability of MNB to handle large-scale text datasets and capture informative features.

A recent study by Han et al. [17] investigated the performance of MNB in multiclass text classification. They compared MNB with other classifiers and found that MNB achieved competitive accuracy and F1-score in classifying text documents into multiple classes. The study highlighted the efficiency and effectiveness of MNB in multiclass text classification scenarios.

These studies demonstrate the effectiveness of MNB in text classification tasks, including document classification, sentiment analysis, and multiclass classification. MNB's simplicity, efficiency, and assumption of feature independence make it a popular choice for text classification applications.

## 3 Problem Definition

The problem addressed in the above work is text classification using a dataset of chat instances. The objective is to build classifiers that can accurately categorize chat instances into one of three classes: positive, negative, or neutral. The dataset consists of 584 instances, with 178 instances labeled as positive, 147 as negative, and 259 as neutral.

The task involves employing three different classifiers: Random Forest, Support Vector Machine (SVM), and Multinomial Naive Bayes (MNB). These classifiers will be trained on the chat dataset to learn patterns and relationships between textual features and their corresponding classes.

Two different text representation techniques will be utilized: Bag of Words (BoW) and TF-IDF (Term Frequency-Inverse Document Frequency). BoW represents each chat instance as a vector of word frequencies, while TF-IDF assigns weights to words based on their frequency in the instance and their rarity in the overall dataset.

The problem is to assess the performance of these classifiers and text representation techniques in accurately classifying chat instances into the predefined positive, negative, and neutral classes. The evaluation will involve measuring various performance metrics, such as accuracy, precision, recall, and F1-score, to determine the effectiveness and robustness of the classifiers in handling the imbalanced class distribution.

The goal of this work is to build reliable and accurate classifiers that can be used for chat classification tasks, such as sentiment analysis or customer feedback analysis. By addressing this problem, insights will be gained into the effectiveness of different classifiers and text representation techniques for text classification tasks, specifically in the context of chat data.

## 4 Proposed Methodology

The proposed methodology for the work involves several steps to build classifiers for chat classification using Random Forest, Support Vector Machine (SVM), and Multinomial Naive Bayes (MNB) algorithms. The methodology also includes the utilization of two text representation techniques: Bag of Words (BoW) and TF-IDF.

(Term Frequency-Inverse Document Frequency). Here is an outline of the proposed methodology:

1. *Data Preprocessing:*
   - Perform data cleaning and normalization, including removing punctuation, stop-words, and special characters.
   - Tokenize the chat instances into individual words or n-grams.
   - Perform stemming or lemmatization to reduce words to their root form, if required.
2. *Extraction:*
   - Implement the Bag of Words (BoW) technique to represent chat instances as vectors of word frequencies.
   - Alternatively, apply the TF-IDF (Term Frequency-Inverse Document Frequency) technique to assign weights to words based on their importance and rarity.

3. *Model Training:*
   - Train Random Forest, SVM, and MNB classifiers on the preprocessed and feature-extracted training data.
   - Optimize the hyperparameters of each classifier using techniques like grid search or cross-validation.
4. *Model Evaluation:*
   - Evaluate the trained classifiers using appropriate metrics such as accuracy, precision, recall, and F1-score.
   - Assess the performance of each classifier on the testing set and compare their results.
5. *Fine-tuning and Model Selection:*
   - Fine-tune the classifiers by adjusting hyper parameters based on the evaluation results.
   - Select the best-performing classifier or ensemble of classifiers based on their performance and robustness.
6. *Final Model Deployment:*
   - Utilize the selected classifier or ensemble to classify new, unseen chat instances into positive, negative, or neutral classes.

The framework for the proposed method is represented in the following block diagram in Fig. 1.

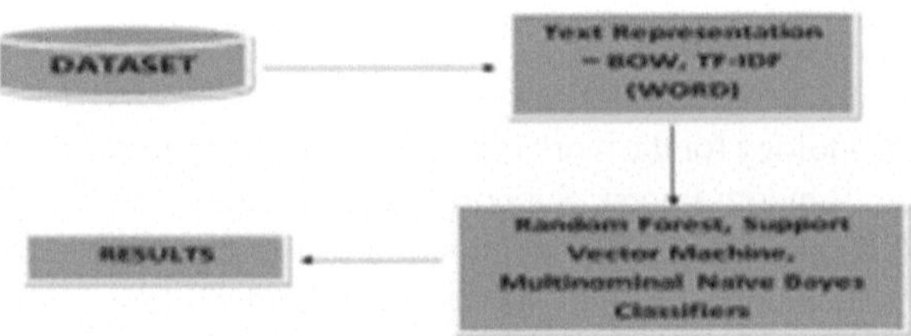


**Fig. 1.** Proposed Methodology

## 5 Experimental Environment

The dataset used for this research is the chat dataset from Kaggle [19]. The dataset used in this work for chat classification consists of 584 instances. The dataset is labeled with three classes: positive, negative, and neutral. The distribution of instances across the classes is as follows:

- Positive: 178 instances
- Negative: 147 instances
- Neutral: 259 instances

The dataset represents chats or textual conversations that are categorized into these three classes based on their sentiment or overall tone. Each chat instance in the dataset is associated with a class label indicating whether it is positive, negative, or neutral in nature (Fig. 2).

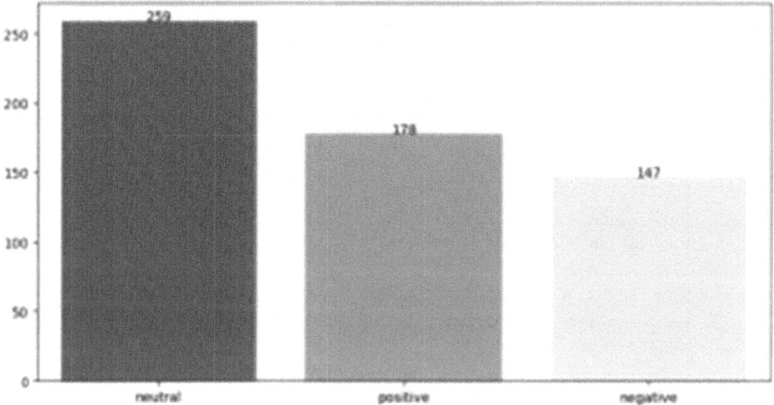


**Fig. 2.** Class Distribution

### 5.1 Models Evaluated

There are six models evaluated for this study.

1. Random Forest _ Bag of Words (RBOW)
2. Random Forest _ TF-IDF (RTI)
3. Support Vector Machine _ Bag of Words (SBOW)
4. Support Vector Machine _ TF-IDF (STI)
5. Multinomial Niave Bayes _ Bag of Words (MNBOW)
6. Multinomial Niave Bayes _ TF-IDF (MNTI)

### 5.2 Evaluation Parameters

The evaluation parameters used are

- Accuracy
- Precision
- Recall
- F1-Score
- Specificity

## 6 Results and Discussions

The confusion matrixes of the classifiers are given in the following figures (Figs. 3, 4, 5, 6 and 7).

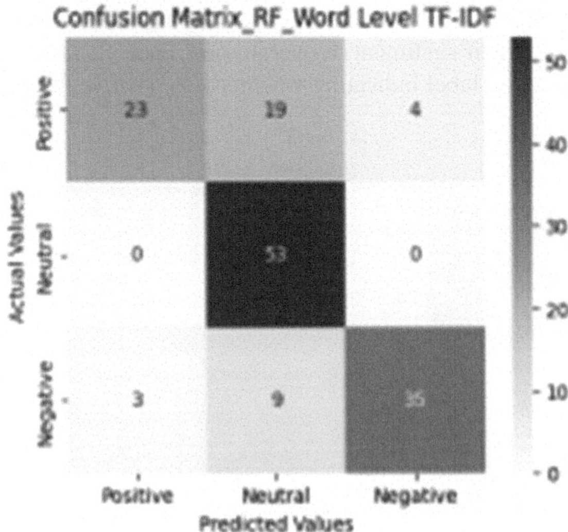


**Fig. 3.** Confusion Matrix of

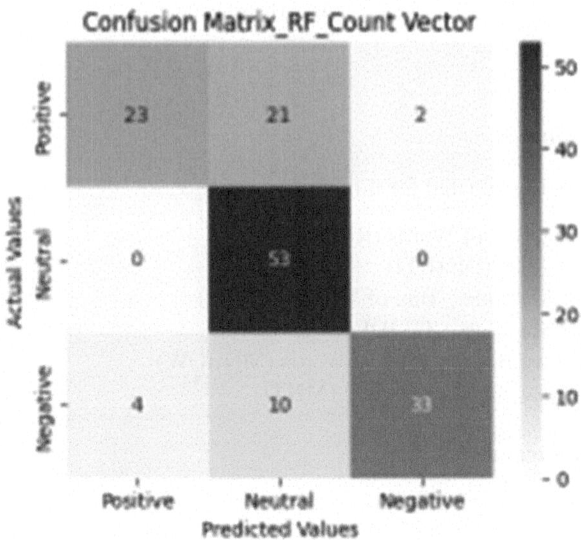


**Fig. 4.** Confusion Matrix of RTI

The results obtained for the classifiers in the chat classification task are as follows in the Table 1.

Accuracy measures the overall correctness of the classifier predictions, with higher values indicating better performance. In this case, SVM with TF-IDF (STI) achieved the highest accuracy of 0.868, followed closely by Random Forest with TF IDF (RTI) at 0.840.

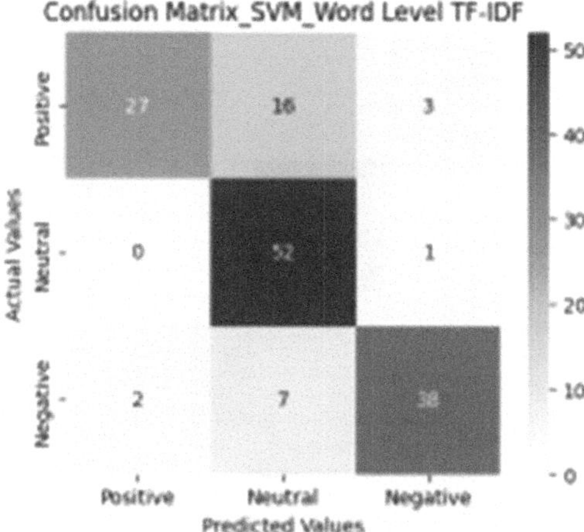


**Fig. 5.** Confusion Matrix of SBOW

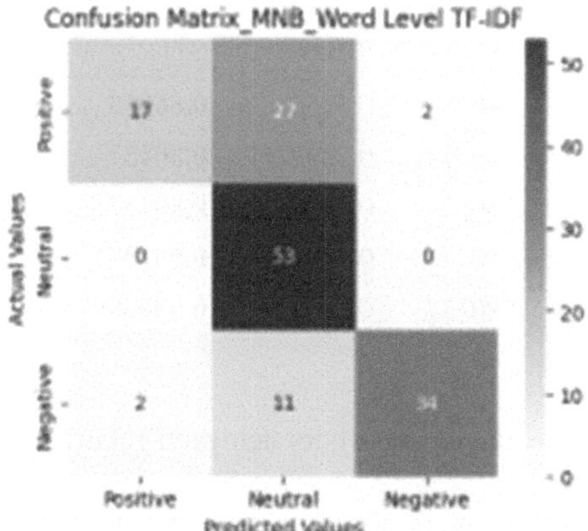


**Fig. 6.** Confusion Matrix of STI

Precision measures the proportion of correctly predicted positive instances out of all instances predicted as positive. Similarly, recall measures the proportion of correctly predicted positive instances out of all actual positive instances. F1-Score is the harmonic mean of precision and recall, providing a balanced measure of their performance. Among the classifiers, SVM with TF-IDF (STI) achieved the highest precision, recall, and F1-Score, indicating its effectiveness in correctly classifying positive instances.

Specificity measures the proportion of correctly predicted negative instances out of all actual negative instances. SVM with TF-IDF (STI) also achieved the highest specificity at 0.901, indicating its ability to accurately classify negative instances.

Overall, SVM with TF-IDF (STI) demonstrated the best performance across multiple evaluation metrics, achieving high accuracy, precision, recall, and F1-Score. Random Forest with TF-IDF (RTI) also performed well, with comparable accuracy and precision

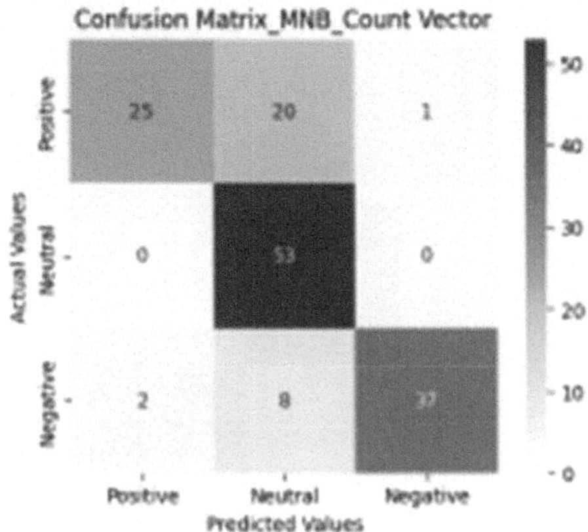


**Fig. 7.** Confusion Matrix of MNBOW

**Table 1.** Results obtained from the classifiers

| | RBOW | RTI | SBOW | STI | MNBOW | MNTI |
|---|---|---|---|---|---|---|
| Accuracy | 0.83105 | 0.840183 | 0.826484 | **0.86758** | 0.858447 | 0.808219 |
| Precision | 0.746575 | 0.760274 | 0.739726 | **0.80137** | 0.787671 | 0.712329 |
| Recall | 0.746575 | 0.760274 | 0.739726 | **0.80137** | 0.787671 | 0.712329 |
| F1-Score | 0.746575 | 0.760274 | 0.739726 | **0.80137** | 0.787671 | 0.712329 |
| Specificity | 0.873288 | 0.880137 | 0.869863 | **0.900685** | 0.893836 | 0.856164 |

scores. Multinomial Naive Bayes classifiers achieved relatively lower scores but still provided reasonable results.

The choice of text representation technique (Bag of Words vs. TF-IDF) did not result in significant performance variations within each classifier type. However, TF IDF generally outperformed Bag of Words in terms of accuracy and other evaluation metrics.

These results provide insights into the performance of different classifiers and text representation techniques for chat classification tasks. They can guide the selection of the most suitable classifier for future chat classification projects, with SVM using TF-IDF representation showing promise in achieving accurate and balanced predictions.

## 7 Conclusion

In this work, we tackled the task of chat classification using three different classifiers: Random Forest, Support Vector Machine (SVM), and Multinomial Naive Bayes (MNB). We explored two text representation techniques, namely Bag of Words (BoW) and TF-IDF (Term Frequency-Inverse Document Frequency). The goal was to accurately classify chat instances into positive, negative, and neutral classes.

The evaluation of the classifiers revealed that SVM with TF-IDF (STI) achieved the highest accuracy, precision, recall, F1-Score, and specificity among all the classifiers. This suggests that SVM with TF-IDF is a robust choice for chat classification tasks, providing accurate predictions for both positive and negative instances.

Random Forest with TF-IDF (RTI) also performed well, closely following SVM with TF-IDF. This indicates that Random Forest is a viable alternative for chat classification, especially when TF-IDF is used as the text representation technique.

Multinomial Naive Bayes classifiers (MNBOW and MNTI) achieved relatively lower scores compared to the other classifiers. While MNB may not have performed as well as SVM and Random Forest in this specific task, it still provided reasonable results, especially when using TF-IDF representation.

The choice of text representation technique showed that TF-IDF consistently outperformed Bag of Words in terms of accuracy and other evaluation metrics across all classifiers. TF-IDF effectively captured the importance of terms in the chat instances, enhancing the classifiers' performance.

In conclusion, this work contributes to the field of chat classification by comparing the performance of different classifiers and text representation techniques. The results suggest that SVM with TF-IDF and Random Forest with TF-IDF are effective approaches for accurately classifying chat instances into positive, negative, and neutral classes.

The outcomes of this work can be valuable for various applications, such as sentiment analysis, customer feedback analysis, or chatbot systems. By understanding the strengths and weaknesses of different classifiers and text representation techniques, practitioners can make informed decisions when building chat classification models for real-world scenarios.

The future enhancement may be to investigate the feasibility of online learning approaches for chat classification, where the models can adapt and update themselves in real-time as new chat instances arrive. This would be particularly useful for chatbot systems that need to continuously learn and adapt to evolving user interactions and language patterns. We can also extend the study to include multilingual chat classification, considering the challenges and opportunities associated with different languages. This could involve exploring language-specific features, utilizing pre trained multilingual embeddings, or adapting the models to handle code-switching scenarios.

## References

1. Harris, Z., Harris, Z., Clark, B., Clark, B.: Bag of words and sentiment analysis. In: The Routledge ndbook of Corpus Linguistics, pp. 510–523(2013)
2. Zhang, M., Zhang, C., Zhang, C.: Textual features for authorship attribution: an empirical study. J. Am. Soc. Inform. Sci. Technol. **61**(2), 369–380 (2010)
3. Sparck Jones, K.: A statistical interpretation of term specificity and its application in retrieval. J. Doc. **28**(1), 11–21 (1972)
4. Wang, S., Manning, C.D.: Baselines and bigrams: simple, good sentiment and topic classification. In: Proceedings of the 50th Annual Meeting of the Association for Computational Linguistics: Short Papers, vol. 2, pp. 90–94 (2012)
5. Yang, X., Gao, Y., Zhou, Z.: An effective feature selection method based on improved information gain ratio and TF-IDF. J. Ambient. Intell. Humaniz. Comput. **9**(5), 1663–1673 (2018)

6. Breiman, L.: Random forests. Mach. Learn. **45**(1), 5–32 (2001)
7. Joachims, T.: Text categorization with support vector machines: learning with many relevant features. In: Machine Learning: ECML-98, pp. 137–142 (1998). https://doi.org/10.1007/BFb0026683
8. Pang, B., Lee, L.: Opinion mining and sentiment analysis. Found. Trends Inf. Retr. **2**(1–2), 1–135 (2008)
9. Zhang, H., Yang, Q.: A survey on multi-label learning. ACM Comput. Surveys (CSUR) **42**(3), 1–34 (2010)
10. Chen, T., Guestrin, C., Xgboost, X.: A scalable tree boosting system. In: Proceedings of the 22nd ACM SIGKDD International Conference on Knowledge Discovery and Data Mining, pp. 785–794 (2020)
11. Yang, Y., Pedersen, J. O., Pedersen, J. O. A comparative study on feature selection in text categorization. In: Proceedings of the Fourteenth International Conference on Machine Learning, pp. 412–420 (1999)
12. Cai, D., He, X., Han, J.: Using SVM to learn information extraction rules. In: Proceedings of the 2004 SIAM International Conference on Data Mining, vol. 13, pp. 104–115 (2004)
13. Huang, J., Ling, C.X., Li, C.: Wrapper feature selection via multilayer perceptron reinforcement. Artif. Intell. **144**(1–2), 195–212 (2003)
14. McCallum, A., Nigam, K.: A comparison of event models for naive bayes text classification. In: AAAI-98 Workshop on Learning for Text Categorization, vol. 41, no. 42, pp. 752–759 (1998)
15. Rennie, J.D., Shih, L., Teevan, J., Karger, D.R.: Tackling the poor assumptions of Naive Bayes text classifiers. ICML **3**, 616–623 (2003)
16. Zhang, H., Zhang, H.: Text classification from labeled and unlabeled documents using EM. Machine Learning: ECML-04, pp. 69–80 (2004)
17. Han, J., Pei, J., Kamber, M.: Data mining: concepts and techniques. Elsevier, vol. 19 (2019). https://www.kaggle.com/datasets/nursyahrina/chat-sentiment-dataset. Accessed Jan 2023

# Correction to: Feature Fusion Based Bayesian Model Detection in Prognosis of Glioma – A Survey

S. Gowthami, K. H. Mohammed Sazzad, M. Nethra, S. Santhya, and A. Arnold Sylevester

**Correction to:**
**Chapter 1 in: R. Geetha et al. (Eds.):**
***Advances in Artificial Intelligence and Machine Learning in Big Data Processing*, CCIS 2203,**
**https://doi.org/10.1007/978-3-031-73068-9_1**

The first name "S." of chapter author's "S. Gowthami" was inadvertently omitted in the published version. This has now been corrected in the updated version.

The updated version of this chapter can be found at
https://doi.org/10.1007/978-3-031-73068-9_1


R. Geetha et al. (Eds.): AAIMB 2023, CCIS 2203, p. C1, 2025.
https://doi.org/10.1007/978-3-031-73068-9_26

# Author Index





R. Geetha et al. (Eds.): AAIMB 2023, CCIS 2203, pp. 327–329, 2025.
https://doi.org/10.1007/978-3-031-73068-9

The manufacturer's authorised representative in the EU is Springer Nature Customer Service Centre GmbH, Europaplatz 3, 69115 Heidelberg, Germany. If you have any concerns regarding our products, please contact ProductSafety@springernature.com

Printed and bound by CPI Group (UK) Ltd, Croydon, CR0 4YY
15/07/2026
02167617-0004